D1272349

CORRELATION PATTERN RECOGNITION

B. V. K. VIJAYA KUMAR

Department of Electrical and Computer Engineering
Carnegie Mellon University
Pittsburgh, PA 15213, USA

ABHIJIT MAHALANOBIS

Lockheed Martin Missiles & Fire Control
Orlando

RICHARD JUDAY

Formely of NASA Johnson Space Center
Longmont

CAMBRIDGE
UNIVERSITY PRESS

CAMBRIDGE UNIVERSITY PRESS
Cambridge, New York, Melbourne, Madrid, Cape Town, Singapore, São Paulo

CAMBRIDGE UNIVERSITY PRESS
The Edinburgh Building, Cambridge CB2 2RU, UK

Published in the United States of America by Cambridge University Press, New York

www.cambridge.org
Information on this title: www.cambridge.org/9780521571036

First published 2005

Printed in the United Kingdom at the University Press, Cambridge

Typeface Times 11/14 pt. *System* Advent 3B2 8.07f [PND]

A catalog record for this publication is available from the British Library

ISBN-13 978-0-521-57103-6
ISBN-10 0-521-57103-0 hardback

The publisher has used its best endeavors to ensure that the URLs for
external websites referred to in this publication are correct and active at the time
of going to press. However, the publisher has no responsibility for the
websites and can make no guarantee that a site will remain live or that the
content is or will remain appropriate.

Contents

Preface

Mathematically, correlation is quite simply expressed. One begins with two functions $f(\bullet)$ and $g(\bullet)$, and determines their correlation as a third function $c(\bullet)$:

$$c(t) \triangleq \int\limits_{-\infty}^{\infty} f(\tau)g^*(t+\tau)\,\mathrm{d}\tau$$

This simplicity is at the core of a rich technology in practical pattern recognition. For unit-energy signals (and images or higher-dimensional signals), the correlation output $c(t)$ achieves its maximum of 1 if and only if the signal $f(\tau)$ matches the signal $g(t+\tau)$ exactly for some t value. Thus, correlation is an important tool in determining whether the input signal or image matches a stored signal or image. However, the straightforward correlation operation (defined by the above equation) does not prove satisfactory in practical situations where the signals are not ideal and suffer any of the many distortions such as image rotations, scale changes, and noise. Over the last 20 years, the basic correlation operation has been improved to deal with these real-world challenges. The resulting body of concept, design methods, and algorithms can be aptly summarized as *correlation pattern recognition* (CPR).

Correlation pattern recognition, a subset of statistical pattern recognition, is based on selecting or creating a reference signal and then determining the degree to which the object under examination resembles the reference signal. The degree of resemblance is a simple statistic on which to base decisions about the object. We might be satisfied with deciding which class the object belongs to, or beyond that we might want more sophisticated information about which side we are viewing the object from – or conversely we might wish our pattern recognition to be quite independent of the aspect from which the object is viewed. Often it is critical to discriminate an object from classes that differ only subtly from the interesting class. Finally, the object may be embedded in (or surrounded by) clutter, some of whose characteristics may be similar to the interesting class. These considerations are at quite different levels, but the correlation algorithms create reference signals such that their correlation against the object produce statistics with direct information for those questions.

One of the principal strengths of CPR is the inherent robustness that results from its evaluating the whole signal at once. The signal is treated in a gestalt – CPR does not sweat the individual details. In contrast, feature-based techniques tend minutely to extract information from piecewise examination of the signal, and then compare the relationships among the features. By comparing the whole image against the template, CPR is less sensitive to small mismatches and obstructions.

For many years, the testing grounds for CPR have mainly been automatic target recognition (ATR) applications where correlation filters were developed to locate multiple occurrences of targets of interest (e.g., images of tanks, trucks, etc.) in input scenes. Clearly, processing speed is of interest in such applications, which has led to much interest in coherent optical correlators because of their ability to yield two-dimensional Fourier transforms (FTs) at the speed of light. However, the input and output devices in optical correlators have not progressed as fast as one would like and it is reasonable to say that today most image correlations are calculated digitally. Over the past few years, there has been a growing interest in the use of correlation filters for biometrics applications such as face recognition, fingerprint recognition, and iris recognition. In general, correlation filters should prove valuable in many image recognition applications.

Correlation can be implemented either in the time domain (space domain for images) or in the frequency domain. Because diffraction and propagation of coherent light naturally and conveniently produce the two-dimensional FT – and do so "at the speed of light" – early applications of coherent optical processing focused on correlation. This frequency domain approach is the reason for the use of the phrase "correlation filters." With the availability of the fast Fourier transform (FFT) algorithm and very high-speed digital processors, nowadays image correlations can be carried out routinely using digital implementations. In this book, we present both digital and optical processing approaches to correlation and have tried to indicate the differences and similarities. For example, in digital correlators, filter values may range more widely than in optical correlators where the optical devices impose constraints (e.g., that transmittance has to be a real value between 0 and 1). Another example is that the optical detectors detect only intensity (a real, positive value) whereas digital methods can freely produce and manipulate complex values. These differences have led to vigorous debates of the comparative advantages of digital and optical correlators and we hope that this energy has carried through to the book itself. We have enjoyed writing it.

Readers who are new to the correlation field may regard the superficial simplicity of the correlation paradigm to be anti-climactic and make no further attempt to grasp the versatility of the correlation pattern recognition techniques. Because the output from a matched filter is the cross-correlation of the received signal with the stored template, often correlation is simply misinterpreted as just matched filtering. We have sought to dispel this myth with a complete treatment of the diverse techniques for designing correlation filters that are anything but simple matched filters. It is well known that the filter theory finds widespread applications in controls, communications, adaptive signal processing, and audio and video applications. From a pattern recognition viewpoint, the same filtering concepts offer substantial benefits such as shift-invariance, graceful degradation, and avoidance of segmentation, not to mention computational simplicity (digitally or optically), and analytical closed-form solutions that yield optimal performance.

In putting together this book, our vision was to provide the reader with a single source that touches on all aspects of CPR. This field is a unique synthesis of techniques from probability and statistics, signals and systems, detection and estimation theory, and Fourier optics. As a result, the subject of CPR is rarely covered in traditional pattern recognition and computer vision books, and has remained elusive to the interested outsider.

The book begins with a practical introduction to CPR, and it ends with the current state of the art in computer-generated correlation filters. It discusses the sometimes seemingly abstract theories (e.g., detection theory, linear algebra, etc.) at the foundation of CPR, and it proceeds to applications. It presents the material necessary for a student to operate a first optical or digital correlator (aiming the level of the material at first-year graduate students in electrical engineering or optics programs). The book is intended to summarize recently published research and to put a usefully current overview of the discipline into the hands of the seasoned worker. In short, to take a line from Stuart L. Meyer, we are writing the book we would like to have owned as we began working in the field.

We believe that one of the main reasons that CPR is not used in more applications is that its practitioner must become familiar with some basic concepts in several fields: linear algebra, probability theory, linear systems theory, Fourier optics, and detection/estimation theory. Most students would not be exposed to such a mix of courses. Thus, Chapters 2, 3, and 4 in this book are devoted to providing the necessary background.

Chapter 2 reviews basic concepts in matrix/vector theory, simple quadratic optimization and probability theory, and random variables. Quadratic optimization will prove to be of importance in many correlation filter designs; e.g., when minimizing the output noise variance that is a quadratic function of the filter being designed. Similarly, basic results from probability theory, random variables, and random processes help us to determine how a filter affects the noise in the input.

As discussed before, correlation is implemented efficiently via the frequency domain. This shift-invariant implementation is based on ideas and results from the theory of linear systems, which is summarized in Chapter 3. This chapter reviews basic filtering concepts as well as the concept of sampling, an important link between continuous images and pixelated images. This chapter also introduces random signal processing, where a random signal is input to a deterministic linear, shift-invariant system.

The usual task of a pattern recognition system is to classify an input pattern into one of a finite number of classes (or hypotheses) and, if underlying statistics are known or can be modeled, we can use the results from detection theory to achieve goals such as minimizing classifier error rates or average cost. Another related topic is estimation theory, where the goal is to estimate an unknown parameter from the observations. One application of estimation is the estimation of a classifier error rate. Chapter 4 summarizes some basic concepts from detection and estimation theory.

Chapters 5 and 6 are aimed at introducing the various correlation filter designs. Chapter 5 introduces the basic correlation filters, which are aimed at recognizing a single image. It starts with the basic notion of matched filters and shows how its output is nothing but a correlation. But then the limitations of the matched filter are discussed and other alternatives such as optimal tradeoff filters (that tradeoff noise tolerance and correlation peak sharpness) are introduced. Performance metrics useful for characterizing correlation filters are introduced. Chapter 5 also introduces some correlation filter variants (e.g., binary phase-only filter) that were introduced because of optical device limitations.

Chapter 6 presents many advanced correlation filters (also called synthetic discriminant function or SDF filters), which are the correlation filters being used in many ATR and biometrics applications. In most of these advanced correlation filter designs, the main idea is to synthesize a filter from training images that exhibit the range of image distortions that the filter is supposed to accommodate. One breakthrough filter is the minimum average correlation energy (MACE) filter, which produces sharp correlation peaks and high discrimination. The MACE filter has been used with good success in ATR and biometrics applications. This and other advanced correlation filters are discussed in Chapter 6.

Chapters 7 and 8 are devoted to optical correlator implementations. Chapter 7 is aimed at introducing some basic optics concepts such as diffraction, propagation, interference, coherence, and polarization. This chapter also introduces the important topic of spatial light modulators (SLMs), which are the optical devices that convert electrical signals to optical signals. Historically, SLMs have been the limiting factors in the speed and capabilities of optical correlators. Nowadays, SLMs originally intended for the display industry are fueling a growth of small laboratory tinkering. For less than $4000, a single color television projector provides three high quality (though slow) modulators of several hundred pixels on a side, along with their necessary drive electronics. Other SLMs and architectures are becoming available whose speeds are substantially higher than the 30 frames per second for conventional broadcast television. Conventional wisdom in optical filter computation does not make appropriate use of these modulators, as is now possible using the recent algorithmic advances. Many of these SLMs are potentially very powerful but are often improperly used. The algorithms now allow us to make productive use of SLM behavior that until very recently would have been regarded as difficult and inferior. These concepts are discussed in Chapter 7.

Chapter 8 provides the mathematical details as well as the algorithms for designing correlation filters that can be implemented on limited-modulation SLMs. Unlike digital designs, these designs must carefully consider the SLM constraints right from the start. Over the past few years, significant mathematical advances (in particular, applying the minimal Euclidean distance [MED] principle) have been made in the design of such limited modulation correlation filters, the topic of Chapter 8.

Finally, Chapter 9 provides a quick review of two correlation filter applications. First is the automatic recognition of targets in synthetic aperture radar (SAR) scenes and the second is the verification of face images. Some MATLAB® code is provided to illustrate the design and application of the correlation filters.

This book would not have been possible without the help of many. At the risk of offending many others who have helped, we would like to acknowledge a few in particular. B. V. K. Vijaya Kumar (BVKVK) acknowledges Professor David Casasent of Carnegie Mellon University (CMU) for introducing him to the topic of optical computers, various colleagues and students for the many advances summarized in this book, the Electrical and Computer Engineering Department at CMU for supporting this effort through a sabbatical leave, and the Carnegie Institute of Technology for the Phil Dowd Fellowship that has accelerated the completion of this book. BVKVK also acknowledges the profound positive influences of his late parents (Ramamurthy Bhagavatula and Saradamba Bhagavatula) and the immense patience and love of his wife Latha Bhagavatula. Abhijit Mahalanobis (AM) would like to acknowledge his mother and late father for their guiding hand, and his wife for her patience in not ceasing to believe in the fact that all good things must come to an end (although this book nearly proved her wrong). Richard Juday wishes to acknowledge the support that NASA's Johnson Space Center provided through a decade and a half of his work in this field, and also the contributions of literally dozens of students, visiting faculty, post-doctoral fellows, and external colleagues. Dr. Stanley E. Monroe has been a particularly steadfast contributor, advisor, critic, and friend to all whose work has touched the Hybrid Vision Laboratory.

The MathWorks, Inc., very kindly provided their state-of-the-art software, MATLAB®, which we have found very useful in developing algorithms and graphics for this book. MATLAB® is a trademark of The MathWorks, Inc., and is used with permission. The MathWorks does not warrant the accuracy of the text in this book. This book's use or discussion of MATLAB® software or related products does not constitute endorsement or sponsorship by The MathWorks of a particular pedagogical approach, or particular use of the MATLAB® software.

1

Introduction

There are many daily pattern recognition tasks that humans routinely carry out without thinking twice. For example, we can recognize those that we know by looking at their face or hearing their voice. You can recognize the letters and words you are reading now because you have trained yourself to recognize English letters and words. We can understand what someone is saying even if it is slightly distorted (e.g., spoken too fast). However, human pattern recognition suffers from three main drawbacks: poor speed, difficulty in scaling, and inability to handle some recognition tasks. Not surprisingly, humans can't match machine speeds on pattern recognition tasks where good pattern recognition algorithms exist. Also, human pattern recognition ability gets overwhelmed if the number of classes to recognize becomes very large. Although humans have evolved to perform well on some recognition tasks such as face or voice recognition, except for a few trained experts, most humans cannot tell whose fingerprint they are looking at. Thus, there are many interesting pattern recognition tasks for which we need machines.

The field of machine learning or pattern recognition is rich with many elegant concepts and results. One set of pattern recognition methods that we feel has not been explained in sufficient detail is that of correlation filters. One reason why correlation filters have not been employed more for pattern recognition applications is that their use requires background in and familiarity with different disciplines such as linear systems, random processes, matrix/vector methods, statistical decision theory, pattern recognition, optical processing, and digital signal processing. This book is aimed at providing such background as well as introducing the reader to state-of-the-art in design and analysis of correlation filters for pattern recognition. The next two sections in this chapter will provide a brief introduction to pattern recognition and correlation, and in the last section we provide a brief outline of the rest of this book.

1

1.1 Pattern recognition

In pattern recognition, the main goal is to assign an observation into one of multiple classes. The observation can be a signal (e.g., speech signal), an image (e.g., an aerial view of a ground scene) or a higher-dimensional object (e.g., video sequence, hyperspectral signature, etc.) although we will use an image as the default object in this book. The classes depend on the application at hand. In automatic target recognition (ATR) applications, the goal may be to classify the input observation as either natural or man-made, and follow this up with finer classification such as vehicle vs. non-vehicle, tanks vs. trucks, one type of tank vs. another type.

Another important class of pattern recognition applications is the use of biometric signatures (e.g., face image, fingerprint image, iris image, and voice signals) for person identification. In some biometric recognition applications (e.g., accessing the automatic teller machine), we may be looking at a verification application where the goal is to see whether a stored template matches the live template in order to accept the subject as an authorized user. In other biometric recognition scenarios (e.g., deciding whether a particular person is in a database), we may want to match the live biometric to several stored biometric signatures.

One standard paradigm for pattern recognition is shown in Figure 1.1. The observed input image is first preprocessed. The goals of preprocessing depend very much on the details of the application at hand, but can include: reducing the noise, improving the contrast or dynamic range of the image, enhancing the edge information in the image, registering the image, and other application-specific processes.

A feature extraction module next extracts features from the preprocessed image. The goal of feature extraction is to produce a few descriptors to capture the essence of an input image. The number of features is usually much smaller than the number of pixels in that input image. For example, a 64×64 image contains 4096 numbers (namely the pixel values), yet we may be able to capture the essence of this image using only 10 or 20 features. Coming up with good features depends very much on the designer's experience in an application domain. For example, for fingerprint recognition, it is well known that features such as ridge endings and bifurcations called minutiae (shown in

Figure 1.1 Block diagram showing the major steps in image pattern recognition

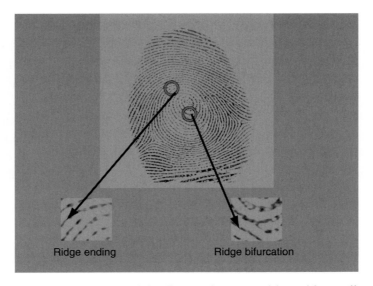

Figure 1.2 Some features used for fingerprint recognition: ridge ending (left) and ridge bifurcation (right)

Figure 1.2) are useful for distinguishing one fingerprint from another. In other pattern recognition applications, different features may be used. For example, in face recognition, one may use geometric features such as the distance between the eyes or intensity features such as the average gray scale in the image, etc. There is no set of features that is a universal set in that it is good for all pattern recognition problems. Almost always, it is the designer's experience, insight, and intuition that help in the identification of good features.

The features are next input to a classifier module. Its goal is to assign the features derived from the input observation to one of the classes. The classifiers are designed to optimize some metric such as probability of classification error (if underlying probability densities are known), or empirical error count (if a validation set of data with known ground truth[1] is available). Classifiers come in a variety of flavors including statistical classifiers, artificial neural-network-based classifiers and fuzzy logic-based classifiers. The suitability of a classifier scheme depends very much on the performance metric of interest, and on what a-priori information is available about how features appear for different classes. If we have probability density functions for various features for different classes, we can design statistical classification schemes. Sometimes, such probability density information may not be available and, instead, we may have sample feature vectors from different classes. In such a

---

[1] A term from remote sensing to denote the correct class of the object being tested.

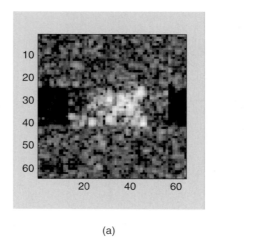
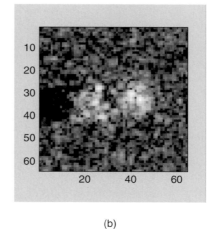

(a)   (b)

Figure 1.3 Synthetic aperture radar (SAR) images of two vehicles, (a) T72 and
(b) BTR70, from the public MSTAR database [3]

situation, we may want to use trainable classifiers such as neural networks. In
this book, we will not discuss these different pattern recognition paradigms.
Interested readers are encouraged to consult some of the many excellent
references [1, 2] discussing general pattern recognition methods.

Another important pattern recognition paradigm is to use the training data
directly instead of first determining some features and performing classification based on those features. While feature extraction works well in many
applications, it is not always easy for humans to identify what the good
features may be. This is particularly difficult when we are facing classification
problems such as the one shown in Figure 1.3, where the images were acquired
using a synthetic aperture radar (SAR) and the goal is to assign the SAR
images to one of two classes (tank vs. truck). Humans are ill equipped to come
up with the "best" features for this classification problem. We may be better off
letting the images speak for themselves, rather than imposing our judgments of
what parts of SAR images are important and consistent in the way a target
appears in the SAR imagery. Correlation pattern recognition (CPR) is an
excellent paradigm for using training images to design a classifier and to
classify a test image.

1.2 Correlation

Most readers are probably familiar with the basic concept of correlation as it
arises in probability theory. We say that two random variables (RVs, the

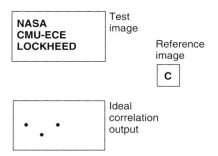

Figure 1.4 Schematic of the image correlation: reference image, test image, and ideal correlation output

concept to be explained more precisely in Chapter 2) are correlated if knowing something about one tells you something about the other RV. There are degrees of correlation and correlation can be positive or negative. The role of correlation for pattern recognition is not much different in that it tries to capture how similar or different a test object is from training objects. However, straightforward correlation works well only when the test object matches well with the training set and, in this book, we will provide many methods to improve the basic correlation and to achieve attributes such as tolerance to real-world differences or distortions (such as image rotations, scale changes, illumination variations, etc.), and discrimination from other classes.

We will introduce the concept of CPR using Figure 1.4. In this figure, we have two images: a reference image of the pattern we are looking for and a test image that contains many patterns. In this example, we are looking for the letter "C." But in other image recognition applications, the reference $r[m, n]$ can be an (optical, infrared, or SAR) image of a tank and the test image $t[m, n]$ can be an aerial view of the battlefield scene. In a biometric application, the reference may be a client's face image stored on a smart card, and the test image may be the one he is presenting live to a camera. For the particular case in Figure 1.4, let us assume that the images are binary with black regions taking on the value 1 and white regions taking on the value 0.

The correlation of the reference image $r[m, n]$ and the test image $t[m, n]$ proceeds as follows. Imagine overlaying the smaller reference image on top of the upper left corner portion of the test image. The two images are multiplied (pixel-wise) and the values in the resulting product array are summed to obtain the correlation value of the reference image with the test image for that relative location between the two. This calculation of correlation values is then repeated by shifting the reference image to all possible centerings of the reference image with respect to the test image. As indicated in the idealized correlation output in Figure 1.4, large correlation values should be obtained at the three locations where the reference matches the test image. Thus, we can locate the targets of interest by examining the correlation output for peaks and determining if those correlation peaks are sufficiently large to indicate the presence of a reference object. Thus, when we refer to CPR in this book, we are not referring to just one correlation value (i.e., one inner product of two arrays), but rather to a correlation output $c[m, n]$ that can have as many pixels as the test image. The following equation captures the cross-correlation process

$$c[m,n] = \sum_k \sum_l t[k,l]r[k+m,l+n] \qquad (1.1)$$

From Eq. (1.1), we see that correlation output $c[m, n]$ is the result of adding many values, or we can say that the correlation operation is an integrative operation. The advantage of such an integrative operation is that no single pixel in the test image by itself is critical to forming the correlation output. This results in the desired property that correlation offers graceful degradation. We illustrate the graceful degradation property in Figure 1.5. Part (a) of this figure shows a full face image from the Carnegie Mellon University (CMU) Pose, Illumination, and Expression (PIE) face database [4] and part (b) shows the correlation output (in an isometric view) from a CPR system designed to search for the image in part (a). As expected, the correlation output exhibits a large value indicating that the test image indeed matches the reference image. Part (c) shows the same face except that a portion of the face image is occluded. Although the resulting correlation output in part (d) exhibits correlation peaks smaller than in part (b), it is clear that a correlation peak is still present indicating that the test image does indeed match the reference object. Some other face recognition methods (that rely on locating both eyes to start the feature extraction process) will not exhibit similar graceful degradation properties.

Another important benefit of CPR is the in-built shift-invariance. As we will show in later chapters, correlation operation can be implemented as a linear, shift-invariant filter (this shift-invariance concept will be made more precise in Chapter 3 on linear systems), which means that if the test image contains the reference object at a shifted location, the correlation output is also shifted by exactly the same amount. This shift-invariance property is illustrated in parts (e) and (f) of Figure 1.5. Part (e) shows a shifted and occluded version of the reference image and the resulting correlation output in part (f) is shifted by the same amount, but the correlation peak is still very discernible. Thus, there is no need to go through the trouble of centering the input image prior to recognizing it.

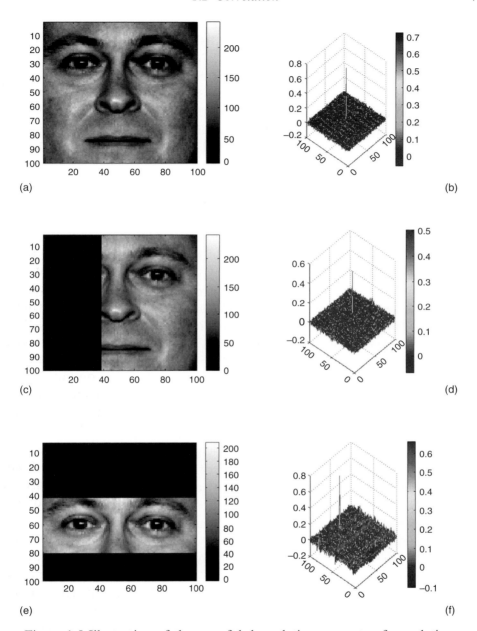

(a)

(b)

(c)

(d)

(e)

(f)

Figure 1.5 Illustration of the graceful degradation property of correlation operation, (a) a full face image from the CMU PIE database [4], (b) correlation output for test image in part (a), (c) occluded face image, (d) correlation output for image in part (c), (e) shifted and occluded face image, and (f) correlation output for image in part (e)

A reasonable question to ask at this stage is why one needs to read the rest of this book when we have already explained using Figure 1.4 and Figure 1.5 the basic concept of correlation and advantages of using correlation.

We need to discuss more advanced correlation filters because the simple scheme in Figure 1.4 works well only if the test scene contains exact replicas of the reference images, and if there are no other objects whose appearance is similar to that of the reference image. For example, in Figure 1.4, the letter "O" will be highly correlated with letter "C" and the simple cross-correlation will lead to a large correlation output for the letter "O" also, which is undesirable. Thus, we need to, and we will, discuss the design of correlation templates that not only recognize the selected reference image, but also reject impostors from other classes. Also the book discusses practical issues of computing correlation using digital methods and optical methods. One way to summarize the contents of this book is that it contains much of the material we wish had been available when starting into CPR.

Another deficiency of the straightforward correlation operation in Eq. (1.1) is that it can be overly sensitive to noise. Most test scenes will contain all types of noise causing randomness in the correlation output. If this randomness is not explicitly dealt with, correlation outputs can lead to erroneous decisions. Also, as illustrated in Figure 1.5, sharp correlation peaks are important in estimating the location of a reference image in the test scene. It is easier to locate the targets in a scene if the correlation template is designed to produce sharp peaks. Unfortunately, noise tolerance and peak sharpness are typically conflicting criteria, and we will need design techniques that optimally trade off between the two conflicting criteria.

The straightforward correlation scheme of Figure 1.4 does not work well if the reference image appears in the target scene with significant changes in appearance (often called *distortions*), perhaps owing to illumination changes, viewing geometry changes (e.g., rotations, scale changes, etc.). For example, a face may be presented to a face verification system in a different pose from the one used at the time of enrolment. In an ATR example based on infrared images, a vehicle of interest may look different when compared to the reference image because the vehicle may have been driven around (and as a result, the engine has become hot leading to a brighter infrared image). A good recognition system must be able to cope with such expected variability. In this book, we will discuss various ways to increase the capabilities of correlation methods to provide distortion-tolerant pattern recognition.

Another important question in connection with the correlation method is how it should be implemented. As we will show later in this book, straightforward implementations (e.g., image–domain correlations as in Figure 1.4) are inefficient, and more efficient methods based on fast Fourier transforms (FFTs) exist. Such efficiency is not just a theoretical curiosity; this efficiency of FFT-based correlations is what allows us to use CPR for demanding applications such as real-time ATR and real-time biometric recognition. This book will provide the theory and details to achieve such efficiencies.

It is fair to say that the interest in CPR is mainly due to the pioneering work by VanderLugt [5] that showed how the correlation operation can be implemented using a coherent optical system. Such an optical implementation carries out image correlations "at the speed of light." However, in practice, we don't achieve such speed owing to a variety of factors. For example, bringing the test images and reference images into the optical correlators and transferring the correlation outputs from the optical correlators for post-processing prove to be bottlenecks, as these steps involve conversion from electrons to photons and vice versa. Another challenge is that the optical devices used to represent the correlation templates cannot accommodate arbitrary complex values as digital computers can. Some optical devices may be phase-only (i.e., magnitude must equal 1), binary phase-only (i.e., only +1 and −1 values are allowed), or cross-coupled where the device can accommodate only a curvilinear subset of magnitude and phase values from the complex plane. It is necessary to design CPR schemes that take into account such implementation constraints if we want to achieve the best possible performance. This book will provide sufficient information for designing *optical* CPR schemes.

1.3 Organization

As discussed in the previous section, CPR is a rather broad topic requiring background in many subjects including linear systems, matrix and vector methods, RVs and processes, statistical hypothesis testing, optical processing, digital signal processing, and, of course, pattern recognition theory. Not surprisingly, it is difficult to find all these in one source. It is our goal to provide the necessary background in these areas and to illustrate how to synthesize that knowledge to design CPR systems. In what follows, we will provide brief summaries of what to expect in the following chapters.

Chapter 2, Mathematical background In this chapter, we provide brief reviews of several relevant topics from mathematics. We first review matrices and vectors, as the correlation templates (also known as *correlation filters*) are designed using linear algebra methods and it is important to know concepts such as matrix inverse, determinant, rank, eigenvectors, diagonalization, etc. This chapter also introduces some vector calculus (e.g., gradient) and illustrates its use in optimization problems that we will need to solve for CPR. As we mentioned earlier, randomness is inevitable in input patterns, and a short review of probability theory and RVs is provided in this chapter. This review includes the case of two RVs as well as more RVs (equivalently, more compactly represented as a random vector).

Chapter 3, Linear systems and filtering theory In this chapter, we review the basic concepts of linear shift-invariant systems and filters. These are important for CPR since most implementations of correlation are in the form of filters, which is why we refer to the correlation templates also as correlation filters (strictly speaking, *templates* refer to image domain quantities whereas *filters* are in the frequency domain). In addition to standard one-dimensional (1-D) signals and systems topics, we review some two-dimensional (2-D) topics of relevance when dealing with images. This is the chapter where we will see that the correlation operation is implemented more efficiently via the frequency domain rather than directly in the image domain. Both optical and digital correlation implementations originate from this frequency domain version. This chapter reviews sampling theory, which is important to understand the connections between digital simulations and optical implementations. Since digital correlators are heavily dependent on the FFT, this chapter reviews the basics of both 1-D and 2-D FFTs. Finally, we review random signal processing, as the randomness in the test images is not limited to just one value or pixel. The randomness in the images may be correlated from pixel to pixel necessitating concepts from random processes, which are reviewed in this chapter.

Chapter 4, Detection and estimation The goal of this relatively short chapter is to provide the statistical basis for some commonly used CPR approaches. First, we derive the optimal methods for classifying an observation into one of two classes. Then, we show that the optimum method is indeed a correlator, if we can assume some conditions about the noise. Another topic of importance is estimation, which deals with the best ways to extract unknown information from noisy observations. This is of particular importance when we need to estimate the error rates from a correlator.

Chapter 5, Correlation filter basics In some ways, this is the core of this book. It starts by showing how correlation is optimum for detecting a known reference signal in additive white Gaussian noise (AWGN). This theory owes its origins to the matched filter (MF) [6], introduced during World War II for radar applications. Next, we show how MFs can be implemented digitally and optically using Fourier transforms (FTs). As MFs cannot be implemented (as they are) on limited-modulation optical devices, we next discuss several variants of the MF including phase-only filters, binary phase-only filters and cross-coupled filters. The focus of this chapter is on correlation filters designed for single images and not for multiple appearances of a target image.

Chapter 6, Advanced correlation filters Matched filters, the main topic of Chapter 5, work well if the test image contains an exact replica of the reference image except possibly for a shift and AWGN. However, in practice the appearance of a target can vary significantly because of illumination changes, changes in the viewing geometries, occlusions, etc. It is desirable that the correlation filters still recognize the target of interest and discriminate it from other objects in the scene. One of the first methods to address this problem is the *synthetic discriminant function* (SDF) filter [7] approach. In the basic SDF filter method, the correlation filter is designed from a linear combination of training images of the target, where the training images are chosen to reflect anticipated appearances of the target. This chapter starts with the basic SDF, but quickly moves on to more advanced correlation filter designs aimed at achieving excellent distortion tolerance, without overly sacrificing discrimination. In this chapter, we discuss selected advanced correlation filters, rather than trying to cover all the correlation filter design methods that have appeared in the literature. More detailed treatments of the correlation filter designs and applications are available elsewhere [8, 9].

Chapter 7, Optical considerations The main catalyst for CPR is the pioneering work by VanderLugt [5], that shows that correlation can be implemented in coherent optical processors by using holographic methods to record the complex-valued matched filters. The goal of this chapter is to provide the background needed for implementing correlation optically. This chapter introduces fundamental concepts in optics such as diffraction, propagation, interference, and polarization. Jones' calculus approach is presented to facilitate the representation as well as manipulation of polarized light. This is then followed by a discussion of the use of spatial light modulators (SLMs) to represent both image domain and frequency domain quantities in optical correlators.

Chapter 8, Limited-modulation filters The matched-filter optical correlator introduced by VanderLugt [5] uses a hologram to represent the filter, which takes on complex values. The use of holograms is not attractive in practical applications, as we may have to change the correlation filters rapidly to match the test image against many different reference images. The input scene can also change quickly needing an SLM, which can be controlled externally. Most SLMs cannot accommodate arbitrary complex values, and in that sense are *limited-modulation* devices. This chapter is aimed at describing how correlation filters can be implemented on limited-modulation SLMs. The output devices in optical correlators are usually intensity detectors, and in that sense differ from digital implementations where the output can be complex. Thus, blindly applying digital designs in optical correlators is not advised, and this chapter provides optimal methods for designing correlation filters that take into account optical system constraints.

Chapter 9, Application of correlation filters This chapter is aimed at providing a couple of application studies for CPR. First is the use of correlation filters for classification of two different types of vehicles from their SAR images of the type shown in Figure 1.3. What makes this problem particularly challenging is that these vehicles must be discriminated from each other and from clutter, although they may be viewed at different aspect angles and probably at different elevations. Another application considered in this chapter is face verification. One way to improve the security of access to physical spaces (e.g., buildings) or virtual spaces (e.g., computers) is to match the live biometric (e.g., face image, fingerprint, or iris image) to a stored biometric of an authorized user. Correlation pattern recognition is proving to be an attractive candidate for this application and this chapter briefly reviews the face verification application.

Our intent in writing this book on CPR is to provide in one place much of the diverse background needed, as well as to provide some discussion of its advantages and limitations. However, our bias is towards the underlying theory and we may not be covering the applications in the detail they deserve. However, we refer the reader to various journals [10, 11] and conferences [12] for application examples.

2

Mathematical background

Correlation filter theory relies heavily on concepts and tools from the fields of linear algebra and probability theory. Matrices and vectors provide succinct ways of expressing operations on discrete (i.e., pixelated) images, manipulating multiple variables and optimizing criteria that depend on multiple parameters. A vector representation also facilitates parallel operations on a set of constants or variables. Thus linear algebra provides powerful tools for digitally synthesizing and analyzing correlation filters.

If the world of interest contained only deterministic (i.e., non-random) signals and images, there would be no need for advanced pattern recognition methods in general and for correlation techniques in particular. In practice, reference images suffer from unpredictable disturbances such as noise, occlusion, illumination changes, rotation, and scale changes. Such unpredictability leads to randomness that can be characterized only by probabilistic models. We also need to understand what happens to such input randomness as it passes through signal processing systems such as correlation filters. Such knowledge will enable us to analyze the response of signal/image processing systems to noisy inputs, and to design systems that will preserve or enhance the desired signals while suppressing unwanted noise. This chapter aims to provide a quick review of the basics of matrix/vector techniques as well as the basics of probability theory and RVs.

While a comprehensive coverage of these fields is beyond the scope of this book, some topics relevant to correlation methods are discussed here for ease of reference. We begin in Section 2.1 with a review of some definitions of matrix algebra, followed by a discussion in Section 2.2 of several useful fundamental properties. Section 2.3 introduces the important concept of eigen-analysis, which is used in Section 2.4 to illustrate optimization methods for quadratic criteria (e.g., minimizing the output noise variance) of utility in correlation filter design. Finally, Section 2.5 provides a brief summary of relevant concepts from probability theory and RVs, including the concept of a random vector containing multiple RVs.

2.1 Matrix–vector notation and basic definitions

A vector is an ordered collection of real or complex scalars arranged either as a column or as a row. We will use a column vector as the default vector. Thus, $\mathbf{x}$ is a column vector containing N scalars x_i, $1 \leq i \leq N$, which are referred to as its *elements* or components. Lower-case, bold roman letters will usually indicate vectors. However, we may occasionally find it convenient to represent a vector compactly in terms of its elements as $\{x_i\}$, i.e.,

$$\mathbf{x} = \{x_i\} = \begin{bmatrix} x_1 \\ x_2 \\ \vdots \\ x_N \end{bmatrix} \tag{2.1}$$

The *transpose* operation, generally denoted by the superscript T, converts a row vector to a column vector and vice versa. Thus, $\mathbf{x}^{\mathrm{T}}$ (pronounced "$\mathbf{x}$ transpose") is the row vector $[x_1\ x_2\ \ldots\ x_N]$. The *conjugate transpose*, or *Hermitian,* denoted by the superscript $+$ (pronounced "$\mathbf{x}$ Hermitian" or "$\mathbf{x}$ conjugate transpose") leads to $\mathbf{x}^{+} = [x_1^*\ x_2^*\ \ldots\ x_N^*]$, where the superscript $*$ denotes the complex conjugate operation.

An $M \times N$ matrix $\mathbf{A}$ has elements arranged in M rows and N columns, i.e.,

$$\mathbf{A} = \begin{bmatrix} a_{11} & a_{12} & \cdots & a_{1N} \\ a_{21} & a_{22} & \cdots & a_{2N} \\ \vdots & \vdots & \ddots & \vdots \\ a_{M1} & a_{M2} & \cdots & a_{MN} \end{bmatrix} \tag{2.2}$$

Here a_{ij}, $1 \leq i \leq M$, $1 \leq j \leq N$, represents the element in the ith row and jth column. Upper-case, bold roman letters will be used to denote matrices. For the sake of convenience, we may express a matrix more compactly as $\mathbf{A} = \{a_{ij}\}$, $1 \leq i \leq M$, $1 \leq j \leq N$. A matrix is said to be *square* when $M = N$. The *transpose* $\mathbf{A}^{\mathrm{T}}$ is obtained by interchanging the rows and columns of $\mathbf{A}$. A matrix $\mathbf{A}$ is *symmetric* if $\mathbf{A} = \mathbf{A}^{\mathrm{T}}$. The *conjugate transpose* $\mathbf{A}^{+}$ is obtained by conjugating all elements of $\mathbf{A}^{\mathrm{T}}$.

Let us now consider some matrices with special structures. A diagonal matrix has non-zero entries only along its main diagonal. A special case of the diagonal matrix is the *identity matrix* (denoted by $\mathbf{I}$), which is a square matrix with 1s along the main diagonal and 0s elsewhere, i.e.,

$$\mathbf{I} = \begin{bmatrix} 1 & 0 & \cdots & 0 \\ 0 & 1 & \cdots & 0 \\ \vdots & \vdots & \ddots & \vdots \\ 0 & 0 & \cdots & 1 \end{bmatrix} \tag{2.3}$$

Another special structure is a *Toeplitz* matrix $\mathbf{T}$, which contains identical elements along any of its diagonals; e.g.,

$$\mathbf{T} = \begin{bmatrix} t_0 & t_1 & t_2 & t_3 \\ t_{-1} & t_0 & t_1 & t_2 \\ t_{-2} & t_{-1} & t_0 & t_1 \\ t_{-3} & t_{-2} & t_{-1} & t_0 \end{bmatrix} \tag{2.4}$$

Another square matrix of some interest is the Vandermonde structure shown below:

$$\mathbf{V} = \begin{bmatrix} 1 & 1 & \cdots & 1 \\ v_0 & v_1 & \cdots & v_{n-1} \\ \vdots & \vdots & \ddots & \vdots \\ v_0^{n-1} & v_1^{n-1} & \cdots & v_{n-1}^{n-1} \end{bmatrix} \tag{2.5}$$

For $v_k = \exp(-j2\pi k/N)$, the $N \times N$ Vandermonde matrix in Eq. (2.5) is the N-point discrete Fourier transform (DFT) matrix of importance in digital signal processing in general, and digital correlation in particular.

In a sense, a discrete image is a matrix with the pixel values as its elements. However in the pattern recognition context, an image is a spatially varying two-dimensional function whereas a matrix is just an array of numbers. The distinction becomes important when we purposefully represent an image as a vector or matrix to facilitate desired manipulations. As an example, consider the 4×4 image shown in Figure 2.1(a) in the form of an array and in the form of a gray-scale image in (b). The same data can be arranged as a diagonal matrix or as a column vector as shown in (c) and (d), respectively.

2.2 Basic matrix–vector operations

We now review some basic vector and matrix operations of use in correlation pattern recognition. The sum of $\mathbf{a}$ and $\mathbf{b}$, two column vectors of length N, is obtained by adding their elements, i.e., $\mathbf{a} + \mathbf{b} = \{a_i + b_i\}$. There are at least two

1	2	2	1
3	1	0	3
3	1	1	3
1	2	2	1

(a)

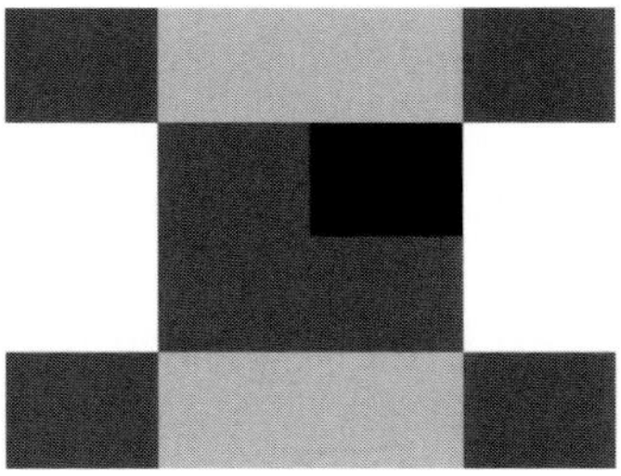

(b)

$$\mathbf{x} = \begin{bmatrix} 1 & & & & & & & & & & & & & & & \\ & 2 & & & & & & & & & & & & & & \\ & & 2 & & & & & & & & & & & & & \\ & & & 1 & & & & & & & & & & & & \\ & & & & 3 & & & & & & & & & & & \\ & & & & & 1 & & & & & & & & & & \\ & & & & & & 0 & & & & & & & & & \\ & & & & & & & 3 & & & & & & & & \\ & & & & & & & & 3 & & & & & & & \\ & & & & & & & & & 1 & & & & & & \\ & & & & & & & & & & 1 & & & & & \\ & & & & & & & & & & & 3 & & & & \\ & & & & & & & & & & & & 1 & & & \\ & & & & & & & & & & & & & 2 & & \\ & & & & & & & & & & & & & & 2 & \\ & & & & & & & & & & & & & & & 1 \end{bmatrix} \qquad \mathbf{x} = \begin{bmatrix} 1 \\ 2 \\ 2 \\ 1 \\ 3 \\ 1 \\ 0 \\ 3 \\ 3 \\ 1 \\ 1 \\ 3 \\ 1 \\ 2 \\ 2 \\ 1 \end{bmatrix}$$

(c)   (d)

Figure 2.1 Representation of a 4×4 image as (a) an array of numbers, (b) gray value pixels, (c) a diagonal matrix, and (d) a column vector

types of vector multiplication. The *inner* product $\mathbf{a}^+\mathbf{b}$ (also referred to as the *dot* product) yields a scalar result and is defined as

$$\mathbf{a}^+\mathbf{b} = \sum_{i=1}^{N} a_i^* b_i \qquad (2.6)$$

The *outer* product $\mathbf{ab}^+$ results in an $N \times N$ matrix given by

$$\mathbf{ab}^+ = \begin{bmatrix} a_1 b_1^* & a_1 b_2^* & \cdots & a_1 b_N^* \\ a_2 b_1^* & a_2 b_2^* & \cdots & a_2 b_N^* \\ \vdots & \vdots & \ddots & \vdots \\ a_N b_1^* & a_N b_2^* & \cdots & a_N b_N^* \end{bmatrix} \tag{2.7}$$

A third product operation, available in MATLAB,[1] is the *array product* operation defined as follows. The result of an array product of two vectors with N elements is another vector with N elements.

$$\mathbf{a} \cdot *\mathbf{b} = [a_1 b_1 \ a_2 b_2 \cdots a_N b_N]^{\mathrm{T}} \tag{2.8}$$

If $\mathbf{A}$ and $\mathbf{B}$ are two matrices of the same size, then $\mathbf{A} + \mathbf{B} = \{a_{ij} + b_{ij}\}$. If $\mathbf{A}$ is an $L \times M$ matrix and $\mathbf{B}$ is an $M \times N$ matrix, then their $L \times N$ matrix product $\mathbf{C} = \mathbf{AB}$ has elements given by

$$c_{ij} = \left\{ \sum_{k=1}^{M} a_{ik} b_{kj} \right\}, 1 \le i \le L, \ 1 \le j \le N \tag{2.9}$$

The number of columns in $\mathbf{A}$ must be equal to the number of rows in $\mathbf{B}$ for the matrix multiplication $\mathbf{AB}$ to be valid. The product matrix $\mathbf{C}$ will have the same number of rows as $\mathbf{A}$ and same number of columns as $\mathbf{B}$. Since a vector can be treated as a matrix with only one column (or row), the multiplication of a matrix with a vector is a special case of Eq. (2.9). In general, matrix multiplication does not commute (i.e., $\mathbf{AB}$ and $\mathbf{BA}$ are not usually equal). In fact, $\mathbf{BA}$ may not even be defined even though $\mathbf{AB}$ is defined.

Just as in the case of vectors, we can define an array product between two matrices $\mathbf{A}$ and $\mathbf{B}$ of the same dimensions. The resulting matrix $\mathbf{C}$ will be of the same dimensions, but the elements of $\mathbf{C}$ will be the products of corresponding elements of $\mathbf{A}$ and $\mathbf{B}$. Another matrix–matrix product is the Kronecker product where multiplying an $M \times N$ matrix by a $K \times L$ matrix results in an $MK \times NL$ matrix. We will not discuss that product any further as we will not need it.

2.2.1 Vector norms and the Cauchy–Schwarz inequality

The *norm* serves as a generalized measure of length of a vector. In general, the p-norm of a vector $\mathbf{x}$ is defined as

---

[1] MATLAB is a registered trademark of MathWorks Inc.

$$\|\mathbf{x}\|_p = \left[\sum_{i=1}^{N}|x_i|^p\right]^{\frac{1}{p}}, \ p \geq 1 \tag{2.10}$$

where p is an integer, and $|\cdot|$ indicates the absolute value of the scalar elements. For $p = 1$, the norm is the sum of absolute values for $p = 2$, it is the well-known Euclidean distance; and for p approaching infinity, it is the maximum value. The *unit* vector is one that satisfies $\|\mathbf{x}\|_p = 1$. The *Cauchy–Schwarz* inequality states that for two vectors $\mathbf{x}$ and $\mathbf{y}$

$$\|\mathbf{x}^+\mathbf{y}\| \leq \|\mathbf{x}\|_2\|\mathbf{y}\|_2 \tag{2.11}$$

with equality holding if and only if $\mathbf{x} = \alpha\mathbf{y}$ where α is any complex constant. The Cauchy–Schwarz inequality also leads to the following inequality [13] where $\mathbf{x}$ and $\mathbf{y}$ are N-dimensional column vectors and $\mathbf{R}$ is an $N \times N$ Grammian matrix, i.e., $\mathbf{R} = \mathbf{A}^\mathsf{T}\mathbf{A}$ for some matrix $\mathbf{A}$.

$$\left(\mathbf{x}^\mathsf{T}\mathbf{R}\mathbf{y}\right)^2 \leq \left(\mathbf{x}^\mathsf{T}\mathbf{R}\mathbf{x}\right)\left(\mathbf{y}^\mathsf{T}\mathbf{R}\mathbf{y}\right) \text{with equality if } \mathbf{x} = \alpha\mathbf{y} \tag{2.12}$$

and

$$\left(\mathbf{x}^\mathsf{T}\mathbf{y}\right)^2 \leq \left(\mathbf{x}^\mathsf{T}\mathbf{R}\mathbf{x}\right)\left(\mathbf{y}^\mathsf{T}\mathbf{R}^{-1}\mathbf{y}\right) \text{ with equality if } \mathbf{x} = \alpha\mathbf{R}^{-1}\mathbf{y} \tag{2.13}$$

The Cauchy–Schwarz inequality is a special case of the *Hölder* inequality [13], which states that if $\mathbf{x}$ and $\mathbf{y}$ contain real, positive elements, then

$$\sum_i x_i y_i \leq \|\mathbf{x}\|_p\|\mathbf{y}\|_q \quad \frac{1}{p} + \frac{1}{q} = 1, \ p > 1 \tag{2.14}$$

with equality if and only if $y_i \propto x_i^{p-1}$. The figures of merit frequently used in this book are based on quadratic performance criteria such as the mean squared error, and correspond to the case where $p = 2$. Therefore, unless indicated otherwise, we will assume the 2-norm for the rest of this book and drop the subscript p for the sake of simplicity.

2.2.2 Linear independence, rank, matrix inverse and determinant

Two vectors $\mathbf{a}$ and $\mathbf{b}$ are said to be *linearly independent* if the only solution to $\alpha\mathbf{a} + \beta\mathbf{b} = \mathbf{0}$ is $\alpha = \beta = 0$. In general, n vectors $\{\mathbf{x}_1, \mathbf{x}_2, \ldots, \mathbf{x}_n\}$ are said to be linearly independent if the only solution to $\sum_{k=1}^{n} \alpha_k\mathbf{x}_k = \mathbf{0}$ is $\alpha_1 = \alpha_2 = \ldots \alpha_n = 0$. If at least one α_k is non-zero, then the n vectors are linearly dependent. The *rank* of a matrix is the number of linearly independent rows (or equivalently, the number of linearly independent columns). Thus a matrix and its transpose have the same rank. If **A** and **B** are square matrices of size $n \times n$ with ranks r_A and r_B respectively, then the rank of their product **AB** can be bounded as follows:

$$\min(r_A, r_B) \geq \text{rank}(\mathbf{AB}) \geq r_A + r_B - n \qquad (2.15)$$

A square matrix is said to be of *full rank* if its rank is equal to the number of its rows (or columns). If **A** is a square matrix of full rank, then **A** is also called *non-singular*, and a unique matrix $\mathbf{A}^{-1}$ (known as its *inverse*) exists such that

$$\mathbf{A}^{-1}\mathbf{A} = \mathbf{A}\mathbf{A}^{-1} = \mathbf{I} \qquad (2.16)$$

The matrix inverse can be used to express the solutions to a set of linear equations. Consider the following system of N variables x_j and M linear equations:

$$
\begin{aligned}
y_i &= a_{11}x_1 + a_{12}x_2 + \cdots + a_{1N}x_N \\
y_2 &= a_{21}x_1 + a_{22}x_2 + \cdots + a_{2N}x_N \\
&\vdots \\
y_M &= a_{M1}x_1 + a_{M2}x_2 + \cdots + a_{MN}x_N
\end{aligned}
\qquad (2.17)
$$

where a_{ij}, $1 \leq i \leq M$, $1 \leq j \leq N$ are scalars. In matrix–vector notation this is equivalent to $\mathbf{y} = \mathbf{Ax}$, where $\mathbf{y} = \{y_i\}$, $\mathbf{x} = \{x_i\}$, and $\mathbf{A} = \{a_{ij}\}$. Matrix **A** and vector **y** are usually known or measurable, and we desire to solve the system of equations in Eq. (2.17) to obtain **x**. For the case when $M = N$ and **A** is a full-rank matrix, the solution is simply given by $\mathbf{x} = \mathbf{A}^{-1}\mathbf{y}$. We will discuss more about finding solutions to a system of equations in Section 2.4.2.

One way to determine $\mathbf{A}^{-1}$ requires the *determinant* of **A**. For a one-by-one matrix (i.e., a scalar) the determinant is simply that scalar. For matrices larger than one-by-one, the definition for the determinant is as follows:

$$\det(\mathbf{A}) = \sum_{j=1}^{N} (-1)^{j+1} a_{1j} \det(\mathbf{A}_{1j}) \qquad (2.18)$$

where $\mathbf{A}_{1j}$ is the $(N-1) \times (N-1)$ matrix obtained by deleting the first row and jth column of **A**. The determinant for the Vandermonde matrix in Eq. (2.5) is given by $\prod_{i<j} (v_i - v_j)$.

Another useful parameter for a square matrix is its *trace*, which is nothing but the sum of all its diagonal elements. Some useful rules concerning matrix transpose, trace, and inverse operations are as follows:

$$(\mathbf{AB})^\mathrm{T} = \mathbf{B}^\mathrm{T}\mathbf{A}^\mathrm{T} \tag{2.19}$$

$$\det(\mathbf{AB}) = \det(\mathbf{A})\det(\mathbf{B}) \tag{2.20}$$

$$\det(\mathbf{A}^\mathrm{T}) = \det(\mathbf{A}) \tag{2.21}$$

$$\left(\mathbf{A}^{-1}\right)^\mathrm{T} = \left(\mathbf{A}^\mathrm{T}\right)^{-1} \tag{2.22}$$

$$\mathrm{tr}(\mathbf{AB}) = \mathrm{tr}(\mathbf{BA}) \tag{2.23}$$

$$\mathrm{tr}(\mathbf{S}^{-1}\mathbf{AS}) = \mathrm{tr}(\mathbf{A}) \tag{2.24}$$

$$\mathbf{x}^\mathrm{T}\mathbf{Ax} = \mathrm{tr}(\mathbf{Axx}^\mathrm{T}) \tag{2.25}$$

The matrix $\mathbf{A}$ is called *unitary* if its conjugate transpose is equal to its inverse, i.e.,

$$\mathbf{A}^{-1} = \mathbf{A}^+ \Rightarrow \mathbf{AA}^+ = \mathbf{A}^+\mathbf{A} = \mathbf{I} \tag{2.26}$$

$\mathbf{A}$ is called *idempotent* if $\mathbf{A}^2 = \mathbf{A}$.

The matrix $\mathbf{A}$ is *orthogonal* if the inverse is equal to its transpose, i.e.,

$$\mathbf{A}^{-1} = \mathbf{A}^\mathrm{T} \Rightarrow \mathbf{AA}^\mathrm{T} = \mathbf{A}^\mathrm{T}\mathbf{A} = \mathbf{I} \tag{2.27}$$

2.2.3 Partitioned matrices

Often, we may find it convenient to express a large matrix in terms of its submatrices. Suppose that matrix $\mathbf{A}$ is partitioned as $\mathbf{A} = \begin{bmatrix} \mathbf{P} & \mathbf{Q} \\ \mathbf{R} & \mathbf{S} \end{bmatrix}$, where $\mathbf{P}$ and $\mathbf{S}$ are square matrices and $\mathbf{P}$ is non-singular. Then the determinant of $\mathbf{A}$ can be obtained from the determinant of smaller matrices as follows:

$$|\mathbf{A}| = \begin{vmatrix} \mathbf{P} & \mathbf{Q} \\ \mathbf{R} & \mathbf{S} \end{vmatrix} = |\mathbf{P}||\mathbf{S} - \mathbf{RP}^{-1}\mathbf{Q}| \tag{2.28}$$

Similarly, we can express the inverse of a partitioned symmetric matrix $\mathbf{A}$ as follows [13]:

$$\mathbf{A}^{-1} = \begin{bmatrix} \mathbf{P} & \mathbf{Q} \\ \mathbf{Q}^\mathrm{T} & \mathbf{S} \end{bmatrix}^{-1} = \begin{bmatrix} \mathbf{P}^{-1} + \mathbf{FE}^{-1}\mathbf{F}^\mathrm{T} & -\mathbf{FE}^{-1} \\ -\mathbf{E}^{-1}\mathbf{F}^\mathrm{T} & \mathbf{E}^{-1} \end{bmatrix} \tag{2.29}$$

where $\mathbf{P}$ and $\mathbf{S}$ are obviously symmetric matrices and where $\mathbf{E} = \mathbf{S} - \mathbf{Q}^T\mathbf{P}^{-1}\mathbf{Q}$ and $\mathbf{F} = \mathbf{P}^{-1}\mathbf{Q}$. Equation (2.29) shows how to determine the inverse of a larger matrix $\mathbf{A}$ in terms of the inverses of smaller matrices, namely $\mathbf{P}$ and $\mathbf{E}$. A special case of interest is when $\mathbf{A}$ is an $n \times n$ matrix and $\mathbf{P}$ is of size $(n-1) \times (n-1)$. Then, $\mathbf{Q}$ in Eq. (2.29) is actually a column vector $\mathbf{q}$, and $\mathbf{S}$ is actually a scalar s. Then $\mathbf{E}$ will equal the scalar $e = s - \mathbf{q}^T\mathbf{P}^{-1}\mathbf{q}$, and $\mathbf{F}$ becomes column vector $\mathbf{f} = \mathbf{P}^{-1}\mathbf{q}$. Then the matrix inverse in Eq. (2.29) can be written as follows:

$$\mathbf{A}^{-1} = \begin{bmatrix} \mathbf{P} & \mathbf{q} \\ \mathbf{q}^T & s \end{bmatrix}^{-1} = \begin{bmatrix} \mathbf{P}^{-1} + 1/e\,\mathbf{f}\mathbf{f}^T & -1/e\,\mathbf{f} \\ -1/e\,\mathbf{f}^T & 1/e \end{bmatrix} = \begin{bmatrix} \mathbf{P}^{-1} & \mathbf{0} \\ \mathbf{0}^T & 0 \end{bmatrix} + \frac{1}{e}\begin{bmatrix} \mathbf{f}\mathbf{f}^T & -\mathbf{f} \\ -\mathbf{f}^T & 1 \end{bmatrix}$$

(2.30)

Sometimes, a non-singular matrix $\mathbf{A}$ of interest (e.g., a correlation matrix to be defined later) is the sum of outer products of n column vectors and we may have determined its inverse $\mathbf{A}^{-1}$. Suppose we get one more observation of vector $\mathbf{x}$ and want to determine the inverse $(\mathbf{A} + \mathbf{x}\mathbf{x}^T)^{-1}$. The following lemma [13] enables us to determine $(\mathbf{A} + \mathbf{x}\mathbf{x}^T)^{-1}$ from $\mathbf{A}^{-1}$:

$$\left(\mathbf{A} + \mathbf{x}\mathbf{x}^T\right)^{-1} = \mathbf{A}^{-1} - \frac{\left(\mathbf{A}^{-1}\mathbf{x}\right)\left(\mathbf{x}^T\mathbf{A}^{-1}\right)}{1 + \mathbf{x}^T\mathbf{A}^{-1}\mathbf{x}}$$

(2.31)

Another matrix inversion lemma of potential use is the following, where $\mathbf{A}$, $\mathbf{B}$, $\mathbf{C}$, and $\mathbf{D}$ are real matrices of size $n \times n$, $n \times m$, $m \times m$ and $m \times n$, respectively:

$$(\mathbf{A} + \mathbf{B}\mathbf{C}\mathbf{D})^{-1} = \mathbf{A}^{-1} - \mathbf{A}^{-1}\mathbf{B}\left(\mathbf{C}^{-1} + \mathbf{D}\mathbf{A}^{-1}\mathbf{B}\right)^{-1}\mathbf{D}\mathbf{A}^{-1}$$

(2.32)

2.3 Eigenvalues and eigenvectors

An eigenvector $\mathbf{q}$ of a matrix $\mathbf{A}$ is such that it changes only in length, but not in direction, when multiplied by the matrix $\mathbf{A}$. Thus if $\mathbf{q}$ and λ are an eigenvector and corresponding eigenvalue of $\mathbf{A}$, respectively, then

$$\mathbf{A}\mathbf{q} = \lambda\mathbf{q}$$

(2.33)

which can also be expressed as

$$(\mathbf{A} - \lambda\mathbf{I})\mathbf{q} = \mathbf{0}$$

(2.34)

where $\mathbf{0}$ is a vector with all zeros. If matrix $(\mathbf{A} - \lambda\mathbf{I})$ was invertible, then the only solution to Eq. (2.34) would be $\mathbf{q} = \mathbf{0}$. For a non-trivial solution of Eq. (2.34) to exist, $(\mathbf{A} - \lambda\mathbf{I})$ must be singular, i.e.,

$$|\mathbf{A} - \lambda\mathbf{I}| = 0$$

(2.35)

Expanding the determinant in Eq. (2.35) yields a polynomial (of degree less than or equal to N when $\mathbf{A}$ is an $N \times N$ matrix) in λ, and that polynomial is known as the *characteristic polynomial* of $\mathbf{A}$. The roots of the characteristic polynomial are the eigenvalues λ_i, which can be back-substituted into Eq. (2.34) to obtain the corresponding eigenvectors $\mathbf{q}_i$. Note from Eq. (2.33) that if $\mathbf{q}$ is an eigenvector of $\mathbf{A}$, then $\alpha \mathbf{q}$ is also an eigenvector of $\mathbf{A}$ when α is a scalar. Thus, eigenvectors are not unique. But unless specifically stated otherwise, we will assume that the eigenvectors are normalized to have unit norm, i.e., $\mathbf{q}^+\mathbf{q} = 1$. The number of non-zero eigenvalues equals the rank of the matrix.

2.3.1 Some properties of eigenvalues and eigenvectors of real, symmetric matrices

In general, the eigenvalues of a matrix may be complex. The *trace* of a square matrix equals the sum of its eigenvalues, and its determinant is equal to the product of its eigenvalues; i.e.

$$\mathrm{tr}(\mathbf{A}) = \sum_{i=1}^{N} a_{ii} = \sum_{i=1}^{N} \lambda_i \tag{2.36}$$

$$\det(\mathbf{A}) = |\mathbf{A}| = \prod_{i=1}^{N} \lambda_i \tag{2.37}$$

The eigenvalues of $\mathbf{A}^{-1}$ are the reciprocals of the eigenvalues of $\mathbf{A}$, but the eigenvectors are the same. Similarly, the eigenvalues of $\mathbf{A}^2$ are the squares of the eigenvalues of $\mathbf{A}$, while the eigenvectors are the same.

We will often be using covariance matrices which are real and symmetric. Real symmetric matrices have several properties that are of use in some correlation filter designs. If $\mathbf{A}$ is real and symmetric, then its eigenvalues λ_i and eigenvectors $\mathbf{q}_i$ can be chosen to be real, i.e., the imaginary part of the eigenvectors of a real symmetric matrix can be chosen to be zero. When normalized to unit length (as is typically done), the eigenvectors form an *orthonormal* set such that

$$\mathbf{q}_i^{\mathrm{T}}\mathbf{q}_j = \begin{cases} 1 & i = j \\ 0 & i \neq j \end{cases} \tag{2.38}$$

An $N \times N$ orthogonal matrix $\mathbf{Q}$ can be obtained by using the eigenvectors as its columns, i.e., $\mathbf{Q} = [\mathbf{q}_1 \ \mathbf{q}_2 \ \ldots \ \mathbf{q}_N]$. It follows that $\mathbf{Q}^{\mathrm{T}}\mathbf{Q} = \mathbf{I}$ and, since $\mathbf{Q}$ has N linearly independent columns, it is invertible and thus $\mathbf{Q}^{-1} = \mathbf{Q}^{\mathrm{T}}$. Also, Eq. (2.33) can be expressed as follows:

$$\mathbf{AQ} = \mathbf{Q\Lambda} \tag{2.39}$$

where $\mathbf{\Lambda}$ is the diagonal matrix with eigenvalues along its diagonal, i.e.,

$$\mathbf{\Lambda} = \begin{bmatrix} \lambda_1 & & & 0 \\ & \lambda_2 & & \\ & & \ddots & \\ 0 & & & \lambda_N \end{bmatrix} \tag{2.40}$$

Pre-multiplying Eq. (2.39) by $\mathbf{Q}^T$ leads to the following diagonalization result:

$$\mathbf{Q}^T\mathbf{AQ} = \mathbf{\Lambda} \tag{2.41}$$

Eq. (2.41) shows a method to diagonalize a real, symmetric matrix. Post-multiplying Eq. (2.39) by $\mathbf{Q}^T$ leads to the following result known as the spectral decomposition theorem:

$$\mathbf{A} = \mathbf{Q\Lambda Q}^T = \sum_{i=1}^N \lambda_i \mathbf{q}_i \mathbf{q}_i^T \tag{2.42}$$

Spectral decomposition theory applies only if the eigenvectors are linearly independent, which is the case for real symmetric matrices. If the eigenvalues of a symmetric matrix $\mathbf{A}$ are zero or positive, then $\mathbf{x}^T\mathbf{Ax} \geq 0$ for all $\mathbf{x} \neq \mathbf{0}$, and $\mathbf{A}$ is called a *positive semi-definite* matrix. If the eigenvalues are strictly positive, then $\mathbf{x}^T\mathbf{Ax} > 0$ for all $\mathbf{x} \neq \mathbf{0}$, and $\mathbf{A}$ is called a *positive definite* matrix. We can similarly define negative definite and negative semi-definite matrices.

2.3.2 Relationship between the eigenvalues and eigenvectors of the inner product matrices and outer product matrices

Consider N real, linearly independent column vectors $\mathbf{x}_1, \mathbf{x}_2, \ldots, \mathbf{x}_N$ each with d elements. The inner product and outer product matrices of this set of vectors are real and symmetric, and the special properties discussed in Section 2.3.1 hold. For special cases where an inner product or outer product matrix may contain complex elements, we will have to deal with its real part and imaginary part separately. When $d \gg N$, it may be more convenient to estimate the eigenvalues and eigenvectors of the large $d \times d$ outer product matrix using the smaller $N \times N$ inner product matrix. This result is useful in an image recognition approach known as the principal component analysis (PCA) method [14, 15].

The $d \times d$ outer product matrix of rank N is given by $\mathbf{M} = \mathbf{X}\mathbf{X}^T$ where $\mathbf{X} = [\mathbf{x}_1 \ \mathbf{x}_2 \ \ldots \ \mathbf{x}_N]$ is a matrix of size $d \times N$. For example, columns of $\mathbf{X}$ may represent N images, with each image of d pixels in it. Clearly, $\mathbf{M}$ is not a full-rank matrix since N is smaller than d. In fact, $\mathbf{M}$ has at most N non-zero and remaining (i.e., at least $d - N$) zero eigenvalues. On the other hand, the smaller full-rank inner product matrix of size $N \times N$ is given by $\mathbf{V} = \mathbf{X}^T\mathbf{X}$ and can be factored as follows:

$$\mathbf{V} = \mathbf{X}^T\mathbf{X} = \mathbf{Q}\boldsymbol{\Lambda}\mathbf{Q}^T \tag{2.43}$$

where $\mathbf{Q}$ is the matrix of eigenvectors and the diagonal elements of $\boldsymbol{\Lambda}$ are the associated eigenvalues of $\mathbf{V}$. Both $\mathbf{Q}$ and $\boldsymbol{\Lambda}$ are of size $N \times N$ and can be easily determined.

To obtain the eigenvectors and eigenvalues of $\mathbf{M}$, we square both sides of Eq. (2.43) and use the fact that $\mathbf{Q}^T\mathbf{Q} = \mathbf{Q}\mathbf{Q}^T = \mathbf{I}$ to obtain

$$(\mathbf{X}^T\mathbf{X})(\mathbf{X}^T\mathbf{X}) = \mathbf{Q}\boldsymbol{\Lambda}^2\mathbf{Q}^T \tag{2.44}$$

Pre-multiplying by $\boldsymbol{\Lambda}^{-1/2}\mathbf{Q}^T$ and post-multiplying by $\mathbf{Q}\boldsymbol{\Lambda}^{-1/2}$ both sides of Eq. (2.44) (where $\boldsymbol{\Lambda}^{-1/2}$ denotes a diagonal matrix with $1/\sqrt{\lambda_i}$ as the ith diagonal element, which is valid since λ_i is real and positive for positive definite matrices) yields

$$\boldsymbol{\Lambda}^{-1/2}\mathbf{Q}^T\mathbf{X}^T(\mathbf{M})\mathbf{X}\mathbf{Q}\boldsymbol{\Lambda}^{-1/2} = \boldsymbol{\Lambda} \tag{2.45}$$

Setting

$$\bar{\mathbf{Q}} = \mathbf{X}\mathbf{Q}\boldsymbol{\Lambda}^{-1/2} \tag{2.46}$$

we see that $\bar{\mathbf{Q}}^T\mathbf{M}\bar{\mathbf{Q}} = \boldsymbol{\Lambda}$. Using Eq. (2.46), it can be shown that $\bar{\mathbf{Q}}^T\bar{\mathbf{Q}} = \mathbf{I}$ and that the columns of $\bar{\mathbf{Q}}$ form an orthonormal set. Thus, $\bar{\mathbf{Q}}^T\mathbf{M}\bar{\mathbf{Q}} = \boldsymbol{\Lambda}$ is equivalent to

$$\mathbf{M} = \bar{\mathbf{Q}}\boldsymbol{\Lambda}\bar{\mathbf{Q}}^T \tag{2.47}$$

This allows Eq. (2.47) to be expressed as follows:

$$\mathbf{M}\bar{\mathbf{Q}} = \bar{\mathbf{Q}}\boldsymbol{\Lambda} \tag{2.48}$$

proving that $\bar{\mathbf{Q}}$ is the eigenvector matrix for $\mathbf{M}$. It also follows that the eigenvalues of $\mathbf{M}$ are the diagonal elements of $\boldsymbol{\Lambda}$, and hence the same as the eigenvalues of $\mathbf{V}$. The eigenvector matrix $\bar{\mathbf{Q}}$ of the larger outer product matrix $\mathbf{M}$ can be obtained from the eigenvector matrix $\mathbf{Q}$ of the smaller inner product matrix $\mathbf{V}$ using Eq. (2.46) [15].

In this section, we provided a brief summary of some properties and results from matrix-vector theory relevant to correlation pattern recognition. More detailed treatments on this subject are available [16–18].

2.4 Quadratic criterion optimization

It is often convenient to express the gradient of a function of many variables using vector notation. Consider a scalar function $y = f(x_1, x_2, \ldots, x_N)$ of N independent variables. This may be also expressed as $y = f(\mathbf{x})$ where $\mathbf{x}$ is a vector of size N with elements x_i. The first order *total differential* of y is given by

$$dy = f_1 dx_1 + f_2 dx_2 + \cdots + f_N dx_N \tag{2.49}$$

where dx_i is the incremental change (differential) in x_i, and $f_i = \partial y / \partial x_i$ is the partial derivative of y with respect to x_i. Denoting the vector of partial derivatives by

$$\mathbf{f} = \frac{\partial \mathbf{y}}{\partial \mathbf{x}} = \begin{bmatrix} f_1 \\ f_2 \\ \vdots \\ f_N \end{bmatrix} \tag{2.50}$$

and the vector of differentials by

$$d\mathbf{x} = \begin{bmatrix} dx_1 \\ dx_2 \\ \vdots \\ dx_N \end{bmatrix} \tag{2.51}$$

the differential of y can be expressed as $dy = \mathbf{f}^T d\mathbf{x}$, or $dy / d\mathbf{x} = \mathbf{f}$. The maxima or minima of y occur when all the partial derivatives are zero, i.e., $\mathbf{f} = \mathbf{0}$.

2.4.1 Derivatives of linear and quadratic functions

If y is a weighted sum of variables, $y = \mathbf{a}^T \mathbf{x}$, $dy/d\mathbf{x} = \mathbf{a}$. The general quadratic form for y is $y = \sum_{i=1}^{N} \sum_{j=1}^{N} a_{ij} x_i x_j$, where a_{ij} are real scalars. This can also be expressed in matrix vector notation as $y = \mathbf{x}^T \mathbf{A} \mathbf{x}$ where $\mathbf{A} = \{a_{ij}\}$ is an $N \times N$ matrix of weights. Again, by determining partial derivatives of y with respect to each variable, it can be shown that $dy/d\mathbf{x} = (\mathbf{A} + \mathbf{A}^T)\mathbf{x}$.[2] If $\mathbf{A}$ is symmetric, $dy/d\mathbf{x} = 2\mathbf{A}\mathbf{x}$.

---

[2] It is assumed that $\mathbf{x}$ is a real vector.

2.4.2 *System of linear equations and least squares*

If the number of equations in the system $\mathbf{A}\mathbf{x} = \mathbf{y}$ is less than the number of unknowns (i.e., $M < N$) we have an *under-determined* system for which there are an infinite number of solutions of the form

$$\mathbf{x} = \mathbf{A}^{\mathrm{T}}\left(\mathbf{A}\mathbf{A}^{\mathrm{T}}\right)^{-1}\mathbf{y} + \left[\mathbf{I} - \mathbf{A}^{\mathrm{T}}\left(\mathbf{A}\mathbf{A}^{\mathrm{T}}\right)^{-1}\mathbf{A}\right]\mathbf{z} \tag{2.52}$$

where $\mathbf{z}$ is any vector of length N. The expression for $\mathbf{x}$ in Eq. (2.52) is analogous to the solution for differential equations. The term $\mathbf{A}^{\mathrm{T}}(\mathbf{A}\mathbf{A}^{\mathrm{T}})^{-1}\mathbf{y}$ is the *particular solution* and $\left[\mathbf{I} - \mathbf{A}^{\mathrm{T}}\left(\mathbf{A}\mathbf{A}^{\mathrm{T}}\right)^{-1}\right]\mathbf{z}$ is the *homogeneous* solution. Pre-multiplying both sides of Eq. (2.52) by $\mathbf{A}$ shows that the particular solution yields the desired result ($\mathbf{A}\mathbf{x} = \mathbf{y}$), while the contribution of the homogeneous solution is always $\mathbf{0}$ for any $\mathbf{z}$. Thus, an infinite number of solutions are possible.

The system is said to be *over-determined* when there are more equations than variables, i.e., $M > N$. In general, a unique solution does not exist for this case. However, a *minimum mean square error* (MMSE) solution can be obtained by minimizing the MSE.

$$\mathrm{MSE} = |\mathbf{y} - \mathbf{A}\mathbf{x}|^2 = \mathbf{x}^{\mathrm{T}}\mathbf{A}^{\mathrm{T}}\mathbf{A}\mathbf{x} - 2\mathbf{x}^{\mathrm{T}}\mathbf{A}^{\mathrm{T}}\mathbf{y} + \mathbf{y}^{\mathrm{T}}\mathbf{y} \tag{2.53}$$

where $\mathbf{A}$ is assumed to be real and symmetric, and where $\mathbf{x}$ and $\mathbf{y}$ are assumed to be real. Setting the gradient of the MSE with respect to $\mathbf{x}$ to $\mathbf{0}$ yields

$$2\mathbf{A}^{\mathrm{T}}\mathbf{A}\mathbf{x} - 2\mathbf{A}^{\mathrm{T}}\mathbf{y} = \mathbf{0} \tag{2.54}$$

or

$$\mathbf{x} = \left(\mathbf{A}^{\mathrm{T}}\mathbf{A}\right)^{-1}\mathbf{A}^{\mathrm{T}}\mathbf{y} \tag{2.55}$$

The solution in Eq. (2.55) is sometimes denoted as $\mathbf{x} = \mathbf{A}^{\dagger}\mathbf{y}$ where the pseudo-inverse $\mathbf{A}^{\dagger} = (\mathbf{A}^{\mathrm{T}}\mathbf{A})^{-1}\mathbf{A}^{\mathrm{T}}$ satisfies $\mathbf{A}^{\dagger}\mathbf{A} = \mathbf{I}$, but $\mathbf{A}\mathbf{A}^{\dagger}$ is not necessarily equal to $\mathbf{I}$.

2.4.3 *Constrained optimization with Lagrange multipliers*

The method of Lagrange multipliers is useful for minimizing a quadratic function subject to a set of linear constraints. Suppose that $\mathbf{B} = [\mathbf{b}_1 \ \mathbf{b}_2 \ \dots \ \mathbf{b}_M]$ is an $N \times M$ matrix with vectors $\mathbf{b}_i$ of length N as its columns, and $\mathbf{c} = [c_1 \ c_2 \ \dots \ c_M]^{\mathrm{T}}$ is a vector of M constants. We wish to determine the real vector $\mathbf{x}$ which minimizes the quadratic term $\mathbf{x}^{\mathrm{T}}\mathbf{A}\mathbf{x}$ while satisfying the linear equations $\mathbf{B}^{\mathrm{T}}\mathbf{x} = \mathbf{c}$. Towards this end, we form the functional

$$\Phi = \mathbf{x}^T\mathbf{A}\mathbf{x} - 2\lambda_1\left(\mathbf{b}_1^T\mathbf{x} - c_1\right) - 2\lambda_2\left(\mathbf{b}_2^T\mathbf{x} - c_2\right) - \cdots - 2\lambda_M\left(\mathbf{b}_M^T\mathbf{x} - c_M\right) \quad (2.56)$$

where the scalar parameters $\lambda^1, \lambda^2, \ldots, \lambda_M$ are known as the Lagrange multipliers. These multipliers allow us to convert a constrained extremum problem into an unconstrained extremum problem. Setting the gradient of Φ with respect to $\mathbf{x}$ to zero yields

$$2\mathbf{A}\mathbf{x} - 2(\lambda_1\mathbf{b}_1 + \lambda_2\mathbf{b}_2 + \cdots + \lambda_M\mathbf{b}_M) = \mathbf{0} \quad (2.57)$$

Defining $\mathbf{m} = [\,\lambda_1\ \lambda_2 \cdots \lambda_M\,]^T$, Eq. (2.57) can be expressed as

$$\mathbf{A}\mathbf{x} - \mathbf{B}\mathbf{m} = \mathbf{0} \quad (2.58)$$

or

$$\mathbf{x} = \mathbf{A}^{-1}\mathbf{B}\mathbf{m} \quad (2.59)$$

Substituting Eq. (2.59) for $\mathbf{x}$ into the constraint equation $\mathbf{B}^T\mathbf{x} = \mathbf{c}$ yields

$$\mathbf{B}^T\mathbf{A}^{-1}\mathbf{B}\mathbf{m} = \mathbf{c} \quad (2.60)$$

The Lagrange multiplier vector $\mathbf{m}$ can now be obtained as

$$\mathbf{m} = \left(\mathbf{B}^T\mathbf{A}^{-1}\mathbf{B}\right)^{-1}\mathbf{c} \quad (2.61)$$

Using Eqs. (2.59) and (2.61), we obtain the following solution to the constrained optimization problem.

$$\mathbf{x} = \mathbf{A}^{-1}\mathbf{B}\left(\mathbf{B}^T\mathbf{A}^{-1}\mathbf{B}\right)^{-1}\mathbf{c} \quad (2.62)$$

2.4.4 Maximizing a ratio of two quadratic terms

Another useful figure of merit is the ratio of two quadratic criteria in which we want the numerator to be large while the denominator should be small. For example, the numerator may represent signal power whereas the denominator may characterize the noise power. Suppose we wish to find $\mathbf{h}$ maximizing the following ratio where matrices $\mathbf{A}$ and $\mathbf{B}$ are assumed to be real and symmetric:

$$J(\mathbf{h}) = \frac{\mathbf{h}^+\mathbf{A}\mathbf{h}}{\mathbf{h}^+\mathbf{B}\mathbf{h}} \quad (2.63)$$

The ratio in Eq. (2.63) is known as the Rayleigh quotient, and to maximize $J(\mathbf{h})$ with respect to $\mathbf{h}$, we set the gradient $\nabla_{\mathbf{h}}J(\mathbf{h})$ to zero as shown below:

$$\nabla_{\mathbf{h}} J(\mathbf{h}) = \nabla_{\mathbf{h}} \left(\frac{\mathbf{h}^+ \mathbf{A} \mathbf{h}}{\mathbf{h}^+ \mathbf{B} \mathbf{h}} \right)$$

$$= \frac{2(\mathbf{h}^+ \mathbf{B} \mathbf{h}) \mathbf{A} \mathbf{h} - 2(\mathbf{h}^+ \mathbf{A} \mathbf{h}) \mathbf{B} \mathbf{h}}{(\mathbf{h}^+ \mathbf{B} \mathbf{h})^2} = \mathbf{0} \qquad (2.64)$$

Simple manipulations yield

$$\mathbf{B}^{-1} \mathbf{A} \mathbf{h} = \left(\frac{\mathbf{h}^+ \mathbf{A} \mathbf{h}}{\mathbf{h}^+ \mathbf{B} \mathbf{h}} \right) \mathbf{h} \qquad (2.65)$$

where $\mathbf{B}$ is assumed to be invertible. Since $J(\mathbf{h}) = \mathbf{h}^+ \mathbf{A} \mathbf{h} / \mathbf{h}^+ \mathbf{B} \mathbf{h}$, we see from Eq. (2.65) that $\mathbf{h}$ must be the eigenvector of $\mathbf{B}^{-1} \mathbf{A}$ with the largest eigenvalue.

Strictly speaking, the above derivation works only if $\mathbf{h}$ is assumed to be real. If $\mathbf{h} = \mathbf{h}_R + j\mathbf{h}_I$, then we must set the gradients with respect to both $\mathbf{h}_R$ and $\mathbf{h}_I$ to zero separately. However, assuming that Cauchy–Riemann conditions hold [19], we can set the gradient with respect to $\mathbf{h}$ to zero by treating $\mathbf{h}$ as real and still get the same answer for the solution vector.

2.5 Probability and random variables

Probability is used to describe uncertain events. There is no need to use probabilistic models to characterize events that are certain (e.g., the fact that the Sun will rise in the east). On the other hand, if we roll a die, we do not know a priori (i.e., beforehand) which of the six faces will show up. All we can say is that it is equally likely that any of the six sides will show up. Probability theory is an attempt to quantify such randomness. In this section, we will provide a brief review of probability theory concepts and results that we will find useful in correlation pattern recognition.

2.5.1 *Basics of probability theory*

Let us go back to the example of the six-sided die. This experiment has a random outcome in that the output can be any one of the six numbers. We define an event as a set containing some of these outcomes. For example, event A can be defined as $\{1, 5\}$, and event B might denote {even numbered outputs} (i.e., $\{2, 4, 6\}$). Event S (also known as the *sample set*) denotes the set of all possible outcomes (in this example, the set $\{1, 2, 3, 4, 5, 6\}$) and the null event $\emptyset$ denotes the null set, i.e., the set with no elements in it.

Probability theory provides a framework to define probabilities of such events. Probability is defined as a real, non-negative number associated with every event. The probability of null event $\varnothing$ is zero and the probability of the sample set S is one, because the set S includes all possible outcomes. We also need a method to determine the probabilities of more complicated events such as $\{A \cap B\}$.

Probability theory axioms We will denote the probability of an event A by $P(A)$. Probability theory is based on the following three axioms.

$P(A)$ is a real number in the closed interval [0, 1]
$P(S) = 1$, the probability of the sample set S is 1.
For mutually exclusive sets A and B, $P(A \cup B) = P(A) + P(B)$

In the example of a six-sided die, all six faces have equal probability of appearing in a single roll, i.e., $P(\{k\}) = 1/6$ for $k = 1, 2, 3, 4, 5, 6$. Thus, event A (i.e., $\{1, 5\}$) has probability 2/6 and event B (i.e., {even numbers 2, 4, and 6}) has probability 3/6. Since these two events are mutually exclusive, the third axiom tells us that the probability of $\{A \cup B\}$ is $1/2 + 1/3 = 5/6$.

We will use the easier notation $\{A + B\}$ to denote $\{A \cup B\}$ and $\{AB\}$ to denote $\{A \cap B\}$. Also, $\{\bar{A}\}$ is used to denote the complement of the event $\{A\}$. Using the above three axioms, the following useful properties can be derived:

$$P(\bar{A}) = 1 - P(A); P(A + B) = P(A) + P(B) - P(AB) \qquad (2.66)$$

The above properties can be extended to more than two events by applying them repeatedly.

Conditional probabilities We can define probabilities of events conditioned on other events. Conditioning an event A on another event B reduces the uncertainty and hence increases our knowledge of the event A. We use $P(A|B)$ to denote the probability of event A conditioned on event B. The conditional probability satisfies all the probability properties.

$$P(A|B) = \frac{P(AB)}{P(B)} \qquad (2.67)$$

In the six-sided die example, consider the event C (number greater than or equal to 3, i.e., 3, 4, 5, or 6) as the conditioning event. Then $P(A|C)$ refers to the probability of event A (i.e., $\{1, 5\}$) conditioned on event C (i.e., $\{3, 4, 5, 6\}$). From Eq. (2.67), $P(A|C) = \dfrac{P(AC)}{P(C)} = \dfrac{P(\{5\})}{P(\{3, 4, 5, 6\})} = \dfrac{1/6}{4/6} = \dfrac{1}{4}$, which agrees with our intuitive notion that $P(A|C)$ should be 1/4 since only one of the four outcomes in event C corresponds to event A. The basic conditional probability definition in Eq. (2.67) can be used to derive the following other identities.

$$P(AB) = P(A) \cdot P(B|A) = P(B) \cdot P(A|B)$$

$$P(A) = P(AB) + P(A\bar{B}) = P(B) \cdot P(A|B) + P(\bar{B}) \cdot P(A|\bar{B}) \qquad (2.68)$$

$$P(A|B) = \frac{P(B|A) \cdot P(A)}{P(B)} \quad P(B|A) = \frac{P(A|B) \cdot P(B)}{P(A)}$$

The last line of equalities in Eq. (2.68) is known as Bayes' rule and allows us to connect $P(A|B)$ to $P(B|A)$. Bayes' rule finds significant use in pattern recognition and parameter estimation, among other applications.

Two events A and B are said to be statistically independent if $P(AB) = P(A)P(B)$. For the three events ($A = \{1,5\}$, $B = \{2,4,6\}$, and $C = \{3,4,5,6\}$) we defined for the six-sided die, $P(A) = 1/3$, $P(B) = 1/2$, $P(C) = 2/3$, and $P(AB) = 0$, $P(AC) = 1/6$, $P(BC) = 1/3$. Based on these probabilities, only the pair of events B and C are statistically independent among the three pairs considered. If A and B are statistically independent, then $P(A|B) = P(A)$ and $P(B|A) = P(B)$.

2.5.2 *Random variables*

While probabilities are easy to understand, they are not so easy to apply in many situations. Suppose we want to model the noise present in an image. This noise can take on a continuum of values at each pixel, and we need a more compact representation than enumerating the probabilities for all possible noise values. *RVs* provide such a compact description.

A random variable (RV) X is defined as a mapping from the events in the random experiment to the real line. For example, we can define an RV X as taking on the real value k in our six-sided die experiment where k is the outcome. This RV takes on only discrete values (namely 1, 2, 3, 4, 5, and 6) and is thus called a *discrete* RV. We will use *continuous* RVs to model outcomes such as noise where we can have a continuum of values. Continuous RVs take on real values in an interval or in sets of intervals.

Cumulative distribution functions The probabilities associated with an RV can be expressed using the cumulative distribution function (CDF).

$$F(x) = \Pr\{X \le x\} \qquad (2.69)$$

where x denotes the value that the RV X is taking on. Often, a subscript is used to indicate explicitly a particular RV. Such subscripts will be omitted when the RV is obvious from the context. Since we assume that all six faces of the die are equally likely, the CDF of this RV is as shown in Figure 2.2.

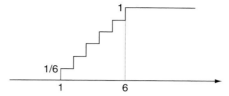

Figure 2.2 Cumulative distribution function (CDF) of the discrete RV modeling the outcome of the rolling of a fair six-sided die

Because of the discrete nature of this RV, its CDF exhibits jumps indicating that those outcomes have a non-zero probability associated with them. In this example, the jumps are of size 1/6 indicating that the probability of each of these outcomes is 1/6. A continuous RV will have a CDF that is continuous, i.e., it has no jumps. The probability that the RV X takes on values in an interval can be easily obtained as

$$\Pr\{x_L < x \le x_R\} = F(x_R) - F(x_L) \tag{2.70}$$

A few other features of a CDF are worth noting. It is non-decreasing, since otherwise we will get negative probabilities. CDF is zero at the left extreme (corresponding to null event $\varnothing$), and equals 1 at the right extreme (corresponding to the sample set S).

2.5.3 Probability density functions

While we can use a CDF to characterize completely an RV, we will find it more convenient to work with the probability density function (PDF), which is the derivative of the CDF, i.e.,

$$f(x) = \frac{\mathrm{d}F(x)}{\mathrm{d}x} \tag{2.71}$$

While the PDF of a continuous RV is well defined, the PDF of a discrete RV contains delta functions owing to the discontinuities in its CDF. Since the PDF is the derivative of the CDF, we can integrate the PDF over an interval to determine the probability that the RV takes on values in an interval, i.e.,

$$\Pr\{x_L < x \le x_R\} = F(x_R) - F(x_L) = \int_{x_L}^{x_R} f(x)\mathrm{d}x \tag{2.72}$$

A few other features of PDFs are worth noting. Since the CDF is a non-decreasing function, its derivative (namely the PDF) is never negative. The total area under any PDF must be 1, since this is the probability that the real-valued RV takes on a value between negative infinity and positive infinity.

Let us look at a few useful PDFs, although there are many more probability density functions that we are not considering. Interested readers should consult some of the excellent probability theory and RVs references [20, 21].

Binomial Consider the tossing of a coin N times. Define the RV X as the number of times heads shows up in these N trials. Clearly, this is a discrete RV taking on values 0, 1, 2, all the way to N. If the probability that heads shows up in a single trial is p and if all the trials are statistically independent, then the probability that X takes on value n is defined by the following binomial distribution:

$$\Pr\{n\} = \binom{N}{n}(p)^n(1-p)^{(N-n)} \quad n = 0, 1, 2, \ldots, N \qquad (2.73)$$

where $\binom{N}{n} = \dfrac{N!}{n!(N-n)!}$

Poisson This is a discrete RV that takes on all non-negative integer values. Consider events that occur at random time instants, such as the arrival of customers at a teller window or the arrival of photons at a photo-detector surface. Let m denote the average number of these events per unit time and let X denote the number of events in a particular unit time. We say that X follows a Poisson distribution provided

$$\Pr\{X = n\} = \frac{m^n}{n!}e^{-m} \quad n = 0, 1, 2, \ldots \qquad (2.74)$$

Uniform Often, all we can say about an RV is that any value in an interval is equally likely. This situation can be modeled by a continuous RV X whose PDF is constant in an interval, and zero outside that interval.

$$f(x) = \begin{cases} 1/(x_R - x_L) & \text{for } x_L \leq x \leq x_R \\ 0 & \text{otherwise} \end{cases} \qquad (2.75)$$

We show the uniform PDF in Figure 2.3(a).

Gaussian A continuous RV X is distributed according to a Gaussian or normal distribution if its PDF is as follows:

$$f(x) = \frac{1}{\sqrt{2\pi\sigma^2}}\exp\left[-\frac{(x-m)^2}{2\sigma^2}\right] \qquad (2.76)$$

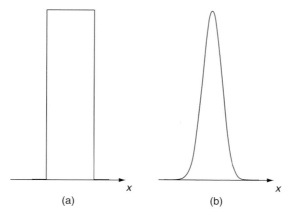

Figure 2.3 (a) Uniform probability density function and (b) Gaussian probability density function

Eq. (2.76) shows that a Gaussian PDF is characterized by two parameters, m (known as its *mean*) and σ^2 (known as its *variance*). The square root of the variance is called the *standard deviation*. We show a Gaussian PDF in Figure 2.3(b). From this figure, we can see that the Gaussian PDF is even symmetric around its mean. While a Gaussian RV can take on any real value, it is more likely to take on values close to its mean m. The smaller the variance, the more narrowly distributed is this set of values around the mean.

We will use Gaussian random variables in several places. The probability of a Gaussian RV taking on values in an interval is obtained by integrating the PDF Eq. (2.76) over that interval. As the needed integral cannot be evaluated in a closed-form manner, it is evaluated numerically and tabulated. The integral of a unit Gaussian (a Gaussian PDF with zero mean and unit variance) is known as the *error function* and many software packages (e.g., MATLAB) contain built-in commands to evaluate the error function (erf).

$$\mathrm{erf}(x) = \frac{2}{\sqrt{\pi}} \int_0^x \mathrm{e}^{-y^2} \mathrm{d}y, \; x \geq 0; \; \mathrm{erf}(\infty) = 1; \; \mathrm{erf}(-x) = -\mathrm{erf}(x);$$

$$\mathrm{erfc}(x) = 1 - \mathrm{erf}(x)$$

(2.77)

where erfc denotes the complementary error function. Other variations of the error function as well as other functions (e.g., $Q(x) \triangleq \frac{1}{\sqrt{2\pi}} \int_x^\infty \mathrm{e}^{-u^2/2} \mathrm{d}u$ is the integral of the unit Gaussian from x to infinity) can be found in the literature, but they can all be related to the error function defined in Eq. (2.77).

Transformation of random variables If we transform the RV X into a new RV Y according to $Y = g(X)$, then the PDF of Y depends on the PDF of X as well as the mapping $g(\cdot)$. We can find the PDF of Y by first finding connection between the CDFs and then taking the derivatives to determine the relation between the two PDFs. Let us first consider the case when the mapping $g(\cdot)$ is one-to-one. Then, the probability that the RV X takes on a value in an elemental interval dx centered at x is the same as the probability that the RV Y takes on a value in an elemental interval dy centered at y, i.e., $f_X(x)|dx| = f_Y(y)|dy|$, given that the sizes of dx and dy are as related through $g(\cdot)$, i.e., $dy = |dg/dx|dx$. From this, the relationship between the two PDFs can be seen to be as follows:

$$f_Y(y) = \left. \frac{f_X(x)}{\left| \frac{dg(x)}{dx} \right|} \right|_{x=g^{-1}(y)} \tag{2.78}$$

We can use this method for other mappings also. For example, if $Y = X^2$, the PDF of Y for negative y values is zero, since the mapping $Y = X^2$ prevents Y from taking on negative values. On the other hand, there are two x values that result in the same positive value of y. When multiple x values lead to the same y value, we add the contributions from each of the x solutions to obtain the complete PDF as follows:

$$Y = X^2 \Rightarrow f_Y(y) = \begin{cases} \frac{f_X(\sqrt{y}) + f_X(-\sqrt{y})}{2\sqrt{y}} & \text{for } y \geq 0 \\ 0 & \text{otherwise} \end{cases} \tag{2.79}$$

Affine transformation Let us consider the affine transformation $Y = aX + b$ where a and b are constants. Using Eq. (2.78) we see that the PDF of Y is related to the PDF of X as follows. (Note: an affine transformation is a linear transform if and only if $b = 0$.)

$$f_Y(y) = \frac{1}{|a|} f_X\left(\frac{y - b}{a} \right) \tag{2.80}$$

If X is a Gaussian RV *with* the PDF given in Eq. (2.76), then Y is also a Gaussian RV as shown below. This is an important result since we will often be interested in what happens to an RV as it passes through a linear system. We now know that Gaussian RVs going into a linear system will come out as Gaussian RVs with perhaps a change in their mean and variance values.

$$f_Y(y) = \frac{1}{|a|} f_X\left(\frac{y-b}{a}\right) = \frac{1}{\sqrt{2\pi a^2 \sigma_x^2}} \exp\left[-\frac{\left(\left(\frac{y-b}{a}\right) - m_x\right)^2}{2\sigma_x^2}\right]$$

$$= \frac{1}{\sqrt{2\pi a^2 \sigma_x^2}} \exp\left[-\frac{(y - [am_x + b])^2}{2a^2 \sigma_x^2}\right]$$

(2.81)

As can be seen from Eq. (2.81), random variable Y is also a Gaussian, but has mean $m_Y = am_X + b$ and variance $\sigma_Y^2 = a^2 \sigma_X^2$. Thus any Gaussian RV X can be mapped into a unit Gaussian (i.e., mean is 0 and variance is 1) RV Y using $Y = aX + b$, with $a = 1/\sigma_X$, and $b = -m_X/\sigma_X$. Thus the probability of a Gaussian RV with arbitrary mean and variance taking on values in an interval can be obtained from the error function (which describes probabilities for the unit Gaussian).

2.5.4 Expectation

Often, we are interested in the average value of an RV X. Such an average value (also known as the *mean m* of the RV) is obtained by multiplying each possible value of the RV by its probability and summing up the results, i.e.,

$$m_X = \int_{-\infty}^{\infty} x f_X(x)\mathrm{d}x = E\{X\}$$

(2.82)

where $E\{X\}$ is known as the *expectation* of X. Expectation can also be defined for functions of X. Thus, we can obtain the expectation of $Y = g(X)$ as follows.

$$m_Y = E\{Y\} = \int_{-\infty}^{\infty} f_Y(y)\mathrm{d}y = E\{g(X)\} = \int_{-\infty}^{\infty} g(x) f_X(x)\mathrm{d}x$$

(2.83)

Eq. (2.83) shows that we don't need to go through the trouble of first performing a transformation of an RV (i.e., $Y = g(X)$) and determining the new PDF of Y to find the average of the RV Y. Instead, we can find the expectation of $g(X)$ using only the PDF of X as in Eq. (2.83).

Moments We earlier defined $E\{X\}$ as the mean of an RV. Higher-order *moments* are defined as the expectation of higher powers of the RV, i.e., the *n*th moment of the RV X is defined as follows:

$$m_n = E\{X^n\} = \int\limits_{-\infty}^{\infty} x^n f(x)\mathrm{d}x \tag{2.84}$$

A smooth PDF can be reconstructed if we know all its moments. But sometimes, all we need are the first two moments (namely the mean and the variance) of the RV. They tell us where the PDF is centered and how broad it is. Also, the popular Gaussian PDF in Eq. (2.76) is completely characterized by its mean m and variance σ^2.

The variance σ^2 of an RV is defined as follows:

$$\begin{aligned}
\sigma^2 = E\left\{(X-m)^2\right\} &= \int (x-m)^2 f(x)\mathrm{d}x \\
&= \int x^2 f(x)\mathrm{d}x - 2m \int x f(x)\mathrm{d}x + m^2 \int f(x)\mathrm{d}x \\
&= E\{X^2\} - 2m^2 + m^2 = E\{X^2\} - m^2
\end{aligned} \tag{2.85}$$

We have omitted the integration limits in the above expression since they are from $-\infty$ to $+\infty$, as should be obvious from the context. Often, mean and variance prove to be adequate descriptors for an RV, although they do not necessarily provide a complete description.

Moments for Gaussian PDF For the Gaussian PDF in Eq. (2.76), we can show that $E\{X\} = m$ and $E\{(X-m)^2\} = \sigma^2$. A Gaussian PDF is one of the few RVs that are completely characterized by their first two moments (namely mean and variance). The Poisson distribution is characterized by a single moment.

Moments for uniform PDF For the uniform PDF in Eq. (2.75), the mean can be easily shown to be $(x_L + x_R)/2$, the mid-point of the interval of support. Its variance can be shown to be $(x_R - x_L)^2/12$. Thus, if we need to simulate a uniform RV with zero mean and unit variance, we must subtract 0.5 from the random numbers uniformly distributed over [0,1] and multiply the results by $\sqrt{12}$.

So far, we have concentrated on a single RV. In practice, we need to consider more than one RV. The next section is devoted to two RVs.

2.5.5 *Two random variables*

In pattern recognition applications we will encounter more than one RV and thus will need tools that can describe and handle multiple RVs. In this sub-section, we will first look at the case of two RVs and introduce the important second-order characterizations such as covariance and correlation coefficient. In Section 2.5.6 we consider multiple random variables, which are best described using a vector formulation. We will review results associated with Gaussian random vectors in some detail as we will be relying on these results. Finally, we will investigate the effects of linear transformations on random vectors.

Suppose we roll two six-sided dice as part of a single random experiment. We can associate two RVs X and Y with that one experiment. X is the number on the first die and Y is the number on the second die. To characterize these two random variables *jointly* (i.e., together), we can use the joint CDF defined below.

$$F(x, y) = \Pr\{X \le x, Y \le y\} \tag{2.86}$$

The joint CDF describes the joint random behavior of the two RVs, not just of each RV by itself, and in that sense is more informative. The joint CDF is non-negative, non-decreasing and must approach 1 as both x and y approach infinity. Similarly, the CDF is zero if either x or y approaches negative infinity.

Joint PDF RVs X and Y can be described using a joint PDF that is related to the joint CDF as below.

$$f(x, y) = \frac{\partial^2 F(x, y)}{\partial x \partial y} \tag{2.87}$$

Since the joint CDF is non-decreasing, the joint PDF must be non-negative. The probability that the two RVs take on values in an area in the (x, y) plane is obtained by integrating the joint PDF (which we will simply refer to as a PDF) over that region. If we integrate the PDF over the entire (x, y) plane, we get 1.

Marginal PDFs We can determine the PDF of just X from knowing the joint PDF. This is done by integrating out the randomness in Y. Resulting density functions are called *marginal* PDFs and we will use subscripts to distinguish marginal PDFs from joint PDFs. Marginal PDFs can be obtained from joint PDFs as below.

$$f_X(x) = \int_{-\infty}^{\infty} f_{X,Y}(x, y) \, \mathrm{d}y \quad f_Y(y) = \int_{-\infty}^{\infty} f_{X,Y}(x, y) \, \mathrm{d}x \tag{2.88}$$

Marginal PDFs describe only what happens to one RV without worrying about the other and in that sense are incomplete. Thus knowing the two marginal

PDFs $f_X(x)$ and $f_Y(y)$ is not same as knowing the joint PDF $f_{X,Y}(x, y)$. An exception however, is when the joint PDF is product separable (often just called separable), i.e.,

$$f_{X,Y}(x, y) = f_X(x) \cdot f_Y(y) \tag{2.89}$$

When the joint PDF is separable as above, the two RVs X and Y are said to be *statistically independent*. Statistical independence is beneficial in many ways. Instead of needing a 2-D joint PDF to describe the associated randomness, all we need are two 1-D marginal PDFs. It also allows us to separate out computations in X and Y. For example, we can determine $E\{X^3 Y^4\}$ as the product of $E\{X^3\}$, the third moment in X, and of $E\{Y^4\}$, the fourth moment in Y. If X and Y are statistically independent, then $E\{f(X)g(Y)\}$ is equal to $E\{f(X)\} \times E\{g(Y)\}$.

Sum of random variables We are often interested in finding the PDF of $Z = (X + Y)$. The CDF of Z is related to the joint PDF of X and Y as follows. By taking the derivative of the CDF of Z with respect to z, we can determine the PDF of Z.

$$F_Z(z) = \Pr\{(X + Y) \le z\} = \int\limits_{-\infty}^{\infty} dy \int\limits_{-\infty}^{(z-y)} dx f_{X,Y}(x, y)$$

$$\Rightarrow f_Z(z) = \frac{dF_Z(z)}{dz} = \int\limits_{-\infty}^{\infty} dy f_{X,Y}(z - y, y) \tag{2.90}$$

For the special case when X and Y are statistically independent, the PDF of their sum is the convolution of the PDFs of X and Y as seen below.

$$f_Z(z) = \int\limits_{-\infty}^{\infty} dy f_{X,Y}(z - y, y) = \int\limits_{-\infty}^{\infty} dy f_X(z - y) f_Y(y) = f_X(z) * f_Y(z) \tag{2.91}$$

where the asterisk denotes the convolution operation, to be discussed in Chapter 3. Thus the PDF of the sum of independent RVs is the convolution of the original PDFs.

Central limit theorem One application of the result in Eq. (2.91) is of particular interest. Suppose Z is the sum of N independent and identically distributed (IID) RVs. By applying Eq. (2.91) repeatedly, we see that the PDF of Z is the N-fold auto-convolution of the PDF of the original RVs. Since convolution is a broadening operation, the result of the N-fold auto-convolution approaches a

Gaussian shape as N increases. The central limit theorem (CLT) states that the CDF of the sum of N IID random variables approaches that of a Gaussian RV as N increases.

The central limit theorem is useful in that it lets us approximate the sums of statistically independent RVs by Gaussian RVs, as long as the number of RVs being summed is large. However, there is no need to invoke CLT if the original RVs themselves were Gaussian RVs. Summing is a linear operation and linear operations on Gaussian RVs lead to Gaussian RVs.

Second-order measures Earlier we defined variance of an RV as its second central (i.e., mean-subtracted) moment. For the case of two RVs X and Y, we can compute the variance for X and Y separately using their marginal PDFs. Unless the two RVs are statistically independent, the two variances do not provide a complete (i.e., joint) second-order characterization. In addition, we need the covariance (cov) defined below.

$$\text{cov}\{X, Y\} = E\{(X - m_X)(Y - m_Y)\} = E\{XY\} - m_X m_Y \qquad (2.92)$$

The covariance describes the joint second-order behavior of the two random variables. If X and Y are statistically independent, we know that $E\{XY\} = E\{X\}E\{Y\} = m_X m_Y$ and, from Eq. (2.92), the covariance is zero. When the covariance of two RVs is zero, they are said to be *uncorrelated*. Thus, statistical independence implies uncorrelatedness. However, uncorrelatedness does not necessarily imply statistical independence as uncorrelatedness deals only with second-order moments whereas statistical independence refers to probability density functions.

A related measure is the *correlation coefficient* (ρ) defined as follows:

$$\rho_{X,Y} = \frac{\text{cov}\{X, Y\}}{\sigma_X \sigma_Y} \qquad (2.93)$$

One can prove that the correlation coefficient can take on values only between -1 and $+1$. If it is zero, the two RVs are uncorrelated. If the correlation coefficient is close to $+1$, then the two RVs track each other closely. If the correlation coefficient is close to -1, the two RVs track each other closely but one is the negative of the other. The larger the absolute value of the correlation coefficient, the more we can tell about one RV from our knowledge of the other RV.

Bivariate Gaussian Two RVs are said to be jointly Gaussian provided that their joint PDF takes on the following form:

$$f_{X,Y}(x,y) = \frac{1}{2\pi\sigma_X\sigma_Y\sqrt{1-\rho^2}}$$

$$\times \exp\left[-\frac{1}{2(1-\rho^2)}\left\{\frac{(x-m_X)^2}{\sigma_X^2} + \frac{(y-m_Y)^2}{\sigma_Y^2} - 2\rho\frac{(x-m_X)(y-m_Y)}{\sigma_X\sigma_Y}\right\}\right]$$

(2.94)

where ρ is the correlation coefficient of X and Y. The bivariate Gaussian requires five parameters (two means, two variances and one correlation coefficient). The marginal PDFs for X and Y can be shown to be univariate Gaussians with the corresponding means and variances.

When the correlation coefficient is zero, the joint PDF in Eq. (2.94) becomes separable indicating that the RVs are statistically independent. Thus, here is an exception to the previous statement that uncorrelatedness does not necessarily imply statistical independence. For jointly Gaussian RVs, statistical independence implies uncorrelatedness, and vice versa. We will later discuss linear transformations that convert correlated RVs into uncorrelated RVs. If we start with Gaussian RVs, such linear transformations will allow us to obtain uncorrelated Gaussian RVs that are also statistically independent.

Conditional PDFs When dealing with two RVs, we can define a conditional PDF as follows, just as we defined a conditional probability before. The symbol $f_{X|Y}(x|y)$ means the PDF of the RV X conditioned on the RV Y, evaluated at x and given that Y took on the value y. Another terminology is that $f_{X|Y}(x|y)$ is the a-posteriori likelihood of X taking on the value x, given that Y was found to have the value y. That is to say, following the measurement of Y we have information on what values of X are likely to occur. Before Y is measured, the PDF for X is said to be the a-priori likelihood.

$$f_{X|Y}(x|y) = \frac{f_{X,Y}(x,y)}{f_Y(y)}$$

(2.95)

Not surprisingly, when X and Y are statistically independent $f_{X|Y}(x|y)=f_X(x)$ and $f_{Y|X}(y|x)=f_Y(y)$. The conditional PDF in Eq. (2.95) is just like any other PDF in that it is non-negative with an area of 1 under the PDF curve, and we can define conditional mean, conditional variance and conditional moments. If X and Y are jointly Gaussian, then their conditional PDFs are also Gaussian.

Bayes' rule Often we wish to reverse the direction of inference afforded by the conditional PDF, and Bayes' rule tells us how to do that. The definition in

Eq. (2.95) allows us to connect conditional PDFs to marginal PDFs as follows. The set of relationships given below are various forms of Bayes' rule that find use in pattern recognition theory.

$$f_{X,Y}(x,y) = f_Y(y)f_{X|Y}(x|y) = f_X(x)f_{Y|X}(y|x)$$

$$\Rightarrow f_{Y|X}(y|x) = \frac{f_Y(y)f_{X|Y}(x|y)}{f_X(x)} = \frac{f_Y(y)f_{X|Y}(x|y)}{\int f_Y(y')f_{X|Y}(x|y')dy'} \qquad (2.96)$$

Suppose X denotes the received signal and Y denotes the transmitted signal. An estimation method known as the maximum a-posteriori method selects the y value that maximizes the a-posteriori PDF $f_{Y|X}(y|x)$ for a given received value y. Bayes' rule lets us express this a-posteriori PDF $f_{Y|X}(y|x)$ in terms of the known conditional PDF $f_{X|Y}(x|y)$ and the marginal a-priori PDF $f_Y(y)$.

Complex random variables Strictly speaking, RVs are real-valued. However, we can construct complex RVs by using one RV for the real part and another for the imaginary part. In particular, a complex RV of specific interest is $Z = X + jY$, where X and Y are statistically independent, zero-mean, Gaussian RVs with the same standard deviation σ. One can show that if the complex Gaussian RV Z is represented by magnitude M and angle θ, (i.e., $Z = Me^{j\theta}$), then M is characterized by a Rayleigh PDF and θ is characterized by a uniform PDF in the interval from $-\pi$ to $+\pi$ and the RVs M and θ are statistically independent. It is also easy to see that the joint PDF $f_{X,Y}(x, y)$ depends only on magnitude $M = \sqrt{X^2 + Y^2}$ and not on angle $\theta = \arctan(Y/X)$, i.e., the complex Gaussian RV Z is circularly symmetric. In fact, if independent RVs X and Y are circularly symmetric (i.e., their joint PDF is a function only of $\sqrt{X^2 + Y^2}$), then they are Gaussian with zero means and equal variances. It is easy to verify that $E\{|Z|^2\} = E\{X^2 + Y^2\} = 2\sigma^2$ whereas $E\{Z^2\} = E\{X^2 - Y^2\} = 0$. Finally, it is useful to realize that the central limit theorem applies to complex RVs also, in the sense that adding many identical and independent complex RVs results in Gaussian complex RVs. This can be seen by applying the central limit theorem to the real part and imaginary part separately. More detailed discussions about complex RVs and complex random processes can be found elsewhere [22].

2.5.6 *Random vectors*

When we want to describe multiple RVs, we will find it more convenient to use vectors. Thus, we can define a random column vector to represent

N RVs by defining $\mathbf{x} = [X_1 \ X_2 \ \cdots \ X_N]^T$ where the superscript T denotes the transpose operation. We denote the joint PDF of the random vector, $f(\mathbf{x})$ and this has all the PDF properties such as being non-negative, and that the total volume under the PDF surface is equal to 1. From this joint PDF, we can derive the marginal PDFs of not only the N component RVs, but also other partial characterizations such as bivariate and trivariate PDFs. We obtain these partial PDFs by integrating out unwanted variables from the joint PDF. These N RVs are statistically independent, provided that we can express the N-variate PDF as the product of N marginal PDFs; i.e.,

$$f(\mathbf{x}) = \prod_{i=1}^{N} f_{X_i}(x_i) \qquad (2.97)$$

Statistical independence will greatly simplify matters by requiring N 1-D functions instead of one N-dimensional function. To see the advantage of independence, suppose we describe each 1-D function by 100 samples and $N = 2$. Then two 1-D PDFs require 200 samples, whereas one 2-D PDF requires $100^2 = 10\,000$ samples. This difference is more dramatic for larger values of N.

Second-order statistics The means of the N RVs can be compactly expressed using the *mean vector* $\mathbf{m}$ defined as:

$$\mathbf{m} = E\{\mathbf{x}\} = [\,E\{X_1\} \quad E\{X_2\} \quad \cdots \quad E\{X_N\}\,]^T \qquad (2.98)$$

The second-order statistics include variances and covariances of the N RVs, which can be captured by the following $N \times N$ *covariance matrix* $\mathbf{C}$.

$$\mathbf{C} = E\left\{ (\mathbf{x} - \mathbf{m})(\mathbf{x} - \mathbf{m})^T \right\}$$

$$= \begin{bmatrix} \sigma_1^2 & \mathrm{cov}\{X_1, X_2\} & \cdots & \mathrm{cov}\{X_1, X_N\} \\ \mathrm{cov}\{X_2, X_1\} & \sigma_2^2 & \cdots & \mathrm{cov}\{X_2, X_N\} \\ \vdots & \vdots & \ddots & \vdots \\ \mathrm{cov}\{X_N, X_1\} & \mathrm{cov}\{X_N, X_2\} & \cdots & \sigma_N^2 \end{bmatrix} \qquad (2.99)$$

where we see that the N by N covariance matrix possesses certain attributes. Diagonal entries of the covariance matrix are indeed the variances of the N RVs, and the off-diagonal entries are the covariances of pairs of RVs. We can see from Eq. (2.99) that the covariance matrix is symmetric since the covariance of X and Y is the same as the covariance of Y and X. If all pair-wise covariances are zero (i.e., all pairs of RVs are uncorrelated), the covariance matrix $\mathbf{C}$ is diagonal with variances as the diagonal elements.

Covariance matrix is positive definite Another important property of the covariance matrix is that it is positive definite; i.e., $\mathbf{a}^T\mathbf{C}\mathbf{a} > 0$ for any $\mathbf{a} \neq 0$. To prove this, let us define a new scalar RV Y as a weighted sum of the N RVs; i.e., $Y = \mathbf{a}^T\mathbf{x} = a_1X_1 + a_2X_2 + \cdots + a_NX_N$. The first two moments of this new RV Y can be computed as follows:

$$m_Y = E\{Y\} = a_1m_1 + a_2m_2 + \cdots + a_Nm_N = \mathbf{a}^T\mathbf{m_X}$$

$$\sigma_Y^2 = E\left\{(Y - m_Y)^2\right\} = E\left\{\left(\sum_{i=1}^{N} a_i(X_i - m_i)\right)^2\right\} \tag{2.100}$$

$$= \sum_{i=1}^{N}\sum_{k=1}^{N} a_ia_k\mathrm{cov}\{X_i, X_k\} = \mathbf{a}^T\mathbf{C_X}\mathbf{a}$$

where we have used subscripts for clarity. Since the variance of any random variable (including that of Y) must be positive, from Eq. (2.100) we see that $\mathbf{a}^T\mathbf{C}\mathbf{a} > 0$ for any non-zero vector $\mathbf{a}$. This is indeed the requirement for the positive definiteness of matrix $\mathbf{C}$. Since the covariance matrix $\mathbf{C}$ is guaranteed to be symmetric and positive definite, its eigenvalues are guaranteed to be real and positive.

Gaussian random vectors N RVs are said to be jointly Gaussian when their joint PDF is of the following form:

$$f(\mathbf{x}) = \frac{1}{\sqrt{(2\pi)^N|\mathbf{C}|}} \exp\left[-\frac{1}{2}(\mathbf{x} - \mathbf{m})^T\mathbf{C}^{-1}(\mathbf{x} - \mathbf{m})\right] \tag{2.101}$$

where $|\mathbf{C}|$ is the determinant of the covariance matrix $\mathbf{C}$. From the above equation, we see that a Gaussian PDF is completely characterized by its mean vector $\mathbf{m}$ with N elements in it, and the symmetric covariance matrix $\mathbf{C}$ with $N(N+1)/2$ distinct entries in it.

When the covariance matrix $\mathbf{C}$ is diagonal, its inverse $\mathbf{C}^{-1}$ is also diagonal and the N-variate PDF in Eq. (2.101) can be decomposed as the product of N marginal PDFs (each a univariate Gaussian), proving once again that for jointly Gaussian RVs, uncorrelatedness and statistical independence are equivalent.

For jointly Gaussian RVs, we can express higher-order moments in terms of variances and covariances. For example, for four zero-mean, jointly Gaussian RVs, we can show the following.

$$E\{X_1X_2X_3X_4\} = E\{X_1X_2\}E\{X_3X_4\} + E\{X_1X_3\}E\{X_2X_4\}$$
$$+ E\{X_1X_4\}E\{X_2X_3\} \tag{2.102}$$

2.5.7 Linear transformations

We have already provided our motivation for wanting to look at affine transformations (sometimes we may refer to them as linear transformations, although strictly speaking linear transformations are a special case of affine transformations) of random vectors. As we will show soon, affine transformations of Gaussian random vectors lead to Gaussian random vectors. We will also show that we can select the affine transformation so that the transformed covariance matrix is diagonal. Thus a properly designed linear transformation can convert a set of correlated Gaussian RVs to a set of uncorrelated (and hence statistically independent) Gaussian RVs.

Suppose that a linear transformation converts N RVs $\{X_1, X_2, \ldots, X_N\}$ into M new RVs $\{Y_1, Y_2, \ldots, Y_M\}$ according to the following relationship:

$$Y_1 = a_{11}X_1 + a_{12}X_2 + \cdots + a_{1N}X_M$$
$$Y_2 = a_{21}X_1 + a_{22}X_2 + \cdots + a_{2N}X_M$$
$$\vdots \qquad\qquad \vdots \qquad\qquad \vdots \tag{2.103}$$
$$Y_M = a_{M1}X_1 + a_{M2}X_2 + \cdots + a_{MN}X_M$$

The linear transformation in Eq. (2.103) can be compactly expressed as $\mathbf{y} = \mathbf{A}\mathbf{x}$ where $\mathbf{A}$ is a matrix with M rows and N columns. The (i, j)th entry of $\mathbf{A}$ is a_{ij}.

Let us first consider the case of $M = N$ where the transformation matrix $\mathbf{A}$ is invertible. The Jacobian (defined as the determinant of the matrix containing $\dfrac{\partial y_i}{\partial x_j}$ as its (i, j) element) of this transformation is $|\mathbf{A}|$, the determinant of the transformation matrix. Thus the PDF of $\mathbf{y}$ can be related to the PDF of $\mathbf{x}$ as follows:

$$f_{\mathbf{Y}}(\mathbf{y}) = \frac{1}{|\mathbf{A}|}f_{\mathbf{X}}(\mathbf{A}^{-1}\mathbf{y}) \tag{2.104}$$

Applying the above result and the N-variate Gaussian PDF in Eq. (2.101), we can see that the transformed random vector is also Gaussian, but with a new mean vector and new covariance matrix.

$$f_Y(\mathbf{y}) = \frac{1}{|\mathbf{A}|} \frac{1}{\sqrt{(2\pi)^N |\mathbf{C_X}|}} \exp\left[-\frac{1}{2}\left(\mathbf{A}^{-1}\mathbf{y} - \mathbf{m_X}\right)^{\mathrm{T}} \mathbf{C_X}^{-1}\left(\mathbf{A}^{-1}\mathbf{y} - \mathbf{m_X}\right)\right]$$

$$= \frac{1}{\sqrt{(2\pi)^N |\mathbf{C_Y}|}} \exp\left[-\frac{1}{2}(\mathbf{y} - \mathbf{m_Y})^{\mathrm{T}} \mathbf{C_Y}^{-1}(\mathbf{y} - \mathbf{m_Y})\right] \tag{2.105}$$

where $\mathbf{m_Y} = \mathbf{A}\mathbf{m_X}$ and $\mathbf{C_Y} = \mathbf{A}\mathbf{C_X}\mathbf{A}^{\mathrm{T}}$

The above result clearly shows that the result of an invertible affine transformation is also a Gaussian vector. If M is smaller than N, all we need is to integrate out the $(N - M)$ unwanted RVs. The resultant PDF will be Gaussian since marginal PDFs of a jointly Gaussian PDF are themselves Gaussian. The new mean vector and the covariance matrix are related to the original mean vector and covariance matrix as in Eq. (2.105).

Whitening transformation A particular linear transformation of interest is the *whitening* transform, which converts a set of RVs with a non-diagonal covariance matrix $\mathbf{C_x}$ to a new set of RVs with a diagonal covariance matrix (and, in fact, the identity matrix). This transformation is based on the property that for symmetric, positive definite matrices such as covariance matrices, the eigenvalues λ_i are real and positive and corresponding eigenvectors $\mathbf{e}_i$ can be selected to be orthonormal, i.e.,

$$\mathbf{C_x}\mathbf{e}_i = \lambda_i\mathbf{e}_i, \; i = 1, 2, \ldots, N \quad \lambda_i > 0 \quad \mathbf{e}_i^{\mathrm{T}}\mathbf{e}_j = \begin{cases} 1 \text{ for } i = j \\ 0 \text{ for } i \neq j \end{cases} \tag{2.106}$$

If all eigenvalues are distinct, corresponding eigenvectors are orthogonal and can be normalized to have unit norm so that the set of eigenvectors is orthonormal. If two eigenvalues are equal, then corresponding eigenvectors can be chosen from a plane and do not have to be orthogonal. However, they can be selected to be orthogonal. One example of the case of repeated eigenvalues is the identity matrix whose eigenvalues are all equal to 1. Clearly, any vector is an eigenvector for the identity matrix, but we can always choose N elemental vectors (e.g., [000100...0]) as the eigenvectors.

Let $\mathbf{E}$ denote a square matrix whose columns are $\mathbf{e}_i$, the normalized eigenvectors, and let $\mathbf{L}$ denote a diagonal matrix whose diagonal entries are the N eigenvalues. Then Eq. (2.106) can be written more compactly as follows:

$$\mathbf{C_X}\mathbf{E} = \mathbf{E}\mathbf{L} \text{ and } \mathbf{E}^{\mathrm{T}}\mathbf{E} = \mathbf{I} \Rightarrow \mathbf{E}^{\mathrm{T}}\mathbf{C_X}\mathbf{E} = \mathbf{L} \text{ and } \mathbf{E}\mathbf{L}\mathbf{E}^{\mathrm{T}} = \mathbf{C_X} \tag{2.107}$$

Let us consider a transformation matrix $\mathbf{A}$ whose rows are the normalized eigenvectors $\mathbf{e}_i$ of the covariance matrix $\mathbf{C_x}$; i.e., $\mathbf{A} = \mathbf{E}^T$. If $\mathbf{y} = \mathbf{Ax}$, then the new covariance is a diagonal matrix as shown below.

$$\mathbf{C_Y} = \mathbf{AC_XA}^T = \mathbf{E}^T\mathbf{C_X}\mathbf{E} = \mathbf{E}^T\mathbf{EL} = \mathbf{L} = \text{Diag}\{\lambda_1, \lambda_2, \ldots, \lambda_N\} \quad (2.108)$$

Thus, using $\mathbf{A} = \mathbf{E}^T$ results in new RVs $\mathbf{y}$ that are uncorrelated since the new covariance matrix is diagonal. If the random vector $\mathbf{x}$ is Gaussian, then $\mathbf{A} = \mathbf{E}^T\mathbf{x}$ is also Gaussian since the transformation is linear. Thus, $\mathbf{A} = \mathbf{E}^T\mathbf{x}$ results in statistically independent Gaussian RVs whose variances are nothing but the eigenvalues of the covariance matrix $\mathbf{C_x}$.

We can also create new RVs that are of unit variance by dividing each new RV by the square root of the ith eigenvalue. Thus, we can obtain statistically independent, Gaussian RVs with unit variance by using the linear transformation $\mathbf{A} = \mathbf{L}^{(-1/2)}\mathbf{E}^T\mathbf{x}$, where $\mathbf{L}^{(-1/2)}$ denotes a diagonal matrix whose ith diagonal entry is $1/\sqrt{\lambda_i}$. This transformation is called a whitening transformation since it creates new uncorrelated RVs with equal variances similar to white light, which has all wavelengths in it. The whitening transform allows us to evaluate the probabilities associated with N-variate Gaussians using univariate Gaussians and hence error functions.

2.6 Chapter summary

In this chapter, we reviewed relevant concepts and results from matrix–vector theory and probability theory and RVs. In this section, we will try to capture the main points of the chapter.

- Images can be represented as full matrices, diagonal matrices, or vectors. Thus many matrix–vector operations such as multiplication, inner product, inverse, etc. will prove useful in correlation pattern recognition.
- The inner product of two vectors is particularly relevant to correlation pattern recognition. The normalized inner product measures the similarity between two vectors. The Cauchy–Schwarz inequality states that the square of the inner product between two vectors is less than or equal to the product of their auto inner products.
- Matrix partitioning can simplify computations such as matrix determinant and matrix inverse, and can also enable recursive updating of matrix inverses.
- Eigenvalues and eigenvectors play a very important role in correlation filter design. For real, symmetric matrices of much interest in CPR, eigenvalues are real and orthonormal eigenvectors can be found. This enables the diagonalization of symmetric matrices.
- For images with $N \times N$ pixels, we are often interested in finding eigenimages, i.e., eigenvectors of the corresponding $N^2 \times N^2$ outer product matrices. As these outer

product matrices are usually determined using d images where $d \ll N^2$, we can find the eigenimages of interest (corresponding to the non-zero eigenvalues) using the $d \times d$ inner product matrix. These eigenimages will depend on the centration of the images being used to estimate the outer product matrix. Thus, care must be taken to register the images prior to extracting eigenimages.

- Correlation filter design often requires the optimization of ratios of quadratic criteria and this optimization leads to solutions in terms of eigenvectors. Similarly, constrained optimization problems can be converted to unconstrained optimization problems using the Lagrange multipliers method.

- Random variables can be characterized by PDFs. Probability density functions of continuous RVs exhibit no jumps whereas PDFs of discrete RVs contain discontinuities. All PDFs are non-negative and must have a total volume under the surface equal to 1.

- Often, we are satisfied with partial characterizations of RVs, such as knowing only their mean and variance. The mean indicates the average value of the RV, whereas the variance tells us how broadly the PDF is spread around this mean value.

- The simultaneous behavior of two RVs requires the specification of a two-dimensional joint PDF. By integrating out the unwanted random variables, we can find the marginal PDFs. If the joint PDF is the product of marginal PDFs, then the RVs are said to be statistically independent.

- The joint second-order behavior of two RVs is captured by their covariance. Two RVs are said to be uncorrelated if their covariance is zero. Statistical independence implies uncorrelatedness, but the converse is not in general true. An exception, however, is the case of jointly Gaussian RVs, where statistical independence and uncorrelatedness are equivalent.

- When dealing with two RVs, we can define a conditional PDF of one RV conditioned on the other RV. A conditional PDF is just like any other PDF in that it is non-negative and has a total volume of 1 under the surface. Bayes' rule allows us to express the a-posteriori PDFs in terms of the a-priori PDFs and to invert the direction of condition.

- It is convenient to use vectors to denote more than two RVs. For random vectors, the second-order statistics are captured by the covariance matrix. The covariance matrix is symmetric and positive-definite.

- Gaussian random vectors can be completely characterized by their mean vectors and covariance matrices. Affine transformations of Gaussian random vectors result in Gaussian random vectors, albeit with new mean vectors and covariance matrices. By using an affine transformation matrix obtained from the orthonormal eigenvectors of the original covariance matrix, we can transform the RVs to become uncorrelated. Since uncorrelatedness implies statistical independence for Gaussian RVs, this affine transformation allows us to convert statistically dependent Gaussian RVs into statistically independent Gaussian RVs.

3

Linear systems and filtering theory

Correlation involves two signals or images. A reference image is correlated with a test image (also called a *scene*) to detect and locate the reference image in the scene. Thus the correlator can be considered as a system with an input (the scene), a stored template or filter (derived from the reference image), and an output (correlation). As we will see in this chapter, such a system is *linear* in the sense that a new input that is a weighted sum of original inputs results in an output that is an identically weighted sum of the original outputs. Thus a correlator can take advantage of the many properties of linear systems. The most important property is that a linear, time-invariant system can be characterized in terms of its frequency response. We use this and other related properties for the synthesis and use of correlation filters with attractive features such as distortion-tolerance and discrimination. In this chapter, we provide a review of some of the useful properties of signals and linear systems.

3.1 Basic systems

Strictly speaking, the signal is denoted $s(\cdot)$, and $s(x)$ is the value of $s(\cdot)$ when the argument value is x. We will occasionally require the strict notation, but usually there is no confusion from writing $s(x)$ to mean "$s(\cdot)$ with x being used as a general value for the argument." Figure 3.1 is a simple block diagram of a system. A system can be characterized as producing an output signal $o(x)$ in response to an input signal $i(x)$. We are using the space variable, x (as opposed to the more commonly used time variable, t) to emphasize our interest in images which are intensities that are functions of two space variables, x and y. A signal can be thought of as the variation of an independent variable (e.g., voltage, gray level of an image) as a function of a dependent variable (e.g., time, spatial coordinates in an image). While signals are usually thought of as one-dimensional (1-D) functions, we can use the theory presented in this

Figure 3.1 Block diagram of a system

chapter with higher-dimensional signals such as images, which can be thought of as 2-D signals (e.g., gray scale as a function of spatial coordinates x and y). In that sense, the input to the system is $i(x, y)$ and the output is $o(x, y)$. For notational brevity, we will refer to these as $i(x)$ and $o(x)$ from now on and show explicitly the two independent variables only where needed. An important sub-class of systems known as *linear, shift-invariant* (LSI) systems can be completely characterized by the system's output for just one particular input, namely a point input at the origin. The resulting output is known as the point spread function (PSF) in 2-D systems, and the impulse response in 1-D systems.

The independent variable used with a signal or an image can be either continuous (e.g., time) or discrete (e.g., pixel number). When the independent variable is continuously varying, we will loosely refer to it as a continuous-time (CT) signal and as a discrete-time (DT) signal if the independent variable is discrete. Thus, a CT signal $i(x)$ is defined for all possible values of its continuous argument x. On the other hand, a DT signal $i[n]$ is defined only for discrete values n of the independent variable. A DT example is the signal representing the Dow Jones daily closing index. This sequence of numbers is defined for only one instant every day and there is no meaning for the closing index for any other time of the day. Similarly, the music signal stored on an audio CD is obtained by taking 44 100 samples for every second of the music signal and only these samples are stored on a CD. Thus, a CD contains a DT signal. The process of converting a CT signal to a DT signal is known as analog-to-digital conversion (ADC), or *sampling*. We will discuss sampling theory in some detail later in this chapter. Most input devices in optical correlators are pixelated and employ sampling. The DT signal stored on the CD is converted to a CT music signal before it is played through the speakers. This process of converting DT signals to CT signals is known as digital-to-analog conversion (DAC).

In the next section, we will establish the notation for some special signals we will be encountering. This will be followed by Section 3.3 which reviews the basics of LSI systems, and discusses the convolution operation that allows us to determine the output of an LSI system for any arbitrary input. Section 3.4 reviews the important concept of Fourier analysis of CT signals, and this is followed by Section 3.5 which reviews the sampling theory. Sampling theory is important to understand well since, although most signals are CT to begin with, anytime we use a digital computer to process them, we have to convert them to DT signals via sampling. Fourier analysis of these DT signals is reviewed in Section 3.6. Finally, Section 3.7 provides a brief review of how to characterize random signals and what happens to them as they pass through linear systems. Knowing what happens to random signals through linear systems enables us to analyze and design correlation filters with the required noise tolerance.

3.2 Signal representation

Physical inputs to physical systems, the systems themselves, and the physical outputs, are all decidedly real. However, the mathematics of the LSI system and the signals – particularly for sinusoidal or nearly sinusoidal signals – is often very conveniently shortened with complex notation. We will often use $A \exp[j(2\pi fx + \phi)]$ to represent $A \cos(2\pi fx + \phi)$. The complex exponential is the *phasor* representing the signal.

We will use x to represent continuous time and n for discrete time. Thus, CT system signals are $i(x)$ and $o(x)$, whereas DT system signals are denoted by $i[n]$ and $o[n]$. Note the notational difference between parentheses and square brackets used for CT and DT signals, although it should normally be obvious from the context whether we are dealing with CT or DT signals. Several basic signal operations are defined in Table 3.1 in terms of a 1-D CT signal $i(x)$. The focus of this book being image correlation, we need to deal with 2-D signals $i(x, y)$. DT images are denoted by $i[n, m]$. Basic signal operations such as shift, scaling, and reflection are applied to 2-D signals in the same way as 1-D signals, except that both x and y must be taken into consideration. Some commonly encountered 1-D CT and DT signals are summarized in Table 3.2 and Table 3.3, respectively. However, it is worth highlighting some special 2-D signals.

Separable signals Often, a 2-D signal can be written as the product of two 1-D signals as below. Such 2-D signals are known as *product-separable,* or commonly, just *separable*, signals.

$$i(x, y) = i_x(x)i_y(y) \quad i[n, m] = i_n[n]i_m[m] \tag{3.1}$$

Separable signals are easier to handle than non-separable, as they require only two 1-D signals or two vectors, instead of one 2-D signal or one matrix. Real images are rarely separable and, even worse, rarely allow compact analytical representations. As a result, we will mostly denote an image as $i(x, y)$ or $i[n, m]$ without any further simplification.

Table 3.1. *Basic signal operations*

Operation	Description
Time shift	Signal $i(x - x_0)$ is $i(x)$ shifted to the right by x_0. If x_0 is positive, then the shift is to the right, and if it is negative, the shift is to the left. For a DT signal $i[n]$, the shift must always be an integer. A signal is *periodic* with period T, if $i(x + nT) = i(x)$ for any integer value of n.
Time scaling	Signal $i(ax)$ denotes the original signal $i(x)$ scaled by a factor a. If $a > 1$, the signal is compressed, whereas if $0 < a < 1$, the signal is dilated.
Reflection	Signal $i(-x)$ denotes a time-reversed or reflected signal. A signal is considered to be an *even signal* when its reflection equals itself, i.e., $i(-x) = i(x)$. A signal is considered to be an *odd signal* if its reflection equals the negative of itself, i.e., $i(-x) = i(x)$. Not every signal has to be either even or odd, though every signal can be expressed as the sum of unique even and odd parts.
Even–odd parts	An arbitrary signal $i(x)$ can be decomposed as the sum of an even part $i_e(x)$ and an odd part $i_o(x)$, as $i(x) = i_e(x) + i_o(x)$. The component $i_e(x) = \frac{i(x)+i(-x)}{2}$ is even, and $i_o(x) = \frac{i(x)-i(-x)}{2}$ is odd.
Energy	The energy of a signal defined as $E = \int_{-\infty}^{\infty} |i(x)|^2 dx$. For periodic signals $i(x)$ with period T, we define an average energy $E_p = \frac{1}{T} \int_{-T/2}^{T/2} |i(x)|^2 dx$. Similar energy and average energy definitions exist for DT signals.

Coordinate transformation An image $i(x, y)$ can be mapped to another image $\hat{i}(x, y)$ by transforming the coordinate system. A particular coordinate transform of interest is the polar transform (PT) from Cartesian coordinates x and y to polar coordinates, namely radius r and angle θ.

$$i(x, y) \rightarrow \hat{i}(r, \theta) \text{ where } r = \sqrt{x^2 + y^2} \text{ and } \theta = \tan^{-1}\left(\frac{y}{x}\right)$$

(3.2)

$$i(x, y) \rightarrow \hat{i}(r, \theta) \text{ where } x = r\cos\theta \text{ and } y = r\sin\theta$$

If the PT of an image is independent of angle θ, then that image is *circularly symmetric*. If the PT is independent of radius r, it is a *radially constant* image. Figure 3.2(a) shows a circularly symmetric image and Figure 3.2(b) indicates its PT. The PT in Figure 3.2(d) is of the radially constant image in Figure 3.2(c). While we have only discussed the polar transform here, there exist other useful coordinate transformations such as the Mellin transform [23] and the log-polar transform [24].

Table 3.2. *Special one-dimensional CT signals*

Signal	Definition	Comments
CT impulse (delta function)	$\delta(x)$	Loosely speaking, this function is zero everywhere except at the origin where it is infinitely large. Multiplying a smooth function by a delta function forces the product to become zero everywhere except at the location of the delta function, i.e., $\int_{-\infty}^{\infty} i(x)\delta(x - x_0)\mathrm{d}x = i(x_0)$, provided $i(x)$ is continuous at $x = x_0$. This is known as the *sifting* property since it picks out the value of $i(\cdot)$ at x_0.
unit step	$u(x) = \begin{cases} 1 \text{ for } x \geq 0 \\ 0 \text{ for } x < 0 \end{cases}$	The unit step is the integral of a delta function, i.e., $u(x) = \int_{-\infty}^{x} \delta(\tau)\mathrm{d}\tau$. It can be used for representing switching systems.
comb function	$\mathrm{comb}_T(x) = \sum_{k=-\infty}^{\infty} \delta(x - kT)$	The comb function is an infinite train of delta functions spaced at uniform intervals of T. Multiplying any signal $i(x)$ by the comb function $\mathrm{comb}_T(x)$ results in a sampled signal that is non-zero only at the sampling instants.
rect function	$r(x) = u(x + 1/2) - u(x - 1/2)$ $= \begin{cases} 1 \text{ for } |x| \leq 1/2 \\ 0 \text{ otherwise} \end{cases}$	The rectangle function $r(x)$, also known as the box function, equals 1 in the interval $[-1/2, 1/2]$ and zero outside. It is easy to verify that the unit rectangle has energy 1.
Sinusoids	$i(x) = A\cos(2\pi f x + \phi)$	A is the *amplitude*, f is the *frequency*, and ϕ is the phase (indicates the relative position of the signal with respect to the origin) of the sinusoid. The period T is related to the frequency as $f = 1/T$. A sinusoid of a particular frequency input to a linear, shift-invariant (LSI) system must lead to an output sinusoid of the *same* frequency. Thus, sinusoids are *eigenfunctions* of LSI systems.

Table 3.2. (*cont.*)

Signal	Definition	Comments
		The amplitude and phase, but not the frequency, of an input sinusoid are altered by an LSI system.
Complex exponentials	$i(x) = A \exp(j2\pi fx)$ $= A[\cos(2\pi fx)$ $+ j\sin(2\pi fx)]$	Because of their close connection with sinusoids, complex exponentials are eigenfunctions of LSI systems. Complex exponentials are periodic signals.
Unit Gaussian	$\text{Gaus}(x) = 1/\sqrt{2\pi} \exp(-x^2/2)$	Often used to describe smoothly tapering apertures and windows. This is the same shape as that of a Gaussian PDF, with zero mean and unit variance.

Table 3.3. *Special one-dimensional DT signals*

Signal	Definition	Comments
Unit DT step	$u[n] = \begin{cases} 1 \text{ for } n \geq 0 \\ 0 \text{ for } n < 0 \end{cases}$	Takes on a value of 1 at the origin and at all positive integer values of n, and a value of 0 at all negative integer values of n.
Unit DT impulse	$\delta[n] = \begin{cases} 1 \text{ for } n = 0 \\ 0 \text{ for } n \neq 0 \end{cases}$ $= u[n] - u[n-1]$	Also known as the DT delta function, or Kronecker delta function. The DT delta function is 1 when $n = 0$, but 0 everywhere else. A DT delta function is the difference between the DT unit step and the DT unit step shifted by 1 to the right.
Sinusoids	$i[n] = A\cos(2\pi fn + \phi)$	Unlike the CT sinusoid the DT sinusoid is not always periodic. In fact, the DT sinusoid is periodic if and only if $2\pi fmN$ is an integers multiple of 2π for some integers m and N, which means that f must be a ratio of integers (e.g., if f is $1/2$, $3/8$, $5/2$, etc., the DT sinusoid is periodic. On the other hand, if f is $\sqrt{2}$, it is not periodic).

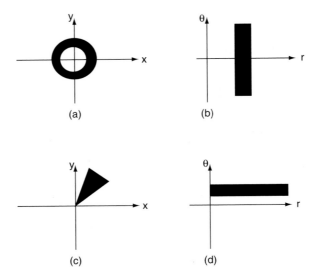

Figure 3.2 (a) A circularly symmetric image, and (b) its polar transform; (c) a radially constant image, and (d) its polar transform

Rectangle function The 2-D rectangle function is 1 inside a rectangular region centered at the origin and 0 outside. It is useful in truncating images and to describe the region of support of images. It is a separable function as shown below:

$$\text{rect}(x, y) = r(x)r(y) = \begin{cases} 1 \text{ if } |x| \leq 1/2 \text{ and } |y| \leq 1/2 \\ 0 \text{ otherwise} \end{cases} \tag{3.3}$$

Circ function For 2-D signals, circular apertures may be more natural than rectangular apertures. The circ function is 1 inside a circle of radius 1 centered at the origin, and 0 outside. It is not a separable function in Cartesian coordinates.

$$\text{circ}(x, y) = \begin{cases} 1 \text{ if } \sqrt{(x^2 + y^2)} \leq 1 \\ 0 \text{ otherwise} \end{cases} \tag{3.4}$$

Unit Gaussian function Both the rect($\cdot$) function and the circ($\cdot$) function are binary in that they take on values of 1 and 0. A useful, circularly symmetric, 2-D function that takes on a continuum of amplitude values is the unit Gaussian centered at the origin. It is separable as shown below:

$$\text{Gaus}(x, y) = \frac{1}{2\pi} \exp\left[-\frac{x^2 + y^2}{2}\right] = \frac{1}{\sqrt{2\pi}} \exp\left[-\frac{x^2}{2}\right] \cdot \frac{1}{\sqrt{2\pi}} \exp\left[-\frac{y^2}{2}\right] \quad (3.5)$$
$$= \text{Gaus}(x) \cdot \text{Gaus}(y)$$

The above definition considers a unit Gaussian (also known as standard Gaussian) that is centered at the origin and has a standard deviation of 1 along both x and y. More general Gaussian functions can be obtained by changing the variables.

3.3 Linear shift-invariant systems

In Section 3.1, we defined a system as producing the output $o(x)$ in response to the input $i(x)$. If both signals are CT, we will refer to that system as a CT system, and if both signals are DT, we consider it to be a DT system. There can be occasions where the input signal is of one type and the output of a different type. We will refer to such systems as hybrid or mixed systems.

Linear, shift-invariant systems offer much in terms of their properties. We will first define what linearity is and what shift-invariance means. This will be followed by a look at the properties of LSI systems. One property of an LSI system of particular interest is that its output can be obtained by *convolving* the input signal and its impulse response (impulse response is the output of the LSI system when the input is a delta function). We will see that the correlation operation is similar to the convolution operation. We will demonstrate the fact that sinusoids are eigenfunctions of LSI systems.[1] While we will use 1-D CT signals, our discussion is easily extended to higher dimensions and to DT systems. We will point out any differences only when they are significant.

Linearity In simple words, linearity requires that weighted summation of inputs should lead to an identically weighted sum of output signals. More rigorously, a linear system must satisfy the following:

If $i_1(x) \rightarrow o_1(x)$ and $i_2(x) \rightarrow o_2(x)$, then $a i_1(x) + b i_2(x) \rightarrow a o_1(x) + b o_2(x)$
for any scalars a, b and any inputs $i_1(x)$ and $i_2(x)$ (3.6)

---

[1] The term *eigenfunction* derives from linear algebra, in which an eigenvector (discussed in Chapter 2) of a matrix is one changed by only a complex factor when multiplied by the matrix. An LSI system may change the phase and magnitude of an input sinusoid, but not the sinusoid's frequency.

If the condition in Eq. (3.6) is not satisfied for even one set of weights or for even one particular signal, then the system is nonlinear. If $a = b = 1$, the above requirement in Eq. (3.6) means that a new input that is the sum of two old inputs results in a new output that is the sum of the corresponding two old outputs. This property is also known as *homogeneity* or the *principle of super-position*. For a linear system, it is easy to show that an all-zero input signal must lead to an all-zero output signal. What this does not mean is that if an input signal is zero over a certain time interval (let us say from t_1 to t_2), then the resulting output signal is also zero over that time interval.

The advantage of linearity is that we can find the output of a system for an input signal by knowing the outputs for some basic signals. We will show in Section 3.4 that we can represent arbitrary signals as a weighted sum of sinusoids. Thus, knowing the outputs of a linear system to input sinusoids is very attractive. Instead of documenting every possible input–output pair for a linear system, we only have to know the outputs for sinusoids. Since sinusoids are eigenfunctions for LSI systems, we need to document only the magnitude and phase response of that system as a function of input frequency.

Shift invariance If shifting the input signal by x_0 results in an output that is shifted by the same amount, then we have a shift-invariant system. In 1-D systems, these are more commonly referred to as time-invariant systems as the independent variable is time. More precisely, if $i(x) \rightarrow o(x)$, then $i(x - x_0) \rightarrow o(x - x_0)$ for any $i(x)$ and any x_0.

Shift-invariance tells us that if we know the output for a particular input, then we know the outputs for every shifted version of that input signal. We will see in the next section that this shift-invariance, coupled with linearity, enables us to characterize an LSI system completely by its *impulse response*.

3.3.1 Impulse response, convolution, and correlation

Consider a unit impulse function $\delta(t)$ or $\delta[n]$ (depending on whether the system is CT or DT, respectively) input signal. Irrespective of whether the system is LSI or not, we will refer to the corresponding output as its *impulse response* ($h(t)$ or $h[n]$). Let us now look at a DT LSI system with impulse response $h[n]$. An example impulse response is shown in Figure 3.3(a). Suppose the input signal to this system is the sequence $i[n]$ shown in Figure 3.3(b). What is the resulting output? This signal comprises three DT delta functions, each weighted by different amounts and each shifted by different amounts as shown

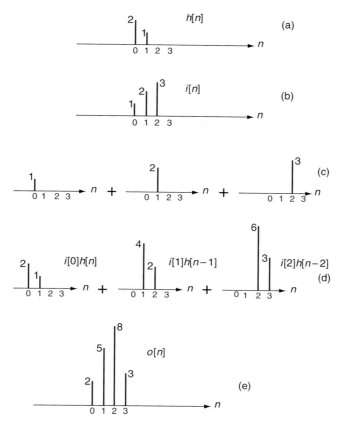

Figure 3.3 (a) The impulse response of a DT system, (b) an example input DT signal, (c) expressed as a sum of weighted, shifted delta functions, (d) output signals for the input-weighted, shifted delta functions, and (e) output signal for the input signal in (b)

in Figure 3.3(c). In general, a DT signal can always be expressed as a weighted sum of shifted delta functions as follows:

$$i[n] = \sum_{k=-\infty}^{\infty} i[k] \cdot \delta[n-k] \tag{3.7}$$

Since the system is shift-invariant, the input $\delta[n-k]$ should lead to the output $h[n-k]$. Since the system is linear, weighting that input signal $\delta[n-k]$ by weight $i[k]$ results in the output $i[k] \cdot h[n-k]$. These output signals are shown in Figure 3.3(d) for the example being considered. Equation (3.7) tells us that the input signal is a sum over k of these weighted, shifted delta function inputs, and, by linearity of the system, the output must be an identical sum over k, i.e.,

$$o[n] = \sum_{k=-\infty}^{\infty} i[k] \cdot h[n-k] = i[n] * h[n] \qquad (3.8)$$

where $i[n] * h[n]$ is used as a shorthand notation for the summation operation in Eq. (3.8). This operation, called the *convolution* between $i[n]$ and $h[n]$, is at the core of LSI systems. We have used both linearity and shift-invariance in obtaining the above result, that the output of the LSI system is the convolution of the input with the impulse response. Figure 3.3(e) shows the output of the LSI system.

It is important to note that in the convolution sum in Eq. (3.8), the sign on k is different in $i[k]$ and $h[n-k]$. If we use the same sign for k, we get a *correlation* $c[n]$, i.e.,

$$c[n] = \sum_{k=-\infty}^{\infty} i[k] \cdot h[n+k] = i[n] \otimes h[n] \qquad (3.9)$$

where we use the symbol $\otimes$ to denote the correlation operation. The convolution in Eq. (3.8) and the correlation in Eq. (3.9) look similar, but produce entirely different results.

Region of support Suppose the two functions being convolved are of finite support. (More precisely, this means $i(x)$ is zero outside $L_i \leq x \leq U_i$, and $h(x)$ is zero outside $L_h \leq x \leq U_h$, with all of L_i, U_i, L_h, and U_h being finite numbers.) Then the convolution of the two, $o(x) = i(x) * h(x)$ is zero outside $L_o \leq x \leq U_o$ where:

$$L_o = L_i + L_h \text{ and } U_o = U_i + U_h \qquad (3.10)$$

Denoting the lengths of the two signals by $A_i = (U_i - L_i)$ and $A_h = (U_h - L_h)$, we see that the output convolution length is at most $A_o = (U_o - L_o) = A_i + A_h$.

The results are slightly different for DT convolution. Assuming that $i[n]$ is zero outside $L_i \leq n \leq U_i$, and $h[n]$ is zero outside $L_h \leq n \leq U_h$, the DT convolution $o[n] = i[n] * h[n]$ is zero outside $L_o \leq n \leq U_o$, where L_o and U_o are as in Eq. (3.10). But for discrete signals, the length of a signal is given by $A_i = U_i - L_i + 1$. Thus, the result for DT convolution length is at most $A_o = (A_i + A_h - 1)$. As an example, convolving a DT sequence of length 64 points with itself would result in an output signal length of at most 127 points. For 2-D CT convolution (which is discussed in the next section), these region-of-support results hold for each axis. The region-of-support results are the same for correlation.

3.3.2 Two-dimensional LSI systems

We can define something similar to the impulse response for 2-D systems. The 2-D counterpart of a 1-D delta function is a 2-D delta function – namely, a point. Let $h[n, m]$ denote the 2-D output signal from a 2-D DT system in response to the input signal $\delta[n, m] = \delta[n] \cdot \delta[m]$. This $h[n, m]$ is also known as the *point spread function* (PSF). Once the PSF is known, the output $o[n, m]$ from an LSI system for any arbitrary input $i[n, m]$ can be obtained as the following 2-D convolution:

$$o[n, m] = \sum_{k=-\infty}^{\infty} \sum_{l=-\infty}^{\infty} i[k, l] \cdot h[n - k, m - l] = i[n, m] ** h[n, m] \qquad (3.11)$$

where $**$ is used to indicate the 2-D convolution.

If both the input image and the PSF are separable, then the resulting output is also separable as shown below. In such separable cases, the 2-D convolution in Eq. (3.11) can be replaced by two 1-D convolutions:

$$
\begin{aligned}
o[n, m] &= \sum_{k=-\infty}^{\infty} \sum_{l=-\infty}^{\infty} i[k, l] \cdot h[n - k, m - l] \\
&= \sum_{k=-\infty}^{\infty} \sum_{l=-\infty}^{\infty} \{i_n[k] \cdot i_m[l]\} \cdot \{h_n[n - k] \cdot h_m[m - l]\} \\
&= \sum_{k=-\infty}^{\infty} \{i_n[k] \cdot h_n[n - k]\} \cdot \sum_{l=-\infty}^{\infty} \{i_m[l] \cdot h_m[m - l]\} = o_n[n] \cdot o_m[m]
\end{aligned}
$$
$$(3.12)$$

where

$$
\begin{aligned}
o_n[n] &= \sum_{k=-\infty}^{\infty} \{i_n[k] \cdot h_n[n - k]\} = i_n[n] * h_n[n] \\
o_m[m] &= \sum_{l=-\infty}^{\infty} \{i_m[l] \cdot h_m[m - l]\} = i_m[m] * h_m[m]
\end{aligned}
$$
$$(3.13)$$

Similarly, we can define the 2-D correlation as follows:

$$c[n, m] = \sum_{k=-\infty}^{\infty} \sum_{l=-\infty}^{\infty} i[k, l] \cdot h[k + n, l + m] = i[n, m] \otimes \otimes r[n, m] \qquad (3.14)$$

where $\otimes \otimes$ is used to indicate 2-D correlation. For separable functions, we show below that the 2-D correlation is equivalent to two 1-D correlations that

Table 3.4. *Selected properties of convolution*

Property	Description
Linearity	$h(x) * [ai_1(x) + bi_2(x)] = \{ah(x) * i_1(x) + bh(x) * i_2(x)\}$
	where a and b are scalars
Shift invariance	$o(x) = i(x) * h(x) \Rightarrow i(x - x_0) * h(x) = o(x - x_0)$
Commutativity	$o(x) = i(x) * h(x) = \int i(\tau) \cdot h(x - \tau)\mathrm{d}\tau = \int i(x - u) \cdot h(u)\mathrm{d}u$
	$= h(x) * i(x)$
Convolution with a shifted delta function	$\delta(x - x_0) * h(x) = h(x - x_0)$

are much easier to compute compared to the one 2-D correlation using the double sum in Eq. (3.14). If $i[k, l]$ and $h[k, l]$ are separable, then

$$
\begin{aligned}
c[n, m] &= \sum_{k=-\infty}^{\infty} \sum_{l=-\infty}^{\infty} i[k, l] \cdot h[k + n, l + m] \\
&= \sum_{k=-\infty}^{\infty} \sum_{l=-\infty}^{\infty} \{i_n[k] \cdot i_m[l]\} \cdot \{h_n[k + n] \cdot h_m[l + m]\} \\
&= \sum_{k=-\infty}^{\infty} \{i_n[k] \cdot h_n[k + n]\} \cdot \sum_{l=-\infty}^{\infty} \{i_m[l] \cdot h_m[l + m]\} \\
&= c_n[n] \cdot c_m[m]
\end{aligned}
\tag{3.15}
$$

where

$$
\begin{aligned}
c_n[n] &= \sum_{k=-\infty}^{\infty} \{i_n[k] \cdot h_n[k + n]\} = i_n[n] \otimes h_n[n] \\
c_m[m] &= \sum_{l=-\infty}^{\infty} \{i_m[l] \cdot h_m[l + m]\} = i_m[m] \otimes h_m[m]
\end{aligned}
\tag{3.16}
$$

For CT systems, the convolution involves 2-D integration and the output is obtained by convolving the input with the PSF $h(x, y)$.

$$
o(x, y) = \int_{-\infty}^{\infty} \mathrm{d}\tau \int_{-\infty}^{\infty} \mathrm{d}\beta \, i(\tau, \beta) h(x - \tau, y - \beta) = i(x, y) ** h(x, y) \tag{3.17}
$$

In this section, we have shown that the convolution of the input signal with its impulse response yields the output signal for an LSI system. We have defined the convolution and the correlation operations. Some selected properties of convolution are shown in Table 3.4.

3.4 Continuous-time Fourier analysis

In this section, we will consider the Fourier analysis of both 1-D and 2-D CT signals, and Section 3.6 will be devoted to Fourier analysis of DT signals. We will first consider Fourier analysis of periodic CT signals and then we will generalize that to non-periodic CT signals. This will be followed by a discussion of the 1-D CT Fourier transform (CTFT). Finally, we will generalize the CTFT to 2-D.

3.4.1 Fourier series of a periodic signal

Consider a periodic signal $i(x)$ with period T; i.e., $i(x+kT)=i(x)$ for any integer k. Fourier showed that, under somewhat general conditions, $i(x)$ can be expressed as a weighted sum of complex exponentials whose frequencies are integer multiples of the fundamental frequency $f_0 = 1/T$, i.e.,

$$i(x) = \sum_{k=-\infty}^{\infty} c_k^i e^{j2\pi k f_0 x} \qquad (3.18)$$

where the kth Fourier series coefficient c_k^i (with the superscript referring to the signal) can be evaluated from one period of the signal as follows:

$$c_k^i = \frac{1}{T} \int_{\langle T \rangle} i(x) e^{-j2\pi k f_0 x} \mathrm{d}x \quad \text{for } k = 0, \pm 1, \pm 2, \ldots \qquad (3.19)$$

where $\langle T \rangle$ indicates any contiguous interval of length T.

Sometimes, we can determine the Fourier series (FS) coefficients just by inspection. For example, the sinusoid $i(x) = A \cos(2\pi f_s x)$ has a fundamental frequency $f_0 = f_s$, the frequency of the sinusoid. Using Euler's relation, the sinusoid can be expressed in terms of complex exponentials as $i(x) = A/2 \left[e^{j2\pi f_0 x} + e^{-j2\pi f_0 x} \right]$, and comparing this expression to the FS expansion in Eq. (3.18), we see that $c_1^i = c_{-1}^i = A/2$ and $c_k^i = 0$ for $k \neq \pm 1$

Equation (3.18) states that a periodic signal is a weighted sum (with weights c_k^i) of complex exponentials at frequencies $kf_0 = k/T$ where k is an integer. The complex exponential for $k=0$ is a constant and is called the DC term.[2] The term for $k=1$ corresponds to the fundamental frequency, the term for $k=2$ corresponds to the second harmonic, and other k values refer to higher-order harmonics. Thus the FS coefficients characterize the periodic signal in terms of

---

[2] It is interesting to see what is really an electrical engineering term from the very early days show up in image processing! The abbreviation DC means *direct current*, of course. An unvarying current is equal to its average, and so the average value of a function has become known as its DC value.

its constituent *frequencies*. Note that while $i(x)$ is in CT and is periodic, its frequency domain description c_k^i is discrete-indexed. Periodicity in one domain (e.g., time) leads to discreteness in the other domain (e.g., frequency).

3.4.2 One-dimensional CTFT

The Fourier series represents periodic signals (with period T) in terms of frequencies that are integer multiples of $f_0 = 1/T$. As the period T increases, the fundamental frequency f_0, and hence the spacing between the harmonics, decreases. In the limit as $T \to \infty$, the signal becomes non-periodic and the spacing between harmonics goes to zero, causing the frequency domain characterization to become continuous. This leads to the CTFT given below:

$$I(f) = \int_{-\infty}^{\infty} i(x) e^{-j2\pi f x} dx \qquad (3.20)$$

where f denotes the frequency. As can be seen in Eq. (3.20), we use lower case to denote signals in time domain and upper case to represent frequency domain functions. The FT relation in Eq. (3.20) is also indicated by $i(x) \leftrightarrow I(f)$. We can find the signal $i(x)$ from its FT $I(f)$ through the inverse Fourier transform (IFT).

$$i(x) = \int_{-\infty}^{\infty} I(f) e^{j2\pi f x} df \qquad (3.21)$$

Note the similarity between the FT and the IFT in Eq. (3.20) and Eq. (3.21), respectively. The only difference is in the sign of the exponent. Thus, from a computational perspective, there is no significant difference between FT and IFT.

Occasionally, we use the radial frequency $\omega = 2\pi f$ rather than the natural frequency f and the resulting FT relations are as follows:

$$I(\omega) = \int_{-\infty}^{\infty} i(x) e^{-j\omega x} dx, \qquad i(x) = \frac{1}{2\pi} \int_{-\infty}^{\infty} I(\omega) e^{j\omega x} d\omega \qquad (3.22)$$

Continuous-time Fourier-transform of a periodic signal Strictly speaking, we cannot define the FT of a periodic signal as it has infinite energy. However, periodic signals lend themselves to a Fourier series expansion as in Eq. (3.18). Since a complex exponential in time has an FT that is a shifted delta function in frequency, we can obtain the FT of a periodic signal as follows:

Table 3.5. *Continuous-time Fourier-transforms of some commonly encountered CT signals*

CT signal (time domain)	CTFT (frequency domain)
$\delta(x)$ (delta function)	1 (constant in frequency)
$\delta(x - x_o)$ (shifted delta function)	$e^{-j2\pi f x_o}$ (linear phase funtion)
1 (constant in time)	$\delta(f)$ (delta function)
$e^{j2\pi f_o x}$ (complex exponential)	$\delta(f - f_o)$ (shifted delta function)
$\mathrm{comb}_T(x) = \sum\limits_{k=-\infty}^{\infty} \delta(x - kT)$	$1/T\ \mathrm{comb}_{1/T}(f) = \frac{1}{T} \sum\limits_{k=-\infty}^{\infty} \delta(f - k/T)$
$\mathrm{rect}\left(\dfrac{x}{T}\right)$	$T\ \dfrac{\sin(\pi f\,T)}{\pi f\,T} = T \cdot \mathrm{sinc}(f\,T)$
	$\mathrm{sinc}(x) \overset{\Delta}{=} \sin(\pi x)/\pi x$ and $\mathrm{sinc}(0) = 1$

$$\mathrm{FT}\{i(x)\} = \mathrm{FT}\left\{ \sum_{k=-\infty}^{\infty} c_k^i e^{j2\pi k f_0 x} \right\} = \sum_{k=-\infty}^{\infty} c_k^i \delta(f - kf_0) \qquad (3.23)$$

Thus, we can use the FT even with periodic signals as long as we allow delta functions. Periodic signals in time lead to delta functions, or line spectra, in frequency. The CTFTs of some special CT signals we will encounter frequently are shown in Table 3.5.

3.4.3 Continuous-time Fourier transform properties

In this subsection, we will look at some of the important properties of 1-D CTFTs. A few of the more obvious ones are listed in Table 3.6 for reference while those requiring clarification are discussed here in more detail. While we did not discuss in detail the properties of the Fourier series, the CTFT properties can be applied to periodic signals by using delta functions in the frequency domain.

Even and odd symmetry When a signal is real and even (i.e., $i(x) = i(-x) = i^*(x)$), its FT is also real and even, i.e., $I(f) = I(-f) = I^*(f)$. For real, even signals the CTFT can be simplified as follows:

$$I(f) = \int_{-\infty}^{\infty} i(x)e^{-j2\pi f x}dx = \int_{-\infty}^{\infty} i(x) \cdot [\cos(2\pi f x) - j\sin(2\pi f x)]dx$$

$$= 2 \int_0^{\infty} i(x) \cdot \cos(2\pi f x)dx = I(-f) \qquad (3.24)$$

Table 3.6. *Some useful properties of the CTFT*

Property	Description
Duality	$i(x) \leftrightarrow I(f) \Rightarrow I(-x) \leftrightarrow i(f)$
Linearity	$FT\{ai_1(x) + bi_2(x)\} = aI_1(f) + bI_2(f)$
Time shifts	$i(x - x_0) \leftrightarrow I(f) \cdot e^{-j2\pi f x_0}$
Time derivative	$di(\dot{x})/dx \leftrightarrow (j2\pi f I(f))$ and $d^n i(x)/dx^n \leftrightarrow (j2\pi f)^n I(f)$
Conjugate symmetry	$i(x)$ is real $\Rightarrow I(f) = I^*(-f) \Rightarrow |I(f)| = |I(-f)|$ and $\theta(f) = -\theta(-f)$
Time reversal	$FT\{i(-x)\} = I(-f)$
	$\therefore$ for real $i(x)$, $FT\{i(-x)\} = I(-f) = I^*(f)$
Modulation	$t(x) = i(x) \cdot h(x) \leftrightarrow I(f) * H(f) = T(f)$
Convolution	$i(x) * h(x) \leftrightarrow I(f) \cdot H(f)$ (See Section 3.4.4)
Correlation	$i(x) \otimes h(x) \leftrightarrow I(f) \cdot H^*(f)$ (See Section 3.4.4)

where we use the fact that $i(x) \cdot \sin(2\pi f x)$ is an odd function of x, and $i(x) \cdot \cos(2\pi f x)$ is an even function since $i(x)$ is even.

On the other hand, if the signal $i(x)$ is an odd function of x, the FT simplifies as follows:

$$I(f) = \int_{-\infty}^{\infty} i(x) e^{-j2\pi f x} dx = \int_{-\infty}^{\infty} i(x) \cdot [\cos(2\pi f x) - j\sin(2\pi f x)] dx$$

$$= -j2 \int_{0}^{\infty} i(x) \cdot \sin(2\pi f x) dx = -I(-f)$$

(3.25)

Thus real, odd signals have an FT that is odd and purely imaginary.

Hartley transform A related transform is the Hartley transform (HT) defined below. We reproduce the FT definition to highlight the difference between the HT and the FT.

$$HT\{i(x)\} = \int_{-\infty}^{\infty} i(x)[\cos(2\pi f x) + \sin(2\pi f x)] dx$$

$$FT\{i(x)\} = \int_{-\infty}^{\infty} i(x)[\cos(2\pi f x) - j\sin(2\pi f x)] dx$$

(3.26)

The FT of a real signal can be complex, but the HT of a real signal is real. This can be an advantage in some applications. The HT appears to be of limited benefit. Perhaps the following two useful properties of FT are the reason for preference for the FT.

Direct current value From the CTFT definition, we can see that $I(0) = \int_{-\infty}^{\infty} i(x)dx$, which equals the total area under the signal. Since this total area is related to the average value of the signal, $I(0)$ is known as the DC value. Similarly, the value $i(0)$ of the signal at origin (i.e., $x = 0$) is equal to $\int_{-\infty}^{\infty} I(f)df$, the total area of the FT. We will use this property in relating correlation values at the origin (also called correlation peaks, presuming that, as is common, the LSI system has been designed to produce large output values there) to the correlation filter.

Parseval's theorem The energy in the time domain is equal to the energy in the frequency domain as proved below. This is a property we will need when relating the correlation output energy to the correlation filter.

$$
\begin{aligned}
\int_{-\infty}^{\infty} |I(f)|^2 df &= \int_{-\infty}^{\infty} df \left[\int_{-\infty}^{\infty} i(x)e^{-j2\pi fx}dx \right] \left[\int_{-\infty}^{\infty} i^*(p)e^{j2\pi fp}dp \right] \\
&= \int_{-\infty}^{\infty}\int_{-\infty}^{\infty} i(x)i^*(p) \left[\int_{-\infty}^{\infty} e^{j2\pi f(p-x)}df \right] dxdp \qquad (3.27) \\
&= \int_{-\infty}^{\infty}\int_{-\infty}^{\infty} i(x)i^*(p)\delta(p-x)dxdp = \int_{-\infty}^{\infty} |i(x)|^2 dx
\end{aligned}
$$

where we use the fact that the IFT of a linear phase function is a shifted delta function (see Table 3.5).

3.4.4 Convolution and correlation using the CTFT

As shown in the following, convolution in the x domain is equivalent to multiplication in the frequency domain.

$$
i(x) \leftrightarrow I(f), \; h(x) \leftrightarrow H(f), \qquad \text{and } o(x) = i(x) * h(x) \leftrightarrow O(f)
$$

$$
\begin{aligned}
O(f) &= \int o(x)e^{-j2\pi fx}dx = \int \left[\int i(\tau)h(x-\tau)d\tau \right]e^{-j2\pi fx}dx \\
&= \int \left[\int i(\tau)e^{-j2\pi f\tau}d\tau \right]h(\beta)e^{-j2\pi f\beta}d\beta = I(f) \cdot H(f) \\
&\therefore \; i(x) * h(x) \leftrightarrow I(f) \cdot H(f)
\end{aligned}
\qquad (3.28)
$$

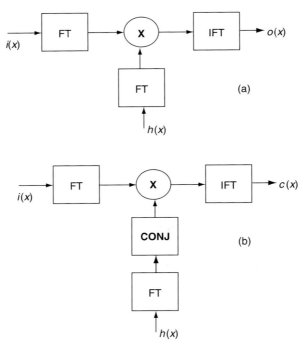

Figure 3.4 Block diagrams showing (a) the computation of convolution using Fourier transforms, and (b) the computation of correlation using Fourier transforms

where in Eq. (3.28) $\beta = (x - \tau)$. We can similarly derive the following property for correlations:

$$c(x) = i(x) \otimes h(x) \leftrightarrow I(f) \cdot H^*(f) \tag{3.29}$$

Correlations and convolutions of two signals can be obtained by performing Fourier transforms on the two signals, multiplying the FTs and then performing an IFT on that product.

Figure 3.4 contains block diagrams that show how convolutions and correlations can be computed using FTs. Note that the only difference between the convolution and the correlation is the complex conjugation step in computing the correlation. If we are not careful in this complex conjugation step (for example, when using MATLAB, you should remember that $\mathbf{x}'$ refers to the conjugate transpose of $\mathbf{x}$), we may end up getting a convolution instead of a correlation. If the function $h(x)$ is real and even, then $H(f)$ is real and even, in which case there is no difference between convolution and correlation. Thus while the two may be identical in some special situations, in general convolution and correlation are very different.

While using the FT to determine convolution and correlation appears to be a rather roundabout process, it turns out to be attractive in two different ways. First, we will see in a later chapter that optical processors can provide FTs (in fact 2-D FTs) rapidly, and this property enables us to compute convolutions and correlations rapidly by taking advantage of the speed of optical FT processors. Secondly, digital computers can compute the FT rapidly using the fast Fourier transform (FFT), and we can compute convolutions and correlations efficiently using this property.

3.4.5 Auto-correlation function peak

When a signal is correlated with itself, the resulting output is called its auto-correlation function (ACF). We will prove here that an ACF always peaks at the origin. Let $c_{ii}(x)$ denote the correlation of $i(x)$ with itself. From Eq. (3.29), we see that $c_{ii}(x)$ and $|I(f)|^2$ form an FT pair. Therefore,

$$|c_{ii}(x)| = \left| \int_{-\infty}^{\infty} |I(f)|^2 e^{j2\pi fx} df \right| \leq \int_{-\infty}^{\infty} |I(f)|^2 |e^{j2\pi fx}| df = \int_{-\infty}^{\infty} |I(f)|^2 df \tag{3.30}$$

$$\Rightarrow |c_{ii}(x)| \leq c_{ii}(0)$$

This property that the ACF always peaks at the origin, coupled with its shift-invariant nature, makes the correlation very useful for locating known reference signals in received signals. For example, consider a simple radar scenario in which a reference signal $r(x)$ is transmitted. This signal is reflected off a target and is returned to the receiver after a time delay of x_0 in the form of a received or input signal $i(x) = r(x - x_0)$. For now, we will ignore problems such as noise, attenuation and interference and concern ourselves only with the time delay. The goal is to determine the time delay (and hence the range) of the target from our knowledge of the reference signal $r(x)$ and the input signal $i(x)$. To determine the time delay, we perform the cross-correlation between $i(x)$ and $r(x)$. Since $i(x) = r(x - x_0)$ and the correlation is a shift-invariant operation, the cross-correlation output will equal $c_{rr}(x - x_0)$, a shifted ACF. Thus, the peak of the output will appear at $x = x_0$ and we can determine the target range by locating the peak of the correlation output. Of course noise and other impairments present in the received signal affect how accurately we will be able to locate this correlation peak.

One word of caution is in order here. Among correlations, only the ACF is guaranteed to have its peak at the origin. There is no such guarantee with cross-correlations unless, as we have done above, we can relate the cross-correlations to auto-correlations. Thus, we cannot blindly assume that all correlation operations will result in peaks at exactly the origin of the input object. When we consider composite correlation filters in later chapters, we will mostly be carrying out cross-correlations and we should be careful not to assume that the peaks automatically correspond to the target location.

3.4.6 Periodic or circular convolution

How can we use FTs to compute convolutions and correlations of *periodic* signals? If only one of the two signals is periodic, the resulting output will also be periodic. If both signals are periodic, the convolution and correlation operations in Eqs. (3.8) and (3.9) cannot be computed since the infinite sum of periodic functions leads to infinity. This is because the convolution or the correlation sum includes infinite periods of the product of the two periodic signals. When both $i[x]$ and $h[x]$ are periodic with period T, we use their *circular* or *periodic* convolution defined as follows:

$$\tilde{o}[x] = i[x] \,\tilde{*}\, h[x] = \sum_{k \in \langle T \rangle} i[k] \cdot h[n - k] \qquad (3.31)$$

where we use the tilde ($\sim$) to indicate the circular nature of the operations and results, and we use $\langle T \rangle$ to denote one contiguous period. The *circular correlation* is similarly defined as

$$c[n] = i[n] \,\tilde{\otimes}\, h[n] = \sum_{k \in \langle T \rangle} i[k] \cdot h[n + k] \qquad (3.32)$$

The main difference between linear and circular convolution is that linear convolution uses the summation from $-\infty$ to $+\infty$ whereas circular convolution uses a summation interval of just one period. Circular convolution and correlation can be similarly defined for CT signals. This difference between circular and linear operations is not just a theoretical curiosity. When we use efficient algorithms such as the FFT to compute convolutions and correlations on a digital computer, we are effectively computing circular convolutions and correlations. It is critical to understand the similarities and differences between the linear and circular operations, and we will look at these differences more closely in Section 3.6 after we introduce the FFT.

3.4.7 Two-dimensional CTFT

So far, we have considered the 1-D CTFT and its properties in detail. When dealing with image correlations, what we need is the 2-D CTFT. We will see in

Chapter 5 that we can assemble coherent optical processors (comprising lasers, spatial light modulators, lenses, and detectors) that can provide the 2-D CTFT at very high speeds. This sub-section is devoted to defining 2-D transforms and examining some sample FT pairs.

The 2-D FT and IFT are defined as follows. Here x and y denote the spatial coordinates whereas u and v denote the corresponding spatial frequencies. We also use $\Leftrightarrow$ to denote 2-D Fourier transform.

$$\text{FT}: \quad I(u,v) = \int_{-\infty}^{\infty}\int_{-\infty}^{\infty} i(x,y)e^{-j2\pi(ux+vy)}\,\mathrm{d}x\mathrm{d}y$$

$$\text{IFT}: \quad i(x,y) = \int_{-\infty}^{\infty}\int_{-\infty}^{\infty} I(u,v)e^{j2\pi(ux+vy)}\,\mathrm{d}u\mathrm{d}v$$

(3.33)

These 2-D transforms involve double integrals and in general are more complicated than carrying out two 1-D integrals. If the signal is separable (i.e., $i(x,y) = i_x(x) \cdot i_y(y)$), then its 2-D FT is also separable and can be computed using only two 1-D transforms.

$$I(u,v) = \iint i_x(x)i_y(y)e^{-j2\pi(ux+vy)}\,\mathrm{d}x\mathrm{d}y$$
$$= \left[\int i_x(x)e^{-j2\pi ux}\mathrm{d}x\right]\left[\int i_y(y)e^{-j2\pi vy}\mathrm{d}y\right] = I_u(u) \cdot I_v(v)$$

(3.34)

There are a few separable signals of interest. However, most 2-D signals are not separable. As we shall see later, even a non-separable 2-D signal is efficiently transformed by the "fast" method, so not all is lost by non-separability.

Two-dimensional CT delta function The 2-D delta function is separable, i.e., $\delta(x,y) = \delta(x)\delta(y)$. Thus, its 2-D CTFT is also separable, i.e., $I(u,v) = 1$ for all u and v.

Two-dimensional comb function A 2-D comb function is the product of 1-D comb functions and leads to the following separable CTFT:

$$\left[\mathrm{comb}_{T_x}(x) \cdot \mathrm{comb}_{T_y}(y)\right] \Leftrightarrow \left[(f_x\mathrm{comb}_{f_x}(u)) \cdot (f_y\mathrm{comb}_{f_y}(v))\right]$$

(3.35)

where $f_x = 1/T_x$ and $f_y = 1/T_y$.

Line functions It is easy to verify the following 2-D CTFT pairs.

$$\delta(x) \Leftrightarrow \delta(v) \quad \delta(y) \Leftrightarrow \delta(u)$$

(3.36)

We see that a line along the x-axis results in a 2-D CTFT that is a line along
the v-axis, whereas a line along the y-axis leads to a CTFT that is a line along the
u-axis. A non-separable function of interest is a tilted line going through the
origin. Its 2-D CTFT is a tilted line in the (u, v) plane that is at right angles to
the line in the (x, y) plane, i.e.,

$$\delta(y - mx) \Leftrightarrow \delta(u + vm) \qquad (3.37)$$

(Incidentally, this is an example where a coordinate transform can give a
desirable property for a particular image quality. Another that we have seen
is that the circ($\cdot$) function becomes separable under the polar transform.)

Two-dimensional rect functions   A rectangular window in space centered at the
origin is the product of 1-D rect($\cdot$) functions. Thus the resulting 2-D CTFT is a
product of sinc($\cdot$) functions in u and v.

$$\left[\text{rect}\left(\frac{x}{T_x}\right) \cdot \text{rect}\left(\frac{y}{T_y}\right)\right] \Leftrightarrow \left[T_x\text{sinc}(uT_x) \cdot T_y\text{sinc}(vT_y)\right] \qquad (3.38)$$

Another 2-D function of interest is the circ($\cdot$) function in Eq. (3.4). As the circ
function is not separable, its 2-D CTFT requires a double integral. Because of
its circular symmetry, we will find it more convenient to determine the 2-D
CTFT of the circ function using the 2-D CTFT properties to be discussed in
the next sub-section.

3.4.8 Two-dimensional CTFT properties

Many properties of the 2-D CTFT are obvious extensions of 1-D CTFT
properties. For example, shifting the 2-D signal $i(x, y)$ by (x_0, y_0) results in
the 2-D CTFT $I(u, v)$ being multiplied by the complex exponential
$e^{-j2\pi(ux_0+vy_0)}$. Other properties such as derivatives and Parseval's relation can
be similarly extended. However, there are a few properties that are unique to
the 2-D CTFT and we will consider these in this sub-section.

Two-dimensional CTFT in polar coordinates   Sometimes, it is more convenient
to use polar coordinates rather than Cartesian coordinates. After PT, (x, y)
becomes (r, θ) and (u, v) becomes (ρ, ϕ). Also, the PTs in the space domain
and in the frequency domain have Jacobians r and ρ. We will denote the polar
transformed functions with a hat.

$$\hat{I}(\rho, \phi) = I(\rho \cos(\phi), \rho \sin(\phi))$$

$$= \int\limits_{-\infty}^{\infty} \int\limits_{-\infty}^{\infty} i(x, y) e^{-j2\pi\rho(x\cos(\phi) + y\sin(\phi))} dxdy$$

$$= \int\limits_{0}^{\infty} rdr \int\limits_{0}^{2\pi} d\theta \hat{i}(r, \theta) e^{-j2\pi\rho r(\cos(\theta)\cos(\phi) + \sin(\theta)\sin(\phi))} \qquad (3.39)$$

$$= \int\limits_{0}^{\infty} rdr \int\limits_{0}^{2\pi} d\theta \hat{i}(r, \theta) e^{-j2\pi\rho r \cos(\theta - \phi)}$$

Circular symmetry The 2-D CTFT of a circularly symmetric image is itself circularly symmetric as shown below. To take advantage of the circular symmetry, it is better to deal with the image and its FT in polar coordinates than in Cartesian coordinates.

$$\hat{I}(\rho, \phi) = \int\limits_{0}^{\infty} \hat{i}(r) rdr \int\limits_{0}^{2\pi} d\theta e^{-j2\pi r\rho \cos(\theta - \phi)} = 2\pi \int\limits_{0}^{\infty} \hat{i}(r) J_0(2\pi r\rho) rdr = \hat{I}(\rho)$$

$$(3.40)$$

where we use the result that $\int_0^{2\pi} e^{-j\beta \cos(\theta)} d\theta = 2\pi J_0(\beta)$ and $J_0(\cdot)$ is the Bessel function of first kind and zeroth order. The 1-D integral relation $2\pi \int_0^\infty \hat{i}(r) J_0(2\pi r\rho) rdr = \hat{I}(\rho)$ between a circularly symmetric image $\hat{i}(r)$ and its circular symmetric 2-D CTFT, $\hat{I}(\rho)$ is known as the *Hankel* transform or *Fourier–Bessel* transform. The inverse Hankel transform relation is the same, i.e.,

$$\hat{i}(r) = 2\pi \int\limits_{0}^{\infty} \hat{I}(\rho) J_0(2\pi r\rho) \rho d\rho \qquad (3.41)$$

Two-dimensional CTFT of circ function As an illustration of the use of the Hankel transform, let us consider the 2-D CTFT of the circ function.

$$FT\left\{ circ\left(\frac{r}{R}\right) \right\} = 2\pi \int\limits_{0}^{R} J_0(2\pi r\rho) rdr$$

$$(3.42)$$

$$= \frac{1}{2\pi\rho} \int\limits_{0}^{2\pi\rho R} \left(\frac{\beta}{\rho}\right) J_0(\beta) d\beta = \frac{R}{\rho} J_1(2\pi\rho R)$$

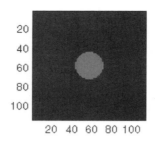

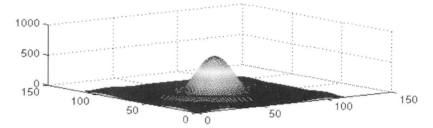

Figure 3.5 (a) A circular aperture, and (b) its 2-D continuous-time Fourier transform

where we use the result that $\int_0^x y J_0(y)\mathrm{d}y = x J_1(x)$ with $J_1(\cdot)$ denoting the Bessel function of first kind and first order. As stated before, the 2-D CTFT of the circularly symmetric circ($\cdot$) function is also circularly symmetric. The function $J_1(x)/x$ is known as an Airy function and it shows up in the diffraction patterns of circular apertures. We show a circular aperture in Figure 3.5(a) and its 2-D CTFT in Figure 3.5(b).

Two-dimensional CTFT of a circular Gaussian Consider a circularly symmetric 2-D Gaussian $i(x,y) = \frac{1}{2\pi\sigma^2}\exp\left(-\frac{x^2+y^2}{2\sigma^2}\right) = \frac{1}{2\pi\sigma^2}\exp\left(-\frac{r^2}{2\sigma^2}\right)$ with a width parameter σ. This 2-D Gaussian is separable and can be expressed as a product of two 1-D Gaussians. We will show below that the 1-D FT of a Gaussian with width parameter σ is a Gaussian with width parameter $1/\sigma$.

$$
\mathrm{FT}\left\{\frac{1}{\sqrt{2\pi\sigma^2}}\exp\left[-\frac{x^2}{2\sigma^2}\right]\right\} = \frac{1}{\sqrt{2\pi\sigma^2}}\int_{-\infty}^{\infty}\exp\left[-\frac{x^2}{2\sigma^2} - j2\pi ux\right]\mathrm{d}x
$$

$$
= \mathrm{e}^{-2\pi^2\sigma^2 u^2}\left\{\frac{1}{\sqrt{2\pi\sigma^2}}\int_{-\infty}^{\infty}\exp\left[-\frac{(x+j2\pi\sigma^2 u)^2}{2\sigma^2}\right]\mathrm{d}x\right\}
$$

$$
= \mathrm{e}^{-2\pi^2\sigma^2 u^2}
$$

$$
\tag{3.43}
$$

where the integral within the { } brackets is the area under a Gaussian probability density function (PDF) and therefore equals 1. Thus, the 2-D CTFT of a 2-D Gaussian is also a 2-D Gaussian, i.e.,

$$\frac{1}{2\pi\sigma^2}\exp\left[-\frac{x^2+y^2}{2\sigma^2}\right] \Leftrightarrow \exp\left[-2\pi^2\sigma^2(u^2+v^2)\right] = \exp\left[-2\pi^2\sigma^2\rho^2\right] \quad (3.44)$$

Central slice theorem Another property that is unique to 2-D CTFT is the relation between image projections and their FT cross-sections and vice versa. The $u=0$ and $v=0$ cross-sections (or slices) in $I(u, v)$ are related to projections onto the y-axis and the x-axis in the image as shown below.

$$I(0, v) = \int_{-\infty}^{\infty}\left[\int_{-\infty}^{\infty} i(x, y)\mathrm{d}x\right]\cdot \mathrm{e}^{-j2\pi vy}\mathrm{d}y = \int_{-\infty}^{\infty} \bar{\imath}_y(y)\cdot \mathrm{e}^{-j2\pi vy}\mathrm{d}y = \mathrm{FT}\{\bar{\imath}_y(y)\}$$

$$I(u, 0) = \int_{-\infty}^{\infty}\left[\int_{-\infty}^{\infty} i(x, y)\mathrm{d}y\right]\cdot \mathrm{e}^{-j2\pi ux}\mathrm{d}x = \int_{-\infty}^{\infty} \bar{\imath}_x(x)\cdot \mathrm{e}^{-j2\pi ux}\mathrm{d}x = \mathrm{FT}\{\bar{\imath}_x(x)\}$$

$$(3.45)$$

where $\bar{\imath}_x(x)$ and $\bar{\imath}_y(y)$ denote the projections of $i(x, y)$ onto the x-axis and y-axis, respectively, and are defined implicitly in the equations. Thus the $v=0$ cross-section in the 2-D FT can be obtained by performing a 1-D FT on the projection of $i(x, y)$ onto the x-axis. Such a result can be generalized for any cross-section through the origin at angle ϕ with respect to the u-axis; i.e.,

$$I(a\cos(\phi), a\sin(\phi)) = \int_{-\infty}^{\infty}\int_{-\infty}^{\infty} i(x, y)\mathrm{e}^{-j2\pi a[x\cos(\phi)+y\sin(\phi)]}\mathrm{d}x\mathrm{d}y$$

$$= \int_{-\infty}^{\infty}\int_{-\infty}^{\infty} i(x', y')\mathrm{e}^{-j2\pi ax'}\mathrm{d}x'\mathrm{d}y' \quad (3.46)$$

(where $x' = x\cos(\phi) + y\sin(\phi)$ & $y' = -x\sin(\phi) + y\cos(\phi)$)

$$= \int_{-\infty}^{\infty} \bar{\imath}_{x'}(x')\mathrm{e}^{-j2\pi ax'}\mathrm{d}x' = \mathrm{FT}\{\bar{\imath}_{x'}(x')\}$$

While the above derivations dealt with projections in the image and cross-sections in its 2-D FT, similar relations can be derived for image cross-sections and 2-D FT projections.

The central slice theorem (CST) has found use in at least two applications. In computer-aided tomography, 1-D projections through a 2-D signal are obtained at various angles. The CST tells us that the 1-D FTs of these projections are nothing but cross-sections of the 2-D FT. By interpolation, we can obtain a "full" 2-D FT from these cross-sections and we can obtain the original 2-D function via a 2-D inverse FT. The same principle can be used in reconstructing three-dimensional volumes from two-dimensional projections.

The second application is in the correlation of an image with a shifted version of itself in order to estimate the shifts. In such an application, we can perform the 1-D correlation of x-projections to determine the shift in x and the 1-D correlation of y-projections to determine the shift in y. Such 1-D correlations of projections can be obtained using cross-sections in the 2-D FT. Similarly, the CST can be used to obtain the 2-D FT when only 1-D FT devices are available. Such 1-D FT devices can be used to provide various cross-sections of the 2-D FT and eventually produce the full 2-D FT.

3.5 Sampling theory

Sampling allows us to convert at least some of the content of a CT signal into a DT signal, which allows simulation and processing of signals using digital computers. The trick is to do this conversion so that there is no signal loss; i.e., we should be able to reproduce the original CT signal faithfully from the DT signal (i.e., samples).

Before digging too deeply into the underlying mathematics, it is important to realize the basic idea of sampling. If a signal is varying slowly in time (e.g., a constant in time), we need only a few samples to describe it. On the other hand, if the signal is varying rapidly, we need many more samples to describe it. We will prove that sampling frequency must exceed twice the highest signal frequency.

Let $i(x)$ be an analog signal with a maximum frequency f_{max}, i.e., $I(f) = 0$ for $|f| > f_{max}$. An example spectrum is shown in Figure 3.6(a). The CT signal is sampled at uniform intervals of T_s to obtain the sampled signal $i_d[n]$, i.e., $i_d[n] = i(nT_s)$. Then $f_s = 1/T_s$ is the sampling frequency. To understand the effects of sampling, we need to see what happens in the frequency domain. Since the sampled signal is equal to the original CT signal at the sampling instants and zero everywhere else, we can consider $\hat{i}(x) = i(x) \cdot \mathrm{comb}_{T_s}(x)$ in place of $i_d[n]$. Using the modulation property in Table 3.6, and the FT of the comb function in Table 3.5, we obtain the following relation between the FTs of the sampled signal and the original signal:

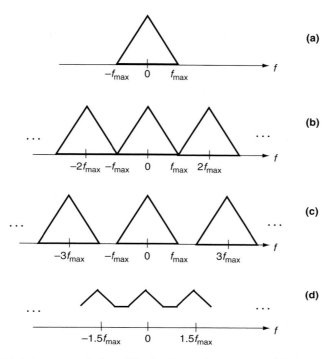

Figure 3.6 (a) An example band-limited signal spectrum, (b) the spectrum of an exactly sampled signal, (c) the spectrum of an over-sampled signal, and (d) the spectrum of an under-sampled signal

$$\mathrm{FT}\{\hat{i}(x)\} = \mathrm{FT}\{i(x) \cdot \mathrm{comb}_{T_s}(x)\} = I(f) * [\mathrm{FT}\{\mathrm{comb}_T(x)\}]$$

$$= I(f) * \left[\frac{1}{T_s} \sum_{k=-\infty}^{\infty} \delta\left(f - \frac{k}{T_s}\right) \right] = \frac{1}{T_s} \sum_{k=-\infty}^{\infty} I\left(f - \frac{k}{T_s}\right) \qquad (3.47)$$

$$= f_s \sum_{k=-\infty}^{\infty} I(f - kf_s)$$

where we use the property that convolving a function with a shifted delta function will simply shift that function to the position of the delta function.

Equation. (3.47) states that the FT of the sampled signal consists of infinite replicas of the original FT, each shifted by integer multiples of $f_s = 1 / T_s$ and each multiplied by the constant $1 / T_s$. If, within the bandlimit of the original signal $i(x)$, $\mathrm{FT}\{i(x)\}$ and $\mathrm{FT}\{\hat{i}(x)\}$ are not identical, the signal is said to be *aliased*. That is, the sampled signal cannot be used to reconstruct faithfully the original signal. (If they are identical, then the original can be reconstructed from information within the band limit. This is because the information in spatial and spectral domains is the same, and we can retrieve the spectral information from the sampled data if the signal is not aliased by the sampling.) If $I(f)$ is not band-limited (i.e., it is not zero for all frequencies above some maximum frequency), then these replicas, each being of infinite width, certainly overlap and the FT of the sampled signal will be different from the original spectrum. Even if the signal is band-limited, aliasing can occur and f_s says how far the replicas are shifted before being added to produce $FT\{\hat{\imath}(x)\}$. If the sampling does not shift the copies of the spectrum by at least the total bandwidth, the copies will "step on each other's toes" in the summation. In such a case, we will not be able to reconstruct the original signal from the sampled signal. Thus, sampling makes sense only when we are dealing with band-limited signals (e.g., see Figure 3.6(a)). If the original CT signal is not band-limited, it should be first passed through a band-limiting filter (more commonly known as an *anti-aliasing* filter) before sampling. Even though such a band-limiting filter may destroy some of the original CT signal information (at frequencies beyond the filter cut-off), we will see later in this sub-section that in practice anti-aliasing filtering is essential prior to sampling.

If the replicas in Eq. (3.47) do not overlap, we can obtain the original CT signal FT by multiplying the sampled signal FT by a suitably designed frequency response. Thus, inputting the properly sampled signal into a *reconstruction filter* with the selected frequency response should result in the original CT signal as its output. Let us now consider three choices for sampling frequency.

Exact sampling   Here $f_s = 2f_{max}$. We show the resulting sampled signal spectrum in Figure 3.6(b). As we can see, the edges of the replicas touch and could result in slightly corrupted information. In this sense, $f_s = 2f_{max}$ does not provide an adequate sampling frequency. To appreciate the potential problem in using exactly $f_s = 2f_{max}$, let us consider the sampling of a sinusoid of frequency f_{max}. Sampling at $f_s = 2f_{max}$ implies that we have two samples per cycle of sinusoid. If these two samples happen to be at the peaks ($+1$ and -1) of the sinusoid, the reconstruction filter will be able to bring back the original CT sinusoids from the samples. On the other hand, if the two samples per cycle happen to be exactly at the zero crossings, the reconstruction filter would simply yield an all zero output that, of course, is not acceptable. Thus using $f_s = 2f_{max}$ is not advisable in practice.

A sampling frequency of f_s should allow us to represent signals with frequencies up to half the sampling frequency, also known as the folding frequency $f_f = f_s/2$. This folding frequency will prove important when considering under-sampling.

Over-sampling   When over-sampling, $f_s > 2f_{max}$. We show in Figure 3.6(c) the FT of the sampled signal when $f_s = 3f_{max}$. As we can see from this figure, the replicas of the original FT do not overlap and thus the original spectrum within the original bandlimits, $-f_{max} \leq f \leq f_{max}$, is preserved. In fact, as long as $f_s > 2f_{max}$, the replicas do not overlap and original information is preserved. This sampling frequency requirement is known as the Nyquist sampling theorem. The original CT spectrum can be obtained by multiplying the spectrum of the sampled signal by any low-pass filter with the following frequency response:

$$H(f) = \begin{cases} 1/f_s \text{ for } |f| \leq f_{max} \\ \text{``Don't care'' for } f_{max} < |f| < (f_s - f_{max}) \\ 0 \text{ for } |f| \geq (f_s - f_{max}) \end{cases} \quad (3.48)$$

Using any filter with frequency response as in Eq. (3.48), we can reconstruct the original CT signal from the sampled signal as follows:

$$i(x) = \text{IFT}\{\hat{I}(f) \cdot H(f)\}$$

$$= \hat{i}(x) * h(x) = \left[\sum_{k=-\infty}^{\infty} i(kT_s)\delta(x - kT_s) \right] * h(x)$$

$$= \sum_{k=-\infty}^{\infty} i(kT_s)h(x - kT_s) \quad (3.49)$$

where $h(x)$ is the impulse response of the reconstruction filter. In deriving Eq. (3.49), we use the result that convolving a function with a shifted delta function simply shifts the original function to the position of the delta function. From Eq. (3.48), we see that the reconstruction low-pass filter has a pass-band ($|f| \leq f_{max}$), a transition band [$f_{max} < |f| < (f_s - f_{max})$], and a stop-band [$|f| \geq (f_s - f_{max})$]. The larger the transition band, the easier it is to implement this low-pass filter in the sense that it requires a lower order (i.e., less complicated) filter. Thus making the sampling frequency f_s larger than the minimum required (i.e., $2f_{max}$), i.e., oversampling, makes the reconstruction filter simpler. In practice, we use a sampling frequency that is at least 10% higher than twice the maximum signal frequency. As an example, audio CDs use a sampling frequency of 44.1 kHz, whereas the maximum audio frequency of interest is about 20 kHz.

Equation (3.49) can be thought of as an interpolation formula in that it shows how to obtain the CT signal from its samples at uniform intervals. If the reconstruction filter is a "brick-wall" low-pass filter with a cut-off frequency $f_s/2$, it satisfies the requirements in Eq. (3.48) and can be used as an interpolation filter. Since the IFT of a rect($\cdot$) function is a sinc($\cdot$) function, we get the following interpolation formula:

$$i(x) = \sum_{k=-\infty}^{\infty} i(kT_s)\text{sinc}(x - kT_s) \quad (3.50)$$

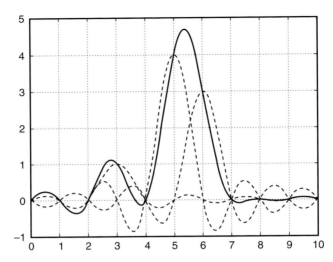

Figure 3.7 An illustration of the sinc interpolation method. In this example, there are only three non-zero signal samples whose contribution to the reconstruction is the three dashed curves. The interpolation result is the solid curve

This sinc interpolation formula is illustrated in Figure 3.7. The sinc($\cdot$) function is 1 at the origin and 0 for all other integer arguments. Thus, the kth term in the summation in Eq. (3.50) leads to a sinc($\cdot$) function that has a value of $i(kT_s)$ at the kth sampling instant, i.e., $x = kT_s$, and zero at all other sampling instants. Summation of all these sinc($\cdot$) functions reproduces the original CT signals. Remember, however, that this interpolation formula is valid only if $f_s > 2f_{max}$. Also, the sinc interpolation formula is not attractive in practice as it suffers from two problems. The first is that it requires an infinite summation, and truncating it to a finite summation leads to unacceptably large errors in reconstruction. The second problem is that the samples must be precisely at $x = kT_s$ and the interpolation is unacceptably sensitive to sampling location errors. As a result, practical interpolation methods use more robust kernels such as spline functions. Sinc interpolation is not the only kernel we can use in building a CT signal from sampled data. All that is required to be a resampling kernel is that a function should be continuous, unity at zero, and zero at all other integers. Sinc($\cdot$) is unique among interpolation kernels in having the most compact transform. Nearest-neighbor, sample-and-hold, and triangular interpolation are interpolation kernels, each of which has wide bandwidth in comparison with sinc($\cdot$); they are illustrated in Figure 3.8.

The reconstructions that these kernels yield for the signal samples of Figure 3.7 are shown in Figure 3.9. Several things are clear. They produce

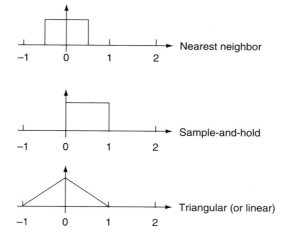

Figure 3.8 Example interpolation kernels

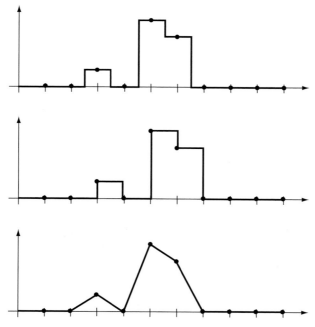

Figure 3.9 Reconstruction from the samples in Figure 3.7 using kernels in Figure 3.8

different reconstructions of the signal from the same sample values, each depends critically on where the samples were drawn, the reconstruction is computationally much simpler than that of the sinc($\cdot$), and their reconstruction's spectrum is wider than that of the sinc($\cdot$) interpolation.

Under-sampling What happens when the sampling frequency is less than twice the maximum signal frequency? We show in Figure 3.6(d) the spectrum of a sampled signal when $f_s = 1.5f_{max}$. As we can see from this figure, the replicas overlap. In fact, the central replica is corrupted for $(f_s - f_{max}) < |f| < f_{max}$ in general and $(0.5f_{max}) < |f| < f_{max}$ for this particular choice of $f_s = 1.5f_{max}$.

A sampling frequency of $f_s = 1.5f_{max}$ could have preserved signal frequencies up to the folding frequency, i.e., $f_f = 0.75f_{max}$. However, we have seen from Figure 3.6(d) that all signal frequencies above $0.5f_{max}$ are in fact corrupted. This is because under-sampling causes a frequency $f_f + \Delta$ (Δ higher than the folding frequency) to appear (i.e., alias) as a signal frequency $f_f - \Delta$ (Δ lower than the folding frequency).

We can prevent aliasing by removing signal frequencies above folding frequency *before* sampling. Thus, if we pass the original CT signal through a CT low-pass filter (usually called an anti-aliasing filter) with a cut-off of $f_f = f_s/2$, then two things happen. Firstly this filter will eliminate all signal frequencies above the folding frequency and thus some signal frequencies will be irrevocably destroyed. On the other hand, the sampling frequency being used will be adequate to sample the low-pass filtered signal and thus no aliasing will occur. The bottom line is that we are able to preserve more signal frequencies ($|f| < f_f$) for the under-sampling case by using the anti-aliasing filter than we would otherwise ($|f| < 2f_f - f_{max}$).

3.5.1 Sampling in two dimensions

When processing images using a digital computer, the images must be digitized. Even with an optical processor, we use digitized images because digital cameras provide digitized images, and because the electronically addressable spatial light modulators (SLMs) that represent the images have a grid structure. Thus, we need to use sampling in 2-D.

The most obvious 2-D sampling strategy is to use a uniform rectangular sampling lattice, i.e., $\hat{i}(x, y) = i(x, y) \cdot \text{comb}_{T_x}(x) \cdot \text{comb}_{T_y}(y)$, where T_x and T_y denote the sampling intervals in x and y, respectively. Because of the separable nature of the rectangular lattice, the spectrum of the sampled signal is related to the original spectrum as follows:

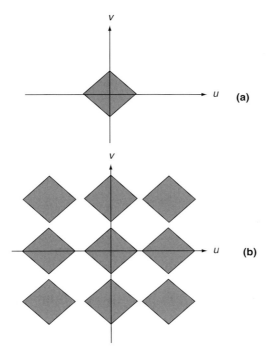

Figure 3.10 (a) An example 2-D band-limited spectrum, (b) spectrum of sampled signal based on rectangular sampling

$$\hat{I}(u, v) = \frac{1}{T_x T_y} \sum_{k=-\infty}^{\infty} \sum_{l=-\infty}^{\infty} I\left(u - \frac{k}{T_x}, v - \frac{l}{T_y}\right) \qquad (3.51)$$

This relation is illustrated pictorially in Figure 3.10. Figure 3.10(a) shows the original spectrum that is band-limited in the (u, v) plane. Figure 3.10(b) shows the spectrum of the sampled signal assuming $(1/T_x) > 2u_{max}$ and $(1/T_y) > 2v_{max}$, where u_{max} and v_{max} denote the maximal u and v frequencies in the signal, respectively.

As long as the sampling intervals in the x and y directions satisfy the above conditions, the spectral replicas do not overlap and the original 2-D CT signal can be reconstructed faithfully using a 2-D low-pass filter. The variety of low-pass filters that can be used for reconstruction in this 2-D case is larger than the set of reconstruction filters for the 1-D sampling case. Of course, we can use a rectangular low-pass filter that is constant when $|u| \leq (1/2T_x)$ and $|v| \leq (1/2T_y)$ and is zero outside. Such a 2-D brick-wall low-pass filter will result in an interpolation method that uses sinc functions in both

x and y directions. If either of the sampling intervals is larger than the maximum allowed, the spectral replicas will overlap leading to aliasing. To avoid the effects of aliasing, it is advisable to pass the original signal through a 2-D anti-aliasing filter with a cut-off of $1/(2T_x)$ along the u-axis and $1/(2T_y)$ along the v-axis. From Figure 3.10(b), we can see that more efficient sampling is possible by using sampling geometries such as a hexagonal lattice.

3.6 Fourier transform of DT signals

While analog processors such as optical computers can handle CT signals and images, digital computers can accommodate only DT signals. In this section, we consider the Fourier transform of DT signals. As mentioned earlier, discreteness in one domain implies periodicity in the other domain. Since digital computers represent both space and time domain signals and their FTs by discrete arrays, both the DT signals and their FTs are implicitly periodic. This implicit periodicity produces circular correlations instead of linear correlations if we are not careful.

3.6.1 Discrete Fourier transform

The N-point discrete Fourier transform (DFT) of a sequence $\{i[n], 0 \leq n \leq N-1\}$ results in an array $\{I[k], 0 \leq k \leq N-1\}$ related to the DT signal as follows:

$$\text{DFT} : I[k] = \sum_{n=0}^{N-1} i[n] e^{-j2\pi \frac{nk}{N}} \quad k = 0, 1, 2, \ldots, (N-1) \quad (3.52)$$

An N-point DFT produces an N-point discrete sequence $I[k]$ in the transform domain from an N-point sequence $i[n]$ in the time domain. As we will show below, increasing N leads to denser sampling in the frequency domain. We can always use a DFT size larger than the number of points in the DT signal by simply concatenating zeros (called *zero-padding*) to the original DT signal.

From Eq. (3.52), we can see that $I[k]$ is periodic with period N, i.e., $I[k] = I[k+N]$. This is because $e^{-j2\pi n\frac{k}{N}} = e^{-j2\pi \frac{nk+N}{N}}$. As we cautioned earlier, using N-point DFTs imposes an implied periodicity of N in the frequency domain as well as in the time domain.

Let us compare the DFT results to the CTFT of the CT signal $i(x) = \sum_{n=0}^{N-1} i[n]\delta(x-n)$. This CT signal $i(x)$ contains the same information as the DT signal $i[n]$. Since the signal $i(x)$ contains delta functions at intervals of 1, its CTFT $I_c(f)$ is periodic with a period of 1 and can be expressed as follows (we use a subscript to distinguish the CTFT $I_c(f)$ from its DT counterpart):

$$
\begin{aligned}
I_c(f) &= \int_{-\infty}^{\infty} \left[\sum_{n=0}^{N-1} i[n]\delta(x-n) \right] e^{-j2\pi f x} \mathrm{d}x \\
&= \sum_{n=0}^{N-1} i[n] \left[\int_{-\infty}^{\infty} \delta(x-n)e^{-j2\pi f x}\mathrm{d}x \right] = \sum_{n=0}^{N-1} i[n]e^{-j2\pi f n}
\end{aligned}
\tag{3.53}
$$

Comparing the N-point DFT in Eq. (3.52) and the CTFT in Eq. (3.53), we can see that the N-point DFT basically samples one period of the CTFT at uniform intervals of $(1/N)$, i.e., $I[k] = I_c(k/N)$, $k = 0, 1, 2, \ldots, (N-1)$. By increasing N, we can obtain denser sampling in the frequency domain.

Note that the first DFT sample (i.e., $I[0]$) corresponds to the zero frequency. Since we are used to spectral plots with zero frequency at the center rather than at the beginning, we usually plot $\{I[N/2], I[(N/2)+1], \ldots, I[N-1], I[0], I[1], \ldots, I[(N/2)-1]\}$. Many FFT software packages will contain a command (e.g., FFTSHIFT in Matlab) that shifts the N-point FFT results so that the DC value is at (or, in the usual case that N is even, adjacent to) the center of the array.

Inverse DFT DT signal $i[n]$ can be obtained from $I[k]$ using the N-point inverse DFT (IDFT) defined below:

$$
\text{IDFT:} \quad i[n] = \frac{1}{N} \sum_{k=0}^{N-1} I[k]e^{j2\pi\frac{nk}{N}} \quad n = 0, 1, 2, \ldots, (N-1)
\tag{3.54}
$$

Note that the DFT in Eq. (3.52) and the IDFT in Eq. (3.54) differ in only two ways. One difference is the $(1/N)$ factor in the IDFT definition, and the second is the sign difference in the exponents. These differences are simple enough for a system or software program designed to carry out the DFT to be easily reconfigured to yield the inverse IDFT.

Fourier-transform of 2-D DT signals For 2-D DT signals, we can define DFT and IDFT as follows (note that we use N and M to denote the extent of the 2-D DT signal in the n and m directions, respectively):

$$\text{2-D DFT:} \quad I[k,l] = \sum_{n=0}^{N-1} \sum_{m=0}^{M-1} i[n,m] e^{-j2\pi\left(\frac{nk}{N} + \frac{ml}{M}\right)},$$

$$k = 0,1,2,\ldots,(N-1) \quad \text{and } l = 0,1,2,\ldots,(M-1)$$

$$\text{2-D IDFT:} \quad i[n,m] = \frac{1}{MN} \sum_{k=0}^{N-1} \sum_{l=0}^{M-1} I[k,l] e^{j2\pi\left(\frac{nk}{N} + \frac{ml}{M}\right)},$$

$$n = 0,1,2,\ldots,(N-1) \quad \text{and } m = 0,1,2,\ldots,(M-1)$$

$$(3.55)$$

A 2-D DFT can be carried out using 1-D DFTs. Firstly, we perform N-point 1-D DFTs along each of the M columns with N points. The results of these 1-D DFTs then replace the original columns. Next, we perform M-point 1-D DFTs along each of the N rows with M points. The resultant DFTs replace the rows. The resulting array is the (N, M)-point 2-D DFT $I[k, l]$. Thus (N, M)-point 2-D DFT can be achieved using N M-point and M N-point 1-D DFTs. With most memory architectures, accessing columns and rows of an array may not always be equally fast. In such a case, the first set of column DFTs is followed by an array transpose operation after which column DFTs are performed.

We will not discuss DFT properties in detail as they are similar to CTFT properties discussed before, except that we should keep in mind the implied periodicity in both domains when using the DFT. For example, multiplying the DFT $I[k]$ by a linear phase factor $e^{-j2\pi\frac{kn_0}{N}}$ is equivalent to a *circular shift n_0* in the time domain.

Circular shift $\text{IDFT}\left\{I[k] \cdot e^{-j2\pi\frac{kn_0}{N}}\right\} = i[(n-n_0)_N]$ where $(\cdot)_N$ denotes modulo N. In other words, circular shift by n_0 is equivalent to creating a periodic DT signal $\tilde{i}[n]$ by repeating the original DT signal $i[n]$ at intervals of N points, shifting the resulting $\tilde{i}[n]$ by n_0 points to the right, and truncating the result to the interval $0 \leq n \leq (N-1)$ to obtain $i[(n-n_0)_N]$. Circular shift can also be thought of as reintroducing the shifted signal to the beginning of the interval as it moves out of the other end. As an example, for $N=8$, $i[(n-2)_8]$ is the following sequence:

$$i\big[(n-2)_8\big] = \{i[6], \ i[7], \ i[0], \ i[1], \ i[2], \ i[3], \ i[4], \ i[5]\} \qquad (3.56)$$

There are other DFT properties such as the correlation and convolution properties that involve circular shifts. We will see in a later sub-section how these circular shifts affect correlation and convolution.

Symmetry properties The DFT has symmetry properties similar to those of the CTFT. For example, real, even time domain signals will have DFTs that are also real and even. However, the symmetry conditions are with respect to $\tilde{i}[n]$ (the periodic version of the original DT signal) and thus must be carefully related to the original DT signal $i[n]$.

If $i[n]$ is real, then $I[k]$ is conjugate symmetric as shown below; i.e., its real part is even and its imaginary part is odd. Thus for real $i[n]$,

$$I[k] = I_R[k] + jI_I[k] = \sum_{n=0}^{N-1} i[n]\mathrm{e}^{-j2\pi\frac{nk}{N}} = \left\{ \sum_{n=0}^{N-1} i[n]\mathrm{e}^{j2\pi\frac{nk}{N}} \right\}^*$$

$$= \left\{ \sum_{n=0}^{N-1} i[n]\mathrm{e}^{-j2\pi\frac{n(N-k)}{N}} \right\}^* = I^*[N-k] = I_R[N-k] - jI_I[N-k] \quad (3.57)$$

$$\therefore\ I_R[k] = I_R[N-k] \quad \text{and} \quad I_I[k] = -I_I[N-k]$$

Other symmetry properties (e.g., the DFT of a real, odd signal is purely imaginary and odd) can be similarly derived except that we need to keep in mind DFT's circular nature.

These symmetry properties can be used to advantage when computing the DFT of real signals. For example, suppose we are interested in computing the N-point DFTs of real sequences $i_1[n]$ and $i_2[n]$ each of length N. Instead of carrying out two separate N-point DFTs, we can get the desired results using only one N-point DFT. This is done by carrying out an N-point DFT on the complex signal $i[n] = i_1[n] + ji_2[n]$. Because of the symmetry property of the DFT, we can obtain the desired DFTs from $I[k] = I_1[k] + jI_2[k]$ as follows:

$$I_1[k] = (I[k] + I^*[N-k])/2 \qquad I_2[k] = (I[k] - I^*[N-k])/j2 \qquad (3.58)$$

3.6.2 Fast Fourier transform

The fast Fourier transform (FFT), introduced in the late 1960s [25], is an efficient algorithm to compute the DFT. It is perhaps safe to say that modern digital signal processing owes its origin to FFT and high-speed digital processors. The FFT has so penetrated digital signal processing that the term "FFT" is sometimes used (incorrectly, in our opinion) where "DFT" is intended. The DFT is a transform; the FFT is an algorithm to implement the DFT. In this sub-section, we discuss the basic principle of the FFT to provide a reasonably high-level understanding of it. For a more thorough understanding of the FFT than is provided here, readers should consult some of the digital signal processing references [26, 27]. While we will focus on 1-D DFTs, our FFT discussion is equally applicable to 2-D DFTs, since we have shown already that a 2-D DFT can be obtained as a series of 1-D DFTs.

Direct DFT We can compute the DFT in a straightforward manner according to Eq. (3.52). For each k, the summation involves N complex multiplications and N additions. We will use the number of multiplications as a measure of the computational complexity.[3] Since we need to determine N DFT samples, the computational complexity of a direct DFT method is of the order of N^2 operations. For short signals, this computational complexity might be acceptable. But for long signals (e.g., a million-point DT signal or equivalently a 1024×1024 image), the computational complexity is of the order of 10^{12} operations which may be prohibitive. The FFT provides a more efficient solution.

Fast Fourier transform The FFT is an efficient DFT computation algorithm based on the "divide and conquer" principle. When N is an integer power of 2 (e.g., 4, 8, 16, etc.), we will show that the FFT algorithm requires only $N \cdot \log_2(N)$ operations to obtain the N-point DFT. Thus if $N = 2^{20}$(about a million), the FFT requires only about 20 million operations, whereas the direct DFT computation requires about a trillion operations. Thus the FFT provides a computational advantage by a factor of about 50 000 in this case. For an N-point DFT, the FFT provides a computational advantage by a factor of $(N^2/N\log_2 N) = (N/\log_2 N)$. As N increases, this FFT efficiency ratio increases. Thus, except for short signals, we should always use the FFT instead of the direct DFT.

We will explain the basics of the FFT algorithm using a 4-point DFT that amounts to determining $\{I[0],\ I[1],\ I[2],\ I[3]\}$ from the 4-point DT signal $\{i[0],\ i[1],\ i[2],\ i[3]\}$. Firstly, we divide the 4-point DT sequence into two 2-point sub-sequences $a[n]$ and $b[n]$ that are the even-indexed and odd-indexed samples of the original sequence as follows:

$$a[0] = i[0], \quad a[1] = i[2] \text{ and } b[0] = i[1], \quad b[1] = i[3] \tag{3.59}$$

Let $A[k]$ and $B[k]$ denote the 2-point DFTs of $a[n]$ and $b[n]$, respectively, i.e.,

$$A[0] = a[0] + a[1], \ A[1] = a[0] - a[1]$$
$$B[0] = b[0] + b[1], \ B[1] = b[0] - b[1] \tag{3.60}$$

These 2-point DFTs can be used to obtain the 4-point DFT $I[k]$:

---

[3] With some modern computer architectures and processors, multiplications are almost as fast as additions, and considering only multiplications as a measure of complexity may not always be correct.

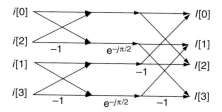

Figure 3.11 A 4-point DFT in terms of two 2-point DFTs

$$I[k] = \sum_{n=0}^{3} i[n]e^{-j2\pi\frac{nk}{4}} = \sum_{n=0,2} i[n]e^{-j2\pi\frac{nk}{4}} + \sum_{m=1,3} i[m]e^{-j2\pi\frac{mk}{4}}$$

(change variables: $n = 2p$ and $m = 2q + 1$) $\hspace{2cm}$ (3.61)

$$= \sum_{p=0}^{1} a[p]e^{-j2\pi\frac{pk}{2}} + e^{-j2\pi\frac{k}{4}} \sum_{q=0}^{1} b[q]e^{-j2\pi\frac{qk}{2}} = A[k] + e^{-j2\pi\frac{k}{4}}B[k]$$

Equation (3.61) shows how the 4-point DFT $I[k]$ can be computed from the two 2-point DFTs $A[k]$ and $B[k]$. Since $A[k]$ and $B[k]$ are 2-point DFTs, k takes on only 0 and 1 values for those, whereas we need $I[k]$ for k values of 0, 1, 2, and 3. This apparent difficulty is easily resolved by noting that 2-point DFTs are periodic with a period of 2 and thus we can express Eq. (3.61) as follows:

$$I[0] = A[0] + B[0], \quad I[2] = A[0] - B[0],$$
$$I[1] = A[1] + e^{-j\frac{2\pi}{4}}B[1], \quad I[3] = A[1] - e^{-j\frac{2\pi}{4}}B[1]$$
$\hspace{4cm}$ (3.62)

This relation is shown pictorially in Figure 3.11, which shows that the 4-point DFT can be obtained from the two 2-point DFTs with only two additional complex multiplications. To extend the above example to N-point DFTs, let us form two $N/2$-point sub-sequences $a[n]$ and $b[n]$ from the even-indexed and odd-indexed samples of the N-point sequence $i[n]$. We will assume N is an integral power of 2. We can relate the N-point DFT $I[k]$ to $N/2$-point DFTs $A[k]$ and $B[k]$ as follows:

$$I[k] = A[k] + e^{-j\frac{2\pi k}{N}}B[k], \quad k = 0, 1, 2, \ldots, (N - 1) \hspace{1cm} (3.63)$$

Thus, going from the two $N/2$-point DFTs to the N-point DFT requires at most N additional complex multiplications. When N is an integer power of 2, each of the $N/2$-point DFTs can in turn be related to two $N/4$-point DFTs. This sub-division can continue until we have 2-point DFTs. It takes $\log_2 N$

stages to sub-divide all the way down to 2-point DFTs, and since each stage requires at most N multiplications, we need at most $N \cdot \log_2 N$ operations to obtain the N-point DFT.

We can carry out the 2-D DFT of an $N \times N$ image using N N-point DFTs along rows and N N-point DFTs along columns. Thus a 2-D DFT requires $2N$ N-point DFTs and assuming that N is an integer power of 2, this requires $2N^2 \log_2 N$ multiplications. This is basically the same number of multiplications we would have had, had we thought of the image as containing N^2 points and performed an N^2-point DFT that requires $N^2 \log_2(N^2) = 2N^2 \log_2(N)$ multiplications.

The FFT method we have described so far (known as a *decimation-in-time* algorithm) is based on dividing the DT signal into even-indexed and odd-indexed sub-sequences. We can also derive an FFT algorithm by dividing the DFT sequence into its even-indexed and odd-indexed sub-sequences. The resulting FFT algorithm is known as a *decimation-in-frequency* algorithm. There exist many other variants of the FFT algorithm that are designed to satisfy different requirements, such as having identical stages so that the same hardware can be used for multiple stages, allowing the generation of the required complex coefficients on line and minimizing the memory access needed. Our coverage of FFT is rather introductory and the interested reader is encouraged to look into the references listed [26–28].

Mixed-radix FFT Does N have to be an integer power of 2 for FFT to work? The answer is yes and no. Yes, because the most efficient form of FFT requires N to be an integer power of 2, so that we can keep halving the DFT size several times until we reach 2-point DFTs. No, because we can design an FFT even when N is not an integer power of 2. For example, a 9-point DFT can be obtained from three 3-point DFTs. The more N can be decomposed into a product of primes, the more efficient the resulting FFT will be.

In practice, we usually zero-pad a sequence until its length is an integer power of 2 before we compute its DFT. For example, it is computationally more attractive to pad one zero at the end of a 31-point sequence and perform a 32-point FFT, rather than carry out a 31-point DFT. One must, however, remember that the two methods yield different frequency samples. A 31-point DFT yields transform samples at frequencies $\{0, 1/31, 2/31, \ldots, 30/31\}$, whereas a 32-point DFT yields frequencies $\{0, 1/32, 2/32, \ldots, 31/32\}$. This difference in frequencies may not be important if these frequency samples are generated in an intermediate stage that is not the final goal (e.g., the correlation output is not affected by the frequency samples used, as long as the FFT size is sufficiently large).

3.6.3 *Correlation and convolution via FFT*

We discussed earlier how CT correlations and convolutions can be computed using CTFT. Similarly, DT convolutions and correlations can be obtained using the DFT. In particular, let $i[n]$ and $h[n]$ be three DT sequences that are zero outside the interval $0 \leq n \leq (N-1)$, and let $I[k]$ and $H[k]$ denote their N-point DFTs. Then the convolution and correlation between the DT signals is related to their DFTs as follows:

$$\text{IDFT}\{I[k] \cdot H[k]\} = \tilde{o}[n] = i[n] \tilde{*} h[n] = \sum_{m=0}^{N-1} i[m] h\big[(n-m)_N\big]$$

$$\text{IDFT}\{I[k] \cdot H^*[k]\} = \tilde{c}[n] = i[n] \tilde{\otimes} h[n] = \sum_{m=0}^{N-1} i[m] h\big[(m+n)_N\big]$$

(3.64)

We will focus on correlation operation although our statements are easily generalized to convolution operation. From Eq. (3.64), we can see that the correlation of the two signals $i[n]$ and $h[n]$ can be obtained as $\text{IDFT}\{[\text{DFT}(i[n])] \cdot [\text{DFT}(h[n])]^*\}$ and thus requires three N-point DFTs. Assuming that N is an integer power of 2, the three FFT operations needed will require about $3N \cdot \log_2 N$ operations. A direct correlation operation will require about N^2 operations. Thus computing convolutions and correlations via FFT is more efficient compared with direct correlation by a factor of about $N/(3 \log_2 N)$. Thus, even for small values such as $N = 128$, using FFTs to compute correlations requires about six times fewer multiplications compared with direct correlation. As N increases, the advantage of the FFT method increases. For example, when dealing with a sequence of about one million points or a 1024×1024 image, the FFT-based correlations will require about 10,000 times fewer multiplications than the direct expressions.

There are a few other ways to make the convolution and correlation operation using the FFT even more efficient. If both signals being convolved or correlated are real, the result will be a real signal. The real nature of the signal imposes certain symmetries on the frequency domain that can be used to reduce the amount of computation by a factor of about 2. Also, often we use the same filter impulse response $h[n]$ with different input signals $i[n]$. In such a case, there is no need to recompute $H[k]$ every time. It can be computed once and reused, reducing the number of FFTs from three to only two.

Circular correlation and convolution Equation (3.64) states clearly that correlations and convolutions obtained via FFT are circular correlations and convolutions rather than the linear versions we desire. Is the difference significant?

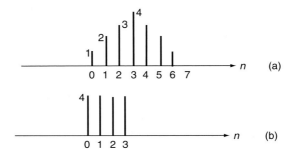

Figure 3.12 (a) Circular convolution using 8-point FFTs, and (b) circular convolution using 4-point FFTs

As we will show through the following example, circular operations can give very different results depending on the size of the DFT employed.

Example Consider the auto-convolution of a DT signal $i[n] = (u[n] - u[n - 4])$ with itself. Since $i[n]$ is a rectangle function with support over the interval $0 \leq n \leq 3$, we can show that auto-convolution results in the following triangular function with support over the interval $0 \leq n \leq 6$:

$$o[n] = i[n] * i[n] = \begin{cases} 4\left(1 - \frac{|n-3|}{4}\right) & \text{for } 0 \leq n \leq 6 \\ 0 \text{ otherwise} \end{cases} \qquad (3.65)$$

The linear convolution result in Eq. (3.65) is a triangle function. Since $i[n]$ is of length 4, the auto-convolution is of length $4 + 4 - 1 = 7$ points. Thus, we will get the linear auto-convolution as long as the DFT size employed is 7 or higher. We show in Figure 3.12(a) the convolution output when we use 8-point FFTs. We also show in Figure 3.12(b) the convolution result obtained by using only 4-point FFTs. This 4-point circular convolution is a rectangle function (of height 4) and not the triangle function expected from a linear convolution.

To understand the result in Figure 3.12(b), let us remember the 4-point DFT of the rectangle function $i[n]$ is $\{I[0] = 4, I[1] = 0, I[2] = 0, I[3] = 0\}$. Multiplying $I[k]$ by itself results in the array $\{16,0,0,0\}$ whose 4-point inverse DFT is $\{4,4,4,4\}$, a rectangle function of height 4.

The above example shows that if we use a long enough DFT size and zero-pad the data (in this example, DFTs of length 7 or higher), the circular convolution will be the same as the linear convolution. We can obtain linear convolution as long as the DFT size employed is at least as large as the extent of the linear convolution. We have shown earlier that if $i[n]$ is zero outside the interval $0 \leq n \leq (N_i - 1)$, and $h[n]$ is zero outside the interval $0 \leq n \leq (N_h - 1)$, the resulting linear convolution $o[n] = i[n] * h[n]$ is zero outside the interval $0 \leq n \leq (N_o - 1)$, where $N_o = N_i + N_h - 1$. Thus, circular convolutions will be the same as linear convolutions as long as we use DFTs of size at least $(N_i + N_h - 1)$.

This difference between circular and linear operations is the reason why we usually need to zero-pad the signals before using FFTs to compute their convolutions. For example, if we want to convolve a signal of size 128 points with another of the same size, it is customary to pad both signals with an extra 128 zeros at the end. The resulting 256-point signals are input to 256-point FFTs to obtain their 256-point DFTs. Their product is then inverse-transformed using a 256-point inverse FFT to obtain a 256-point circular convolution of the original two signals. However, as the linear convolution of the original signals is only of length 255 (i.e., $128 + 128 - 1$), circular convolution will be the same as the desired linear convolution.

Another way to understand the relation between the linear and circular convolutions and correlations is that the circular results are nothing but aliased versions of the linear results; i.e.,

$$\tilde{o}[n] = \sum_m o[n - mN] \text{ and } \tilde{c}[n] = \sum_m c[n - mN] \qquad (3.66)$$

where N is the size of the DFT employed. As long as N is larger than the extent of the linear convolution or correlation, the central replica in the circular version will be the same as the linear version and can be used without any problem. If N is smaller than this minimum DFT size required, circular versions will equal aliased versions of linear correlations and convolutions.

In principle, both circular convolution and correlation are aliased versions of their linear counterparts. However, there is an important practical difference. Convolution of two signals usually results in a broader output, in that linearly convolving a signal of length N with itself results in an output of length $(2N - 1)$ points. Thus, when doing convolution, we usually have no choice but to employ a DFT of sufficiently large size to avoid aliasing.

This is not necessarily the case for correlations, particularly auto-correlations. Typically, auto-correlations of a signal are very sharp and narrow functions and aliasing may not be as troublesome as in the convolution case, since the linear auto-correlations have relatively little energy outside a small region centered at the peak. In such a case, we may be able to use a smaller-size DFT than is required for convolutions. We illustrate this difference in Figure 3.13. Figure 3.13(a) shows a 256-point linear convolution of a pseudo-random sequence of length 128 with itself and its 256-point auto-correlation is shown in Figure 3.13(b). The 128-point circular convolution of this random

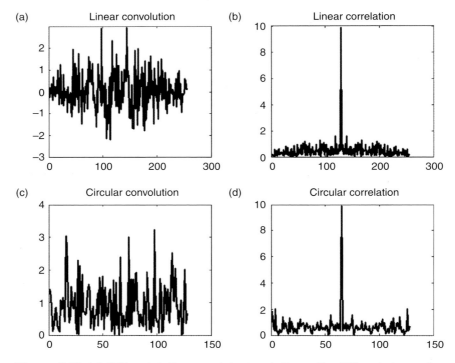

Figure 3.13 (a) 256-point linear auto-convolution of a 128-point pseudo-random sequence, (b) its 256-point linear auto-correlation, (c) its 128-point circular auto-convolution, and (d) its 128-point circular auto-correlation

sequence is shown in Figure 3.13(c). Note that the linear and circular auto-convolutions differ significantly. The 128-point circular correlation is shown in Figure 3.13(d) and note the similarity between the linear and circular auto-correlations.

3.6.4 Overlap-add and overlap-save methods

Suppose we want to convolve a long sequence with a short sequence. We can pad the short sequence with a large number of zeros and the long sequence with a small number of zeros, so that both zero-padded sequences are of the same length and their FFT can be performed. For example, when we want to convolve a 900-point sequence $i[n]$ with a 100-point sequence $h[n]$, the convolution result $o[n]$ will be of length 999 points and we will need to use 1024-point FFTs. This requires that $i[n]$ be padded with 124 zeros and $h[n]$ be padded with 924 zeros. A direct convolution would require about 100 000 multiplications, whereas FFT-based convolution using the three 1024-point

FFTs would require about 30,000 multiplications. This can be made even more efficient by using smaller FFTs and not padding the shorter sequence with so many zeros. Two ways to achieve this increased efficiency are the *overlap-add* and *overlap-save* methods. While we will introduce these in the context of 1-D convolution, similar efficiencies can be achieved with correlations as well as 2-D operations.

Overlap-add method Suppose $h[n]$ is zero outside the interval $0 \leq n \leq (N_h - 1)$. In the overlap-add method of convolution, the long input sequence $i[n]$ is divided into non-overlapping segments of length $L = (N + 1 - N_h)$ as follows:

$$i^k[n] = \begin{cases} i[kL + n] \text{ for } 0 \leq n \leq (L-1) \\ 0 \text{ otherwise} \end{cases} \quad \text{for } k = 0, 1, 2, \ldots \quad (3.67)$$

Linear convolution of $h[n]$ with $i^k[n]$ leads to an output of length $(N_h + L - 1) = N$ points, and thus the N-point circular convolution of $h[n]$ with $i^k[n]$ is the same as the linear convolution $h[n]$ with $i^k[n]$. Since L points separate adjacent input segments, L points also separate the resulting convolution segments. These overlapping convolution segments can be added to obtain the complete convolution output:

$$\tilde{o}^k[n] = i^k[n] \circledast h[n] \Rightarrow o[n] = \sum_k \tilde{o}^k[n - kL] \quad (3.68)$$

The key is to separate the input segments just enough so that each N-point circular convolution leads to N-point long output convolution segments that can be added to get the complete output. Since the convolution segments are separated by $L < N$ points, they overlap and hence the name "overlap-add" for this method.

For the numerical example considered before, $N_h = 100$, and thus it makes sense to use either 128-point, 256-point or 512-point FFTs. For a choice of $N = 256$, we need $L = N + 1 - N_h = 256 + 1 - 100 = 157$, and thus we will segment the 900-point input sequence $i[n]$ into six segments each of length 157 points and separated by 157 points. The last segment will have several zeros at the end. Each of the 157-point input segments is padded with 99 zeros to make them of length 256 points, and the 100-point sequence $h[n]$ is padded with 156 zeros to make it of length 256 points. We perform a 256-point FFT on $h[n]$ to obtain $H[k]$. This FFT needs to be carried out only once. The six input segments are then input to 256-point FFTs, and the resulting DFTs are multiplied by $H[k]$ and inverse-transformed to obtain the six convolution segments. These segments are properly aligned and summed to obtain the final convolution output. Thus we need a total of thirteen 256-point FFTs which requires about 13(256 [$\log_2 256$]) $\approx$ 26 000 multiplications. This is slightly less than the 30 000 multiplications needed for the implementations using 1024-point FFT.

Thus segmenting the input sequence results in less computational complexity than using a large enough FFT to fully accommodate the long input sequence. In some situations, the input sequence arrives in real time and the only possible way of carrying out the needed convolution is to segment the input into reasonable lengths. Such segmentation enables us to obtain output segments without waiting for the complete input to become available.

Overlap-save method The overlap-add method is based on dividing the long input sequence into non-overlapping segments of length L, such that the resulting N-point circular convolutions can be overlapped and summed to yield the final convolution result. In the *overlap-save* method of convolution, the long input sequence is divided into overlapping segments of length N, where N is the size of the FFT being used. Linear convolution of $h[n]$ with each of these N-point sequences leads to an output of length $(N + N_h - 1)$ points. However, N-point circular convolutions yield outputs that are only N points long, and in fact will differ from the correct linear convolutions in $(N_h - 1)$ places. These incorrect circular convolution points are simply discarded and the rest are used. By making sure that the input segments are overlapping, we can obtain the convolution points that had to be discarded in the previous segment. This is the basic idea of the overlap-save method of convolution.

In the overlap-save method, we divide the long input sequence into overlapping segments where the separation L between adjacent input segments is chosen to be $L = (N + 1 - N_h)$. The circular convolution of $i^k[n]$ with $h[n]$ of length N_h differs from their linear convolutions in the first $(N_h - 1)$ points, leading to $L = (N + 1 - N_h)$ circular convolution points agreeing with the linear convolution:

$$\bar{o}^k[n] = i^k[n] \circledast h[n] = \sum_{m=0}^{N_h-1} h[m] i^k[(n-m)_N]$$

$$= \begin{cases} \sum_{m=0}^{N_h-1} h[m] i^k[n-m] & \text{for } n \geq (N_h - 1) \\ \sum_{m=0}^{n} h[m] i^k[n-m] + \sum_{m=n+1}^{N_h-1} h[m] i^k[N+n-m] & \text{for } n < (N_h - 1) \end{cases}$$

$$\text{(3.69)}$$

Equation (3.69) shows that the first $(N_h - 1)$ points in the circular convolution results differ from linear convolution and thus should be discarded.

In our example, once again L equals 157 and we will need six input segments. $(N_h - 1)$ is 99, adjacent input segments overlap by 99 points. The first input segment will have 99 starting zeros and the first 157 points of the input sequence. The resulting convolution output is of length 256 points out of which the first 99 points are discarded. The second input segment overlaps with the first segment in the first 99 points, and thus picks the long input sequence values from point 59 to point 314, inclusive. This second segment is convolved with $h[n]$ the same way, and once again the first 99 points in the result are discarded and the remaining 157 are concatenated with the saved 157 from the first convolution result. The third input segment picks points from 216 to 461, inclusive, and this process continues. The computational complexities of the overlap-add and overlap-save methods are essentially the same and either can be used.

3.7 Random signal processing

The concept of random variables was discussed in Chapter 2. While random variables model the uncertainty of the outcome of an experiment, they do not capture the *time-varying* nature of a signal or image that is random. To model the noisy nature of a signal, we need not only randomness, but also time variation. Random processes provide such models. It is also important to learn how random processes are affected by LSI systems.

We will focus on CT random processes in this section, but our results and observations can be easily generalized to DT random processes. The main difference to keep in mind is that, for DT processes, frequency domain descriptions are periodic, whereas they are not necessarily so for CT random processes.

In a simplistic way, we can think of a random process $X(t)$ as a collection of temporal signals where the uncertainty arises from the fact that we do not know beforehand which of these signals will be selected. While we will use 1-D signals, random processes can be generalized to higher dimensions. There are at least two ways we can characterize a random process.

The first random process characterization method is by specifying the set of signals $\{X(t, \xi)\}$ where ξ represents the random outcome of an experiment such as choice of when to begin a measurement, flip of a coin, decay of a radionuclide, etc. (X is here capitalized as a random variable, not as a frequency domain variable. We also come close to getting into trouble here with respect to the functional nomenclature when we write $\{X(t, \xi)\}$ instead of $\{X(\cdot , \cdot)\}$; pay close attention.) Thus, a different signal in time can be selected depending on the outcome of the experiment. The entire collection $X(t, \xi)$, is known as the

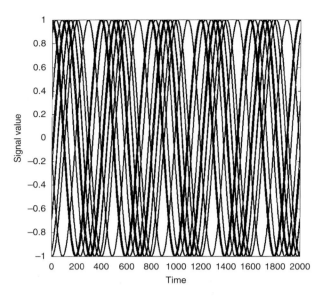

Figure 3.14 An example of the ensemble of sample realizations associated with a sinusoid with unit amplitude, fixed frequency, and uniformly distributed phase

ensemble. If $\xi = \xi_0$, then the randomness disappears and $X(t, \xi_0)$ is just a deterministic signal and is called a *sample realization* of the random process. On the other hand, if $t = t_0$, then $\{X(t_0, \xi)\}$ has no time variation associated with it and is just a random variable, since ξ still has randomness associated with it. Finally $\{X(t_0, \xi_0)\}$ is a deterministic constant since it has no time variation and no randomness.

A more common and more useful random process characterization is by specifying a functional form for a time signal that has some random parameters associated with it. For example, consider a sinusoidal signal with unit amplitude, fixed frequency, but a random phase that is uniformly distributed between $-\pi$ and $+\pi$. This randomness in phase results in an ensemble of sinusoids all with unit amplitude and the same frequency, but with different phases as shown in Figure 3.14. However, it is more compact to describe this random process using the analytical expression in Eq. (3.70).

$$X(t) = \cos[2\pi f_0 t + \theta] \tag{3.70}$$

where θ is uniformly distributed in $[-\pi, \pi]$. We can see in Figure 3.14 that at a given time instant, we have a random variable with values between -1 and $+1$. Each sinusoid corresponds to a particular phase and thus a sample realization. The entire collection of sinusoids is the ensemble.

3.7.1 Random process characterization

How do we characterize a random process? At a fixed time instant t_0, $X(t_0)$ is a random variable and we can use a PDF to describe its randomness. Thus, we will need one PDF $f_{X(t)}(x(t))$ for every time instant t. But that is not a complete description of the random process, as these first-order PDFs cannot tell us the joint random behavior of the two random variables at two time instants (e.g., t_1 and t_2). For that, we will need the bivariate PDFs of the two random variables $X(t_1)$ and $X(t_2)$ for any possible t_1 and t_2. Bivariate PDFs do not provide the complete story either, since we will need trivariate PDFs that describe the random process at any three time instants. In fact, we need the n-variate PDFs of the n random variables $\{X(t_1), X(t_2), \ldots, X(t_n)\}$ for any n and any choice of time instant $t_1, t_2, \ldots, t_n$. As can be easily seen, such a characterization of a random process using n-variate PDFs is impractical. We will most often use functional forms (e.g., Eq. (3.70)) with random parameters to describe random processes. In such cases, we need to characterize only a few random variables.

Stationarity The ensemble of sample realizations in Figure 3.14 appears to have a type of shift-invariance associated with it. If we move the time origin to the right or left, the first-order PDF describing the random process is unaffected by such a shift. More precisely, a random process is said to be *strict sense stationary* (SSS) provided the n-variate PDF of the n random variables $\{X(t_1 + \tau), X(t_2 + \tau), \ldots, X(t_n + \tau)\}$ is independent of τ for any n, any time instants $t_1, t_2, \ldots, t_n$ and any τ.

Thus, for $n = 1$, we see that the first-order PDFs of an SSS process are the same for all time instants. Thus, all the first-order characterizations (e.g., mean, variance, third moment, etc.) are independent of the location. For an SSS process, bivariate PDFs associated with time instants $(t_1 + \tau)$ and $(t_2 + \tau)$ should not depend on τ, which means that the bivariate PDFs depend only on the difference $(t_2 - t_1)$ between the two time instants. This time instant difference is also known as the *lag*.

Strict sense stationarity, while important, is a stringent requirement and sometimes we will settle for the less demanding *weak sense stationarity* (WSS), which we will introduce in the next sub-section.

Ergodicity A random process is an ensemble of sample realizations. However, in practice, we get to observe only one sample realization. Can we determine the random process parameters from only one sample realization? Using a single sample realization, we can compute its *time averages* (e.g., average of the signal, second central moment of the signal in time, etc.). If we had the entire ensemble available, we could have determined *ensemble averages* (e.g., means and variances of random variables at every time instant). We say that a random process is *ergodic* if its time averages equal its ensemble averages.

A random process must be stationary if it is to be ergodic. If the process is not stationary, its ensemble averages change with time and then we have no way of claiming that fixed-time averages equal time-varying ensemble averages. We will assume that the random processes of interest here are ergodic.

3.7.2 Second-order characterizations

As discussed before, characterizing a random process completely is often impossible. However, we are often satisfied with just first-order and second-order characterizations. Accordingly, we define the mean $m_x(t) = E\{X(t)\}$ which is the mean of the PDF of the random variable $X(t)$. If a process is stationary, $m_x(t)$ and variance are constant in time since the first-order PDF does not change with time.

Auto-correlation function When we consider two time instants, t_1 and t_2, we have two random variables, $X(t_1)$ and $X(t_2)$. Each random variable has its own mean and variance. For complete second-order characterization, we also need to specify the ACF of the random process defined below:

$$R_X(t_1, t_2) = E\{X(t_1) \cdot X(t_2)\} \tag{3.71}$$

For stationary processes, the ACF depends only on lag $= (t_2 - t_1)$, i.e.,

$$R_X(\tau) = E\{X(t) \cdot X(t + \tau)\} = E\{X(t + \tau) \cdot X(t)\} = R_X(-\tau) \tag{3.72}$$

We proved in Eq. (3.72) that the ACF is even symmetric. By subtracting the means before forming the product, we get the auto-covariance function. For zero-mean random processes, auto-correlation and auto-covariance functions are identical:

$$\begin{aligned} C_X(t_1, t_2) &= E\{[X(t_1) - m_X(t_1)] \cdot [X(t_2) - m_X(t_2)]\} \\ &= R_X(t_1, t_2) - m_X(t_1) \cdot m_X(t_2) \end{aligned} \tag{3.73}$$

If a process is stationary, then $C_X(\tau) = R_X(\tau) - m_X^2$.

Weak sense stationarity A random process is said to be WSS provided its mean is a constant in time and its ACF is a function of τ, the lag. Weak sense stationarity is also often called wide sense stationarity. It is a much less demanding property than SSS. Every SSS process is WSS, but the converse is not necessarily true. An exception is the Gaussian process to be described in the next section, where we will see that a WSS Gaussian process is also SSS. If a process is WSS and of zero-mean, the ACF at the origin, i.e., $R_X(0)$, is equal to the variance of the random process. We will assume from now on that the random process of interest is WSS unless it is stated to be otherwise.

An example of an appealing WSS process is the *white noise* process characterized by zero mean and a delta-function ACF, i.e.,

$$R_X(\tau) = N_0 \delta(\tau) \tag{3.74}$$

The above ACF implies that the random process $X(t)$ at time instant t and $X(t + \tau)$ at time instant $(t + \tau)$ are uncorrelated since $R_X(\tau)$ is zero for $\tau \neq 0$. Thus, white noise is the fastest varying random process. The broader the ACF, the slower will be the temporal variation of the random process. For most zero-mean random processes, the ACF tapers towards zero as τ goes to infinity.

Power spectral density The ACF is a second-order characterization of a random process. For a WSS process, the power spectral density (PSD) $S_x(f)$ is another second-order characterization. The PSD is defined as the FT of the ACF, i.e.,

$$S_X(f) = \int_{-\infty}^{\infty} R_X(\tau) e^{-j2\pi f \tau} d\tau \qquad R_X(\tau) = \int_{-\infty}^{\infty} S_X(f) e^{j2\pi f \tau} df \tag{3.75}$$

The PSD is a function of frequency f, whereas the ACF of a WSS process is a function of τ, the lag. Since the FT is invertible, ACF and PSD are equivalent second-order descriptors for a WSS random process.

We established earlier that the ACF of a real random process is real and even. Using that property in Eq. (3.75), we see that the ACF and the PSD of a real process can be simplified as follows, indicating that the PSD is also real and even:

$$S_X(f) = 2 \int_0^{\infty} R_X(\tau) \cos(2\pi f \tau) d\tau = S_X(-f)$$

$$\tag{3.76}$$

$$R_X(\tau) = 2 \int_0^{\infty} S_X(f) \cos(2\pi f \tau) df = R_X(-\tau)$$

For the white noise process with the delta function ACF as in Eq. (3.74), the PSD is equal to the constant N_0 at all frequencies. This constant PSD is similar to white light containing all spectral components, and hence the name white noise.

In general, PSD is a convenient way to understand the average spectral content of a random process. If we know the ACF, its FT yields the PSD. On the other hand, if we have only sample realizations of a random process, we can estimate its PSD as follows.

Let $x_T(t)$ denote a sample realization of the random process truncated to a time window of length T. Let $x_T(f)$ denote the FT of $x_T(t)$. Since the selection of a sample realization is random, $x_T(f)$ is also random and we can define its ensemble average. Then the Weiner–Khinchin theorem states that, for many random processes, the expectation of FT magnitude squared approaches the PSD as the truncation window T goes to infinity.

$$\lim_{T \to \infty} \left[\frac{E\left\{|X_T(f)|^2\right\}}{T} \right] = S_X(f) \tag{3.77}$$

Equation (3.77) suggests a method for estimating the PSD of a random process from one sample realization. We truncate this sample realization to length T, compute the magnitude squared of its Fourier transform and divide it by T. In the limit as T goes towards infinity, this time average should approach the PSD.

Because of the FT relation in Eq. (3.75), the total area under the PSD curve can be seen to equal $R(0)$, which is the average power of the random process. For a zero-mean process, this is the same as the variance of the random process, i.e.,

$$E\left\{(X(t))^2\right\} = R_X(0) = \int_{-\infty}^{\infty} S_X(f)\mathrm{d}f = \mathrm{var}\{X(t)\} + m^2 \tag{3.78}$$

Auto-correlation function properties Can any arbitrary function be an ACF? No, for a function to be a valid ACF, it is required that the ACF must exhibit three properties.

1. The ACF must be even symmetric, i.e., $R_x(\tau) = R_x(-\tau)$
2. The ACF must have its peak at the origin as demonstrated below. We utilize below the fact that the PSD is real and positive (we will prove this assertion in Section 3.7.4.)

$$\begin{aligned} |R_X(\tau)| &= \left| \int S_X(f) e^{j2\pi f \tau} \mathrm{d}f \right| \le \int S_X(f) \left| e^{j2\pi f \tau} \right| \mathrm{d}f \\ &= \int S_X(f) \mathrm{d}f = R_X(0) \Rightarrow R_X(0) \ge |R_X(\tau)| \end{aligned} \tag{3.79}$$

3. The ACF must be such that its FT (i.e., the PSD) must be non-negative. Thus, while an ACF can be a sinc function (with a non-negative FT), it cannot be a rectangle function (with a sinc function FT that can be negative).

Typically, ACF peaks at the origin and decays towards zero as we move away from the origin. However, if the random process is periodic, its ACF is also periodic. Also, if the random process has a non-zero mean m_x, then the ACF will asymptotically decay to $(m_x)^2$.

3.7.3 *Gaussian processes*

One random process of particular interest is the Gaussian process. One way to define a Gaussian random process is by requiring that the set of n random variables $\{X(t_1), X(t_2), \ldots, X(t_n)\}$ be jointly Gaussian for any n and any choice of time instants. We know that a Gaussian random vector is completely specified by its mean vector and its covariance matrix. Thus, the n-variate Gaussian PDF corresponding to these n time instants is completely character- ized by the n-element mean vector:

$$\mathbf{m} = [\, E\{X(t_1)\} \quad E\{X(t_2)\} \quad \cdots \quad E\{X(t_n)\}\,]^{\mathrm{T}} \tag{3.80}$$

and the following covariance matrix $\mathbf{C}$ with n rows and n columns:

$$\mathbf{C} = E\left\{(\mathbf{x} - \mathbf{m})(\mathbf{x} - \mathbf{m})^{\mathrm{T}}\right\} = E\{\mathbf{x}\mathbf{x}^{\mathrm{T}}\} - \mathbf{m}\mathbf{m}^{\mathrm{T}} = \mathbf{R} - \mathbf{m}\mathbf{m}^{\mathrm{T}}$$

where

$$\mathbf{R} = \begin{bmatrix} E\{X(t_1)X(t_1)\} & E\{X(t_1)X(t_2)\} & \cdots & E\{X(t_1)X(t_n)\} \\ E\{X(t_2)X(t_1)\} & E\{X(t_2)X(t_2)\} & \cdots & E\{X(t_2)X(t_n)\} \\ \vdots & \vdots & \ddots & \vdots \\ E\{X(t_n)X(t_1)\} & \cdots & \cdots & E\{X(t_n)X(t_n)\} \end{bmatrix} \tag{3.81}$$

The covariance matrix $\mathbf{C}$ is related easily to the correlation matrix $\mathbf{R}$ as shown in Eq. (3.81). Since a Gaussian random vector is completely characterized by $\mathbf{m}$ and $\mathbf{C}$ (or equivalently $\mathbf{m}$ and $\mathbf{R}$, a Gaussian random process is completely specified by its mean function $m_x(t)$ and its ACF $R_x(t_1, t_2)$.

Weak sense stationary Gaussian processes If the Gaussian process is WSS, then $m_x(t)$ is independent of t and $R_x(t_1, t_2)$ depends only on $\tau = (t_1 - t_2)$. Thus the n-variate Gaussian PDF obtained from such a WSS random process will be the same whether we use $\{t_1, t_2, \ldots, t_n\}$ or $\{t_1 + T), (t_2 + T), \ldots, (t_n + T)\}$ for the n time instants for any T. This is because the mean vector $\mathbf{m} = m_x[1\ 1\ \ldots\ 1]^T$ does not depend on the choice of time instants, and the correlation matrix $\mathbf{R}$ does not depend on T, as shown below:

$$\mathbf{R} = \begin{bmatrix} R_X(0) & R_X(t_2 - t_1) & \cdots & R_X(t_n - t_1) \\ R_X(t_1 - t_2) & R_X(0) & \cdots & R_X(t_n - t_2) \\ \vdots & \vdots & \ddots & \vdots \\ R_X(t_1 - t_n) & R_X(t_2 - t_n) & \cdots & R_X(0) \end{bmatrix} \tag{3.82}$$

where we can see that $\mathbf{R}$ does not depend on T. Thus, the n-variate PDF is unaffected by shifting the origin for any n. This means that a WSS Gaussian process is also SSS. We also note that because of the even symmetric nature of the ACF, $\mathbf{R}$ is a symmetric matrix.

Linear transformation Linear transformation of a Gaussian random process $X(t)$ results in another Gaussian process $Y(t)$. Consider the linear transformation $Y(t) = \int h(t, p)\, X(p)\, \mathrm{d}p$. This integral can be thought of as the limit (or generalization) of a weighted sum (where the weights are given by $h(t, p)$) of the random process $X(t)$.

Thus each of the random variables in the set $\{Y(t_1), Y(t_2), \ldots, Y(t_n)\}$ can be thought of as a different weighted sum of the random variables obtained from the Gaussian random process $X(p)$. Since $X(p)$ is Gaussian, weighted sums of samples of $X(p)$ are also Gaussian (see Section 2.5.7 where we showed that affine transformations of Gaussian random vectors must lead to Gaussian random vectors). Thus, any set of n RVs obtained from $Y(t)$ must be jointly Gaussian proving that $Y(t)$ must be a Gaussian random process.

Uniform sampling The correlation matrix $\mathbf{R}$ expression in Eq. (3.82) is valid for any WSS process, not just a Gaussian WSS process. Note that the main diagonal of $\mathbf{R}$ contains the same value, namely $R_x(0) = E\{X^2(t)\}$, the average power of the random process. Suppose the n time instants are at uniform intervals of T, i.e., $t_i = t_0 + iT$, $1 \leq i \leq n$. Then the correlation matrix $\mathbf{R}$ in Eq. (3.82) takes on the following:

$$\mathbf{R} = \begin{bmatrix} R_X(0) & R_X(T) & \cdots & R_X((n-1)T) \\ R_X(T) & R_X(0) & \cdots & R_X((n-2)T) \\ \vdots & \vdots & \ddots & \vdots \\ R_X((n-1)T) & R_X((n-2)T) & \cdots & R_X(0) \end{bmatrix} \quad (3.83)$$

The **R** matrix in Eq. (3.83) is symmetric and Toeplitz, i.e., all entries in a diagonal are equal. The computational complexity and storage requirement associated with the symmetric, Toeplitz matrix in Eq. (3.83) is significantly smaller than those of the symmetric correlation matrix in Eq. (3.82).

3.7.4 Filtering of random processes

We have shown that linear transformation of a Gaussian process leads to a Gaussian process. Thus, Gaussian processes filtered by LSI systems are also Gaussian. More important, however, is what happens to the PSD of a random process as it passes through an LSI system. In this sub-section, we will derive the important result that the PSD at the output of a filter with frequency response $H(f)$ is its input PSD multiplied by $|H(f)|^2$.

Consider the system in Figure 3.1 and let us assume that the system is LSI with impulse response $h(t)$ and the input signal is a sample realization of the random process $x(t)$. The resulting output process is $Y(t)$. The input random process $X(t)$ is an ensemble of sample realizations $X(t, \xi)$, where each sample realization is nothing but a deterministic signal. Each input sample realization $X(t, \xi)$ results in an output signal $Y(t, \xi)$ and an ensemble of these output signals constitutes the output random process $Y(t)$. Let us now see what happens to the mean and the second-order characteristics as they go through the LSI system.

Output mean Since the system is LSI, its output signal $Y(t, \xi)$, is the convolution of the input signal $X(t, \xi)$ with the impulse response; i.e.,

$$Y(t, \xi) = X(t, \xi) * h(t) = \int_{-\infty}^{\infty} h(p)X(t-p, \xi)\mathrm{d}p \quad (3.84)$$

We will drop the variable ξ, which denotes the sample realization, as this parameter will be common to all random processes. Then the mean of the output process $y(t)$ is as follows:

$$m_Y(t) = E\{Y(t)\} = \int\limits_{-\infty}^{\infty} h(p)E\{X(t-p)\}\mathrm{d}p$$

$$= \int\limits_{-\infty}^{\infty} h(p)m_X(t-p)\mathrm{d}p = h(t) * m_X(t) \tag{3.85}$$

We see that the mean of the output from an LSI system is the convolution of the input mean with the impulse response of the system. If the input mean is a constant in time (e.g., the mean of a WSS process), the output mean is also a constant as shown below:

$$m_Y(t) = m_X \int h(t)\mathrm{d}t = m_X H(0) \tag{3.86}$$

where $H(0)$ denotes the filter's frequency response at zero frequency. It should be no surprise that the output mean is the input mean multiplied by $H(0)$, the filter's DC gain.

ACF of the output process: The ACF of $y(t)$ can be related to the ACF of $X(t)$ and $h(t)$:

$$\begin{aligned}
R_Y(t_1, t_2) &= E\{Y(t_1)Y(t_2)\} \\
&= E\left\{\left[\int h(p)X(t_1-p)\mathrm{d}p\right]\left[\int h(q)X(t_2-q)\mathrm{d}q\right]\right\} \\
&= \int\int \mathrm{d}p\mathrm{d}q\, h(p)h(q)E\{X(t_1-p)X(t_2-q)\} \\
&= \int\int \mathrm{d}p\mathrm{d}q\, h(p)h(q)R_X(t_1-p, t_2-q)
\end{aligned} \tag{3.87}$$

If $X(t)$ is WSS, then its ACF depends only on the lag and, as we show below, the ACF of the corresponding $Y(t)$ is also a function only of the lag:

$$\begin{aligned}
R_Y(t_1, t_2) &= \int\int \mathrm{d}p\mathrm{d}q\, h(p)\, h(q)R_X(t_1-p, t_2-q) \\
&= \int\int \mathrm{d}p\mathrm{d}q\, h(p)\, h(q)R_X(t_2-t_1+p-q) = R_Y(t_2-t_1) \tag{3.88}
\end{aligned}$$

$$\Rightarrow R_Y(\tau) = \int\int \mathrm{d}p\mathrm{d}q\, h(p)\, h(q)R_X(\tau+p-q)$$

Thus, if $X(t)$, the input to an LSI system is WSS, then its output $Y(t)$ is also WSS.

Power spectral density Using Eq. (3.88) and the fact that the PSD is the FT of the ACF, we can derive the following relation between the input PSD $S_x(f)$, filter transfer function $H(f)$, and the output PSD $S_y(f)$:

$$S_Y(f) = \int R_Y(\tau)e^{-j2\pi f\tau}d\tau$$

$$= \int \left[\int\int dpdq R_X(\tau + p - q)h(p)h(q)\right]e^{-j2\pi f\tau}d\tau$$

$$= \int\int\int dpdqd\beta h(p)h(q)R_X(\beta)e^{-j2\pi f(\beta-p+q)} \quad \text{where } \beta = (\tau + p - q)$$

$$= \left[\int h(q)e^{-j2\pi fq}dq\right]\left[\int h(p)e^{j2\pi fp}dp\right]\left[\int R_X(\beta)e^{-j2\pi f\beta}d\beta\right]$$

$$= S_X(f)H(f)H^*(f) = S_X(f)|H(f)|^2$$

$$(3.89)$$

where we assume that the impulse response $h(t)$ is real and hence the frequency response $H(f)$ is conjugate symmetric. Note that the output PSD depends on the input PSD and the *magnitude* (but not the phase) of the filter frequency response $H(f)$. In contrast, a filter's output due to a deterministic input signal depends both on the magnitude and the phase of the filter $H(f)$. This difference in an LSI system's response to deterministic signals and random processes allows us to design filters that retain the signals while suppressing noise. We can select the filter magnitude to suppress noise and select the filter phase for the unsuppressed frequencies to enhance the signal.

We can now prove that $S_x(f)$ is non-negative and justify its being called a PSD. Consider a filter whose frequency response $H(f)$ is 1 in a narrow bandpass centered at f_0 and 0 elsewhere, i.e.,

$$|H(f)| = \begin{cases} 1 \text{ for } f_0 - \frac{\Delta f}{2} \leq |f| \leq f_0 + \frac{\Delta f}{2} \\ 0 \text{ otherwise} \end{cases} \qquad (3.90)$$

The average power out of the filter is given as follows:

$$E\left\{(Y(t))^2\right\} = R_Y(0) = \int\limits_{-\infty}^{\infty} S_Y(f)\mathrm{d}f = \int\limits_{-\infty}^{\infty} S_X(f)|H(f)|^2\mathrm{d}f$$

$$= 2 \int\limits_{f_0-\frac{\Delta f}{2}}^{f_0+\frac{\Delta f}{2}} S_X(f)\mathrm{d}f \cong \Delta f S_X(f_0) \tag{3.91}$$

$$\Rightarrow S_X(f_0) \cong \frac{E\left\{(Y(t))^2\right\}}{\Delta f}$$

where we assume that the input PSD $S_x(f)$ is relatively constant in a narrow band centered at f_0. Equation (3.91) indicates that the PSD at frequency f_0 is the ratio of the filter output power to the filter bandwidth, and that is the reason why $S_x(f)$ is known as the *power spectral density*. Equation (3.91) also proves that the PSD is proportional to the average power and thus must be non-negative.

We have established what happens to the first-order and second-order statistics of a random process as it passes through an LSI system. These results apply to any random process, not just a Gaussian process. If the process is indeed Gaussian, all we need are the statistics of the first two orders. If the process is not Gaussian, the analysis will be more challenging.

3.8 Chapter summary

This chapter has provided a review of signals and systems that is relevant for designing and implementing correlations. Here, we will summarize some of the key concepts introduced in this chapter.

- Linear, shift-invariant systems can be completely characterized using impulse response (in 1-D) and the point spread function (in 2-D). For an arbitrary input, the output from an LSI system is the convolution of that input with the system's impulse response.
- Convolution and correlation operations are very similar and thus a filter can be used to obtain a correlation by conjugating the filter's frequency response. However, the results of correlation and convolution are very different.
- Frequencies of sinusoids and complex exponential inputs remain unaffected by an LSI system. Thus one way to characterize an LSI system is through its frequency response, which describes its amplitude and phase response as a function of frequency.
- Fourier series allow us to describe CT periodic signals as a weighted sum of complex exponentials with frequencies that are integer multiples of the fundamental frequency. The continuous-time Fourier transform is a generalization of the Fourier

series for non-periodic CT signals. The 2-D FT can be computed using 1-D FTs sequentially.

- We can sample CT signals at uniform intervals to obtain a DT signal, and we can reconstruct the original CT signal from these samples provided that the sampling frequency exceeds twice the maximum signal frequency. When a smaller sampling frequency than this minimum is used, aliasing occurs. It is important to use analog anti-aliasing filters prior to such under-sampling.

- Fourier analysis of discrete-time signals can be performed using the discrete Fourier transform. The DFT is efficiently implemented using the fast Fourier transform algorithm. When N is a power of 2, the computational complexity of an N-point FFT is in the order of $N \log_2 N$ operations.

- Convolution in time or space domain is equivalent to multiplication in the frequency domain, and vice versa. Correlation is the same, except that the FT of the second function must be conjugated prior to forming the product of the FTs. Computing convolutions and correlations via the FFT is more efficient than their direct computation.

- Using FFTs for convolutions and correlations results in circular correlations and convolutions instead of their linear counterparts. This means that we either need to zero-pad the signals and use sufficiently large FFT sizes, or we have to segment the inputs so that we can assemble linear convolutions and correlations from the obtained circular versions, using overlap-save or overlap-add methods.

- Random processes allow us to combine randomness with time variation. Characterization of a random process requires that we specify its joint PDFs for all possible combinations of time instants. If these PDFs are unaffected by a shift in the time origin, the process is said to be strict sense stationary (SSS). A weaker form of stationarity is the wide sense stationarity in which only the first-and second-order characterizations are unaffected by such a shift of the origin. Strict sense stationarity implies WSS, but the converse is not true in general. Once again, the exception is the Gaussian process, for which SSS and WSS are equivalent.

- Second-order statistics of a random process are characterized by its auto-correlation function, which is the correlation of the two RVs corresponding to the random process at two time instants. For WSS processes, the ACF depends only on the lag, i.e., the difference between the two time instants. Not every function can be a valid ACF. An ACF must be symmetric, must have maximum value at the origin, and its FT must be non-negative. A particular random process of special interest is the white noise process for which the ACF is a delta function centered at the origin.

- The FT of the ACF of a WSS process is known as its power spectral density. The PSD is non-negative and describes the average power per unit bandwidth as a function of the frequency. When a WSS process is passed through an LSI system, the output PSD is the input PSD multiplied by the magnitude squared of the system frequency response. Since the filter phase has no effect on the output PSD of the noise but does affect the output due to a deterministic signal, we can design filters to suppress the noise while allowing deterministic signals to pass through.

4

Detection and estimation

In Chapter 2, we discussed the basics of probability theory, which helps us to model the randomness in signals (called *noise*) and thus allows us to extract the desired signal from unwanted noise. An example source of noise is the thermal noise induced in the voltage across a resistor by the motion of electrons. Similarly, when light is incident on a photodetector, the number of electrons released is random (although the mean is proportional to the incident light intensity), and this randomness leads to noise or uncertainty in the signal. When a signal is corrupted by such random noise, it is often important to extract or restore the original signal from the noisy version; this is known as signal restoration. In other instances, our task is to classify the signal as the noisy version of one of a few possible signals. This task of detecting the signal class is known as detection (or classification) and it is one of the foci of this chapter. It is not surprising that detection theory has a bearing on pattern recognition. A generalization of the notion of detection is estimation, where we try to estimate a parameter (which can assume a value in an interval rather than in a discrete set) from a noisy signal. Estimation theory is also relevant in tasks such as evaluating a correlator; e.g., estimating the error rate from a classifier.

We will first consider in Section 4.1 the case of detecting between two classes and extend this in Section 4.2 to multiple classes. In Section 4.3, we introduce the notion of parameter estimation and discuss its application to error probability estimation. Our coverage in this chapter will be very basic and readers wanting to learn more about detection and estimation theory are advised to consult further references [29, 30].

4.1 Binary hypothesis testing

Let us consider the case of distinguishing (signal + noise) from just noise. Such an additive noise model is useful in applications such as radar where we have two hypotheses: H_0 denoting the absence of a target in the radar path, and H_1 indicating the presence of a target. Let $\mathbf{s}$ denote the reference signal (with N samples in it) transmitted by the radar and let $\mathbf{r}$ denote the received signal. Ignoring the time delay between the transmitter and the receiver and any attenuation in the received signal, we can model the relationship between the transmitted and received signals as follows:

$$\begin{aligned} H_0 : \quad & \mathbf{r} = \mathbf{n} \\ H_1 : \quad & \mathbf{r} = \mathbf{s} + \mathbf{n} \end{aligned} \tag{4.1}$$

where $\mathbf{n}$ denotes the additive noise vector with N elements. While there are other noise models (e.g., multiplicative, non-overlapping, etc.) the additive noise model is useful in many situations. The additive noise model is also simple to analyze and is thus a good starting point to learn detection theory. The objective of binary hypothesis testing is to select the hypothesis (i.e., to decide between whether the target is present or absent) yielding the lowest error probability. What we know are the reference signal $\mathbf{s}$ and the statistics of $\mathbf{n}$.

To determine the minimum probability of error in the detection method, we first need to characterize the probability density functions (PDFs) of the received signal vector $\mathbf{r}$ for the two hypotheses. Let $f_{\mathbf{n}}(\cdot)$ denote the N-variate PDF of the noise vector $\mathbf{n}$. Then the PDFs of the received vector for the two hypotheses are as follows:

$$\begin{aligned} H_0 : \quad & f_0(\mathbf{r}) = f_{\mathbf{n}}(\mathbf{r}) \\ H_1 : \quad & f_1(\mathbf{r}) = f_{\mathbf{n}}(\mathbf{r} - \mathbf{s}) \end{aligned} \tag{4.2}$$

Probability of error Let P_0 and P_1 denote the a-priori probabilities of the two hypotheses. To determine the detection method yielding the smallest probability of error, we first note that a detection error can occur in two different ways. If hypothesis H_0 is true, but H_1 gets selected, that error is called a *false alarm*, since the detection method falsely indicates the presence of a target. On the other hand, if H_1 is true, but H_0 gets chosen, we have a *miss* since the detector missed a target. The probability of error P_e is obtained by weighting the probability of false alarm (P_{FA}) and the probability of miss (P_M) by the corresponding a-priori probabilities P_0 and P_1 and adding them up, i.e.,

$$P_e = P_0 P_{FA} + P_1 P_M \tag{4.3}$$

where $P_M = P(H_0|H_1) = \int \cdots \int_{R_0} f_1(\mathbf{r})d\mathbf{r} = \int \cdots \int_{R_0} f_{\mathbf{n}}(\mathbf{r} - \mathbf{s})d\mathbf{r}$

and $P_{FA} = P(H_1|H_0) = \int \cdots \int_{R_1} f_0(\mathbf{r})d\mathbf{r} = \int \cdots \int_{R_1} f_{\mathbf{n}}(\mathbf{r})d\mathbf{r}$

where $P(H_0 \mid H_1)$ denotes the probability that the detector selects hypothesis H_0 whereas hypothesis H_1 is true. The detector assigns each received vector **r** to one of two hypotheses, i.e., the detector partitions the N-dimensional space of received vectors into two *decision regions*, namely R_0 (for vectors resulting in the selection of H_0) and R_1 (leading to H_1). These two regions *partition* the space of received vectors, i.e., the two regions have no common vectors (otherwise, a received vector would have to be assigned to both hypotheses) and together they fill up the whole space (otherwise, there would be some received vectors that cannot be assigned to either hypothesis).

4.1.1 Minimum probability of error detection

Since the detector partitions the signal space into R_1 and R_0, the integral of either PDF over one of those regions is equal to (1 − the integral of that PDF over the other region). Using this, we can rewrite the probability of error in terms of just one of these decision regions as follows:

$$
\begin{aligned}
P_e &= P_0 \left[\int_{R_1} f_0(\mathbf{r})d\mathbf{r} \right] + P_1 \left[\int_{R_0} f_1(\mathbf{r})d\mathbf{r} \right] \\
&= P_0 \left[\int_{R_1} f_0(\mathbf{r})d\mathbf{r} \right] + P_1 \left[1 - \int_{R_1} f_1(\mathbf{r})d\mathbf{r} \right] \\
&= P_1 + \int_{R_1} [P_0 f_0(\mathbf{r}) - P_1 f_1(\mathbf{r})]d\mathbf{r}
\end{aligned}
\tag{4.4}
$$

Since we want to minimize the integral in Eq. (4.4), we should assign to region R_1 (i.e., to hypothesis H_1) all those vectors **r** for which the term $[P_0 f_0(\mathbf{r}) - P_1 f_1(\mathbf{r})]$ is negative. Similarly, we should assign to region R_0 (i.e., assign to hypothesis H_0) all those received vectors **r** for which the term $[P_0 f_0(\mathbf{r}) - P_1 f_1(\mathbf{r})]$ is positive. Such a detection scheme will yield the smallest probability of error. Thus, the minimum probability of error is obtained by the following decision scheme:

$$
[P_0 f_0(\mathbf{r}) - P_1 f_1(\mathbf{r})] \geq 0 \Rightarrow H_0 \quad \text{and} \quad [P_0 f_0(\mathbf{r}) - P_1 f_1(\mathbf{r})] < 0 \Rightarrow H_1
$$

$$
\therefore \; l(\mathbf{r}) = \frac{f_1(\mathbf{r})}{f_0(\mathbf{r})} \geq \frac{P_0}{P_1} = \eta \Rightarrow H_1 \quad \text{and} \; l(\mathbf{r}) < \eta \Rightarrow H_0
\tag{4.5}
$$

where $l(\mathbf{r})$, the ratio of PDFs, is known as the *likelihood ratio*. Thus the minimum probability of error is achieved by comparing the likelihood ratio $l(\mathbf{r})$ to a threshold η, which is the ratio of the a-priori probabilities, P_0 and P_1.

The detection scheme in Eq. (4.5) minimizes the probability of error defined in Eq. (4.3), which treats the two types of errors (namely, false alarms and misses) as if they are equally costly. In some situations, we may want to assign different costs or weights to the two types of errors. For example, if H_0 denotes the absence of an aircraft and H_1 denotes the presence of it in an air traffic control situation, a false alarm is much more acceptable than a miss. In such a case, the metric to be minimized must weigh the P_{FA} lower than P_M, resulting in an optimal detector different from that in Eq. (4.5), which uses equal costs for the two types of error. In a biometrics application, H_0 may represent an impostor and H_1 may denote an authentic. In such a case $P(H_0|H_1)$ denotes the probability of false rejection and $P(H_1|H_0)$ is the probability of false acceptance and, depending on the security level required, one type of error may be more costly than the other. We can show that, with the new costs, we will still have a detector that compares the likelihood ratio to a threshold, but the new threshold depends not just on the a-priori probabilities, but also on the relative costs.

We can use any monotone mapping (e.g., logarithm) on both sides of the inequality in Eq. (4.5) without any loss of optimality. Thus, an equivalent detection method is to compare the log-likelihood ratio to the logarithm of the threshold, i.e.,

$$\ln[l(\mathbf{r})] \geq \ln\left[\frac{P_0}{P_1}\right] \Rightarrow H_1 \quad \text{and} \ln[l(\mathbf{r})] < \ln\left[\frac{P_0}{P_1}\right] \Rightarrow H_0 \qquad (4.6)$$

In the next sub-section, we use the log-likelihood ratio test in Eq. (4.6) for the case of Gaussian PDFs to illustrate the trade-offs involved in the detector design. Although we focus on the case of Gaussian PDFs, the log-likelihood ratio test in Eq. (4.6) is optimal for any choice of PDF. The attractiveness of Gaussian PDFs is that they result in closed-form expressions for decision tests and resulting error probabilities. Thus, sometimes they are invoked even when it is clear from the context (e.g., when using light intensity) that they may not be appropriate.

4.1.2 *Binary hypotheses testing with Gaussian noise*

Let us now consider the special case where the noise vector $\mathbf{n}$ in Eq. (4.1) is a Gaussian random vector with a zero mean vector and covariance matrix $\mathbf{C}$. For the N-variate Gaussian PDF, the logarithm of the likelihood ratio in Eq. (4.5) becomes

$$\ln[l(\mathbf{r})] = \ln\left[\frac{f_1(\mathbf{r})}{f_0(\mathbf{r})}\right] = \ln\left[\frac{f_\mathbf{n}(\mathbf{r}-\mathbf{s})}{f_\mathbf{n}(\mathbf{r})}\right]$$

$$= \ln\left[\frac{\dfrac{1}{\sqrt{(2\pi)^N|\mathbf{C}|}}\exp\left[-\dfrac{1}{2}(\mathbf{r}-\mathbf{s})^\mathsf{T}\mathbf{C}^{-1}(\mathbf{r}-\mathbf{s})\right]}{\dfrac{1}{\sqrt{(2\pi)^N|\mathbf{C}|}}\exp\left[-\dfrac{1}{2}\mathbf{r}^\mathsf{T}\mathbf{C}^{-1}\mathbf{r}\right]}\right] \tag{4.7}$$

$$= \mathbf{s}^\mathsf{T}\mathbf{C}^{-1}\mathbf{r} - \frac{1}{2}\mathbf{s}^\mathsf{T}\mathbf{C}^{-1}\mathbf{s} = \mathbf{a}^\mathsf{T}\mathbf{r} - b, \quad \text{where } \mathbf{a} = \mathbf{C}^{-1}\mathbf{s}, \quad \text{and } b = \frac{1}{2}\mathbf{s}^\mathsf{T}\mathbf{C}^{-1}\mathbf{s}$$

where we use the fact that $\mathbf{s}^\mathsf{T}\mathbf{C}^{-1}\mathbf{r}$ is a scalar and thus $\mathbf{s}^\mathsf{T}\mathbf{C}^{-1}\mathbf{r} = \mathbf{r}^\mathsf{T}\mathbf{C}^{-1}\mathbf{s}$. (Also note that $\mathbf{C}$ is symmetric.) We see from Eq. (4.7) that the log-likelihood ratio is a linear function of the received vector $\mathbf{r}$, and since the PDF of $\mathbf{r}$ is Gaussian, the PDFs of the log-likelihood ratio are also Gaussian. Using Eqs. (4.6) and (4.7), we can simplify the minimum probability of error detection scheme to be as follows:

$$\psi(\mathbf{r}) = \mathbf{a}^\mathsf{T}\mathbf{r} - b \geq \ln\left[\frac{P_0}{P_1}\right] \implies H_1$$

$$\text{and } \psi(\mathbf{r}) = \mathbf{a}^\mathsf{T}\mathbf{r} - b < \ln\left[\frac{P_0}{P_1}\right] \implies H_0 \tag{4.8}$$

Geometric structure The optimal detector structure in Eq. (4.8) lends itself to a geometric interpretation. Consider the case where the two hypotheses are equally likely, i.e., $P_0 = P_1 = 1/2$. Then the boundary between the two decision regions (called the *decision boundary*) corresponds to the set of $\mathbf{r}$ vectors that satisfy $\mathbf{a}^\mathsf{T}\mathbf{r} - b = 0$. This means that the decision boundary is normal to the vector $\mathbf{a}$. From Eq. (4.7), we also see that $\mathbf{r} = (1/2)\mathbf{s}$ is always a point on the decision boundary, independent of what $\mathbf{C}$ is. Under the assumption of zero-mean noise, the PDF of $\mathbf{r}$ has zero mean under H_0, whereas the PDF of $\mathbf{r}$ has mean $\mathbf{s}$ under H_1. Thus, the decision boundary bisects the vector that connects the means of the two PDFs.

Suppose that the noise covariance matrix $\mathbf{C}$ is proportional to $\mathbf{I}$. Equal probability contours in this case will be circles centered at the class means. Then the decision boundary corresponds to $\mathbf{r}$ vectors satisfying $\mathbf{s}^\mathsf{T}\mathbf{r} - (1/2)\mathbf{s}^\mathsf{T}\mathbf{s} = 0$. From this, we can see that the decision boundary is normal to $\mathbf{s}$, the vector that connects the two means. In this case, the decision boundary is a perpendicular

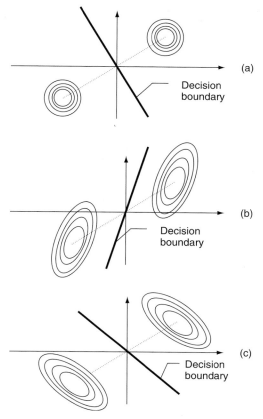

Figure 4.1 Minimum probability of error decision boundaries for two classes with zero-mean Gaussian noise components with a correlation coefficient that is (a) zero, (b) positive, and (c) negative

bisector of the line that connects the two means. This is shown in Figure 4.1(a) for the case where $N = 2$.

When the noise components are not uncorrelated, i.e., $\mathbf{C}$ is not proportional to the identity matrix $\mathbf{I}$, the decision boundary is normal to $\mathbf{a} = \mathbf{C}^{-1}\mathbf{s}$, not $\mathbf{s}$. As a result, the decision boundary will be a tilted bisector of the line that connects the two means. Figure 4.1(b) and (c) show the decision boundaries when the two noise components have positive and negative correlation coefficients, respectively. We show in this figure the equal probability contours for the two PDFs, and it can be seen that the decision boundaries are tilted so as to minimize the overlap between the two PDFs.

Probability of error In general, it is difficult to obtain a compact expression for the probability of error of the optimal binary hypothesis detector. However, for the special case of a Gaussian noise vector with $\mathbf{C} = \sigma^2 \mathbf{I}$, the detector simplifies to the following:

$$\psi(\mathbf{r}) = \frac{1}{\sigma^2}\left(\mathbf{s}^T\mathbf{r} - \frac{1}{2}\mathbf{s}^T\mathbf{s}\right) \geq \ln\left[\frac{P_0}{P_1}\right] \Rightarrow H_1$$

$$\text{and } \psi(\mathbf{r}) = \frac{1}{\sigma^2}\left(\mathbf{s}^T\mathbf{r} - \frac{1}{2}\mathbf{s}^T\mathbf{s}\right) < \ln\left[\frac{P_0}{P_1}\right] \Rightarrow H_0 \tag{4.9}$$

$$\therefore \eta = \mathbf{s}^T\mathbf{r} \geq \frac{E_s}{2} + \sigma^2 \ln\left[\frac{P_0}{P_1}\right] = T \Rightarrow H_1 \quad \text{and } \eta < T \Rightarrow H_0$$

where $E_s = \mathbf{s}^T\mathbf{s}$ is the energy in the signal vector. To determine the error probabilities, we need the PDFs of the test statistic $\eta = \mathbf{s}^T\mathbf{r}$ for the two hypotheses. Since η is the result of an affine operation (inner product plus a bias) on $\mathbf{r}$, it too is a Gaussian random variable. Its mean and variance are as follows for the two hypotheses:

$$E\{\eta|H_0\} = \mathbf{s}^T E\{\mathbf{r}|H_0\} = \mathbf{s}^T E\{\mathbf{n}\} = 0$$

$$E\{\eta|H_1\} = \mathbf{s}^T E\{\mathbf{r}|H_1\} = \mathbf{s}^T E\{(\mathbf{s} + \mathbf{n})\} = \mathbf{s}^T\mathbf{s} = E_s$$

$$\text{var}\{\eta|H_1\} = \text{var}\{\eta|H_0\} = E\left\{(\mathbf{s}^T\mathbf{r})(\mathbf{s}^T\mathbf{r})^T|H_0\right\} \tag{4.10}$$

$$= \mathbf{s}^T E\{\mathbf{r}\mathbf{r}^T|H_0\}\mathbf{s} = \mathbf{s}^T E\{\mathbf{n}\mathbf{n}^T\}\mathbf{s} = \sigma^2\mathbf{s}^T\mathbf{s} = \sigma^2 E_s$$

In Figure 4.2, we show the Gaussian PDFs of the test statistic η for the two hypotheses. While the variances of the two hypotheses are the same, the means are different. The optimal threshold T is related to the noise variance as well as the a-priori probabilities. If the two hypotheses are equally likely, then the optimal threshold T is $E_s/2$, the mid-point between the two means. If the a-priori probabilities are unequal, the optimal threshold moves towards the less likely PDF resulting in a smaller error for the more likely hypothesis.

We need to determine the shaded areas in order to determine the error probabilities. For the case of equally likely hypotheses, we can determine the error probabilities from the shaded areas. We can see from Figure 4.2 that moving the threshold to either the right or left of the mid-point increases the total shaded area and thus the P_e when the a-priori probabilities are equal. We see from Figure 4.2 that the two types of error probability (namely P_{FA} and P_M) are equal for the case of equal a-priori probabilities, and that they depend on

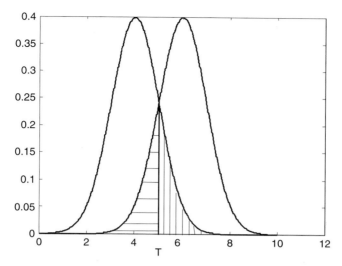

Figure 4.2 Error probabilities for the binary hypothesis case with two Gaussians with equal variances. One shaded area equals the probability of false alarm and the other equals the probability of a miss

(E_s/σ^2), the ratio of signal energy to the noise variance, also known as the input *signal-to-noise ratio* (SNR).

$$P_{FA} = \Pr\left\{\eta \geq \frac{E_s}{2} \Big| H_0\right\} = \frac{1}{\sqrt{2\pi\sigma^2 E_s}} \int_{E_s/2}^{\infty} \exp\left[-\frac{\eta^2}{2\sigma^2 E_s}\right] d\eta$$

$$= \frac{1}{\sqrt{\pi}} \int_{\sqrt{E_s}/(2\sqrt{2}\sigma)}^{\infty} \exp\left[-x^2\right] dx = \frac{1}{2} \operatorname{erfc}\left(\frac{\sqrt{E_s}}{2\sqrt{2}\sigma}\right)$$

Because of symmetry,

$$P_{FA} = P_M \Rightarrow P_e = \frac{1}{2}(P_{FA} + P_M)$$

$$= \frac{1}{2} \operatorname{erfc}\left(\frac{\sqrt{E_s}}{2\sqrt{2}\sigma}\right) = \frac{1}{2} \operatorname{erfc}\left(\sqrt{\frac{SNR}{8}}\right) \qquad (4.11)$$

where erfc($\cdot$), the complementary error function, is defined as follows:

$$\operatorname{erfc}(y) \triangleq \frac{2}{\sqrt{\pi}} \int_{y}^{\infty} \exp(-x^2) dx \qquad (4.12)$$

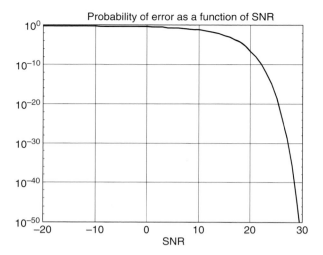

Figure 4.3 Minimum error probability as a function of input SNR (in dB) for the case of two equally likely hypotheses where noise components have zero means, constant variance, and are uncorrelated

Since erfc($\cdot$) in Eq. (4.12) decreases monotonically from 1 to 0 as its argument y increases from 0 to infinity, the error probability in Eq. (4.11) monotonically decreases from 1/2 to 0 as the SNR increases from 0 to infinity. Figure 4.3 shows this dependence except that SNR is shown in dB (i.e., SNR in dB is $10 \log_{10}(E_s/\sigma^2)$).

4.1.3 Receiver operating curves

Suppose we decide to use a threshold other than the one yielding the lowest error probability. This may be done for a variety of reasons including not knowing the a-priori probabilities, or wanting to see the tradeoff between P_{FA} and P_M. If the threshold T in Figure 4.2 moves to the right from the mid-point (i.e., towards the PDF for H_1), then P_{FA} decreases, but P_M will increase. If the threshold moves to the left, P_M will decrease, but P_{FA} will increase.

It is more common to use the probability of detection $P_D = 1 - P_M$ instead of P_M. The plot of P_D versus P_{FA} is known as the *receiver operating curve* (ROC) and an example ROC is shown in Figure 4.4. To understand how this ROC is obtained, we provide below expressions for P_D and P_{FA} as a function of threshold T. By letting T increase from negative infinity to positive infinity, we decrease P_{FA} monotonically from one to zero. Similarly, P_D also decreases monotonically from one to zero as T is increased. That results in the shape of the ROC in Figure 4.4.

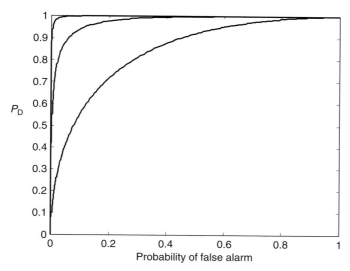

Figure 4.4 Example receiver operating curves (ROC) for different SNR values

$$P_{\text{FA}} = \Pr\{\eta \geq T | H_0\} = \frac{1}{\sqrt{2\pi\sigma^2 E_{\text{s}}}} \int_T^{\infty} \exp\left[-\frac{\eta^2}{2\sigma^2 E_{\text{s}}}\right] d\eta$$

$$= \frac{1}{\sqrt{\pi}} \int_{T/\left(\sigma\sqrt{2E_{\text{s}}}\right)}^{\infty} \exp\left[-x^2\right] dx = \frac{1}{2} \text{ erfc}\left(\frac{T}{\sigma\sqrt{2E_{\text{s}}}}\right)$$

$$\tag{4.13}$$

$$P_{\text{D}} = \Pr\{\eta \geq T | H_1\} = \frac{1}{\sqrt{2\pi\sigma^2 E_{\text{s}}}} \int_T^{\infty} \exp\left[-\frac{(\eta - E_{\text{s}})^2}{2\sigma^2 E_{\text{s}}}\right] d\eta$$

$$= \frac{1}{\sqrt{\pi}} \int_{(T-E_{\text{s}})/\left(\sigma\sqrt{2E_{\text{s}}}\right)}^{\infty} \exp\left[-x^2\right] dx = \frac{1}{2} \text{ erfc}\left(\frac{T - E_{\text{s}}}{\sigma\sqrt{2E_{\text{s}}}}\right)$$

Eq. (4.13) shows that P_{D} and P_{FA} depend not only on the threshold T, but also on the signal energy E_{s} and noise variance σ^2. In Figure 4.4, we show the ROCs for different input SNR values. We see that for low SNR values, the ROC approaches the $P_{\text{D}} = P_{\text{FA}}$ straight line. In fact, the $P_{\text{D}} = P_{\text{FA}}$ straight line is the worst possible binary detection performance in that it cannot distinguish one hypothesis from the other at all. For high SNR values, the ROC approaches a step function that represents the best possible detection performance, in that we can obtain high P_D values for small P_{FA} values. The area under the ROC, known as the *power* of the detector, is a good indicator of how good the test is at distinguishing one hypothesis from the other. We can see from Figure 4.4 that the power increases with SNR, with the lowest power being 0.5 and the highest power being 1.0.

4.2 Multiple hypotheses testing

Often, we need to decide to which of the C classes an input pattern belongs. When C is greater than 2, we will refer to such a problem as the multiple hypotheses testing problem. In particular, the goal is to determine a scheme that assigns the received vector $\mathbf{r}$ to one of the C hypotheses, such that the overall probability of error is as small as it can be. The received vector $\mathbf{r}$ is modeled as follows:

$$H_i: \quad \mathbf{r} = \mathbf{s}_i + \mathbf{n}, \quad i = 1, 2, \ldots, C \tag{4.14}$$

where $\mathbf{s}_i$ denotes the signal for the ith hypothesis and $\mathbf{n}$ denotes the additive noise vector. In this section, we will show how to minimize the probability of error under the model in Eq. (4.14). The resulting detector structure is known as the *maximum a-posteriori probability* (MAP) classifier.

4.2.1 MAP classifier

Let R_i denote the ith decision region, – i.e., the set of received vectors $\mathbf{r}$ that get assigned to hypothesis H_i. Also, let P_i denote the a-priori probability of hypothesis H_i. Then the overall probability of error P_e is the complement of P_c, the probability of correct classification, and can be expressed as follows:

$$P_e = 1 - P_c = 1 - \left[\sum_{i=1}^{C} P_i \cdot \Pr\{\mathbf{r} \in R_i | H_i\} \right] = 1 - \left[\sum_{i=1}^{C} P_i \cdot \int \cdots \int_{R_i} f_i(\mathbf{r}) d\mathbf{r} \right] \tag{4.15}$$

where $f_i(\mathbf{r})$ denotes the PDF of the received vector under hypothesis H_i. For the model in Eq. (4.14), this PDF is related to the noise PDF $f_n(\mathbf{r})$ as follows:

$$f_i(\mathbf{r}) = f_{\mathbf{n}}(\mathbf{r} - \mathbf{s}_i), \quad i = 1, 2, \ldots, C \tag{4.16}$$

From Eq. (4.15), minimizing P_e is equivalent to maximizing P_c. This requires that the decision regions R_i be selected so as to maximize the sum $P_c = \left[\sum_{i=1}^{C} P_i \cdot \int \cdots \int_{R_i} f_i(\mathbf{r}) d\mathbf{r} \right]$. Every received vector $\mathbf{r}$ must be assigned to one and only one of the C decision regions R_i. As a result, every vector $\mathbf{r}$ contributes to only one term in this sum and we should assign $\mathbf{r}$ to that region R_i which results in the largest contribution to the sum, i.e.,

$$P_i \cdot f_i(\mathbf{r}) \geq P_j \cdot f_j(\mathbf{r}), \quad j = 1, 2, \ldots, C \Rightarrow \mathbf{r} \in R_i \tag{4.17}$$

From Bayes' rule (see Chapter 2) we see that the detection scheme in Eq. (4.17) is equivalent to assigning a received vector $\mathbf{r}$ to the hypothesis that maximizes the a-posteriori probability $\Pr\{H_i|\mathbf{r}\} = P_i \cdot f_i(\mathbf{r})/f(\mathbf{r})$. Thus a minimum probability of error is achieved by the MAP classifier.

4.2.2 Additive Gaussian noise model

Let us consider the model in Eq. (4.14) with the additive noise being governed by an N-variate Gaussian PDF with zero-mean vector and covariance matrix $\mathbf{C}$. We will also assume that all C hypotheses are equally likely. In such a case, the MAP classifier assigns received vector $\mathbf{r}$ to the hypothesis H_i corresponding to maximum $f_i(\mathbf{r})$. From Eq. (4.16), $f_i(\mathbf{r})$ are all Gaussian PDFs with the same covariance matrix, but different mean vectors $\mathbf{s}_i$. By using the logarithm, the MAP classifier simplifies to assigning the vector $\mathbf{r}$ to hypothesis H_i with the smallest *Mahalanobis*[1] *distance* d_i defined as follows:

$$d_i^2(\mathbf{r}, \mathbf{s}_i) = (\mathbf{r} - \mathbf{s}_i)^{\mathrm{T}} \mathbf{C}^{-1} (\mathbf{r} - \mathbf{s}_i) \tag{4.18}$$

Before we go back to the MAP classifier discussion, a few remarks about the Mahalanobis distance are in order. For the special case where $\mathbf{C} = \mathbf{I}$, the Mahalanobis distance is the same as the Euclidean distance. If the noise components are uncorrelated (i.e., $\mathbf{C}$ is diagonal), but have different variances, the Mahalanobis distance weights the squared distance in each component by the inverse of its variance, before summing them to find the overall distance. Thus unlike the Euclidean distance, Mahalanobis distance weights the more noisy components less, and in that sense is a more appropriate indicator of the separation between two classes.

---

[1] This "Mahalanobis" here refers to Dr. P. C. Mahalanobis, a famous Indian statistician from the early 1900s.

We have stated that for the case of zero-mean Gaussian noise with equal a-priori probabilities, the MAP classifier is equivalent to minimizing the Mahalanobis distance in Eq. (4.18). On the surface, Eq. (4.18) appears to suggest a quadratic classifier resulting in quadratic (e.g., parabolas, pairs of parallel lines, hyperbolas, etc.) decision boundaries. However the $\mathbf{r}^{\mathrm{T}}\mathbf{C}^{-1}\mathbf{r}$ term common to all $d_i(\mathbf{r}, \mathbf{s}_i)$ is independent of i, and thus the MAP classifier (which minimizes this Mahalanobis distance) is really a *linear* classifier since it assigns $\mathbf{r}$ to the hypothesis that maximizes the following *linear discriminant function* (LDF):

$$\psi_i(\mathbf{r}) = \mathbf{a}_i^{\mathrm{T}}\mathbf{r} - b_i \tag{4.19}$$

where

$$\mathbf{a}_i = \mathbf{C}^{-1}\mathbf{s}_i \quad \text{and} \quad b_i = \frac{1}{2}\mathbf{s}_i^{\mathrm{T}}\mathbf{C}^{-1}\mathbf{s}_i \quad i = 1, 2, \ldots, C$$

The decision boundary separating R_i from R_j is a linear one (e.g., line in 2-D, plane in 3-D, hyperplane in higher dimensions) characterized by $(\mathbf{a}_i - \mathbf{a}_j)^{\mathrm{T}}\mathbf{r} = (b_i - b_j)$. It is easy to verify that the mid-point $(\mathbf{s}_i + \mathbf{s}_j)/2$ between the two means is always part of this decision boundary as long as the a-priori probabilities are equal. If the a-priori probabilities of the hypotheses were not equal, then the decision boundary would move away from this bisecting point towards the less likely hypothesis. We can also see that the decision boundary is normal to the vector $(\mathbf{a}_i - \mathbf{a}_j) = \mathbf{C}^{-1}(\mathbf{s}_i - \mathbf{s}_j)$. For the special case where $\mathbf{C} = \mathbf{I}$, the decision boundary can be seen to be orthogonal to the vector $(\mathbf{s}_i - \mathbf{s}_j)$ connecting the two means.

4.2.3 Error probability for 2-class case

Let us now consider the case where we have just two classes and the noise is additive and Gaussian with zero mean and covariance matrix $\mathbf{C}$. Then the minimum error probability is achieved by the following classifier:

$$l(\mathbf{r}) = (\mathbf{s}_1 - \mathbf{s}_2)^{\mathrm{T}}\mathbf{C}^{-1}\mathbf{r} + \frac{1}{2}\left(\mathbf{s}_2^{\mathrm{T}}\mathbf{C}^{-1}\mathbf{s}_2 - \mathbf{s}_1^{\mathrm{T}}\mathbf{C}^{-1}\mathbf{s}_1\right) \geq 0 \Rightarrow \mathbf{r} \in H_1$$

and $\tag{4.20}$

$$l(\mathbf{r}) < 0 \Rightarrow \mathbf{r} \in H_2$$

Since $l(\mathbf{r})$ is a linear function of $\mathbf{r}$, it is a Gaussian RV with the following means and variances for the two hypotheses:

$$E\{l(\mathbf{r})|H_1\} = (\mathbf{s}_1 - \mathbf{s}_2)^{\mathrm{T}}\mathbf{C}^{-1}E\{\mathbf{r}|H_1\} + \frac{1}{2}\left(\mathbf{s}_2^{\mathrm{T}}\mathbf{C}^{-1}\mathbf{s}_2 - \mathbf{s}_1^{\mathrm{T}}\mathbf{C}^{-1}\mathbf{s}_1\right)$$

$$= (\mathbf{s}_1 - \mathbf{s}_2)^{\mathrm{T}}\mathbf{C}^{-1}\mathbf{s}_1 + \frac{1}{2}\left(\mathbf{s}_2^{\mathrm{T}}\mathbf{C}^{-1}\mathbf{s}_2 - \mathbf{s}_1^{\mathrm{T}}\mathbf{C}^{-1}\mathbf{s}_1\right)$$

$$= \frac{1}{2}\left[\mathbf{s}_1^{\mathrm{T}}\mathbf{C}^{-1}\mathbf{s}_1 - 2\mathbf{s}_2^{\mathrm{T}}\mathbf{C}^{-1}\mathbf{s}_1 + \mathbf{s}_2^{\mathrm{T}}\mathbf{C}^{-1}\mathbf{s}_2\right]$$

$$= \frac{1}{2}(\mathbf{s}_1 - \mathbf{s}_2)^{\mathrm{T}}\mathbf{C}^{-1}(\mathbf{s}_1 - \mathbf{s}_2) = \frac{1}{2}d_{1,2}^2$$

$$E\{l(\mathbf{r})|H_2\} = (\mathbf{s}_1 - \mathbf{s}_2)^{\mathrm{T}}\mathbf{C}^{-1}E\{\mathbf{r}|H_2\} + \frac{1}{2}\left(\mathbf{s}_2^{\mathrm{T}}\mathbf{C}^{-1}\mathbf{s}_2 - \mathbf{s}_1^{\mathrm{T}}\mathbf{C}^{-1}\mathbf{s}_1\right)$$

$$= (\mathbf{s}_1 - \mathbf{s}_2)^{\mathrm{T}}\mathbf{C}^{-1}\mathbf{s}_2 + \frac{1}{2}\left(\mathbf{s}_2^{\mathrm{T}}\mathbf{C}^{-1}\mathbf{s}_2 - \mathbf{s}_1^{\mathrm{T}}\mathbf{C}^{-1}\mathbf{s}_1\right)$$

$$= -\frac{1}{2}\left[\mathbf{s}_1^{\mathrm{T}}\mathbf{C}^{-1}\mathbf{s}_1 - 2\mathbf{s}_2^{\mathrm{T}}\mathbf{C}^{-1}\mathbf{s}_1 + \mathbf{s}_2^{\mathrm{T}}\mathbf{C}^{-1}\mathbf{s}_2\right]$$

$$= -\frac{1}{2}(\mathbf{s}_1 - \mathbf{s}_2)^{\mathrm{T}}\mathbf{C}^{-1}(\mathbf{s}_1 - \mathbf{s}_2) = -\frac{1}{2}d_{1,2}^2$$

$$\mathrm{var}\{l(\mathbf{r})|H_1\} = \mathrm{var}\{l(\mathbf{r})|H_2\} = \mathrm{var}\left\{(\mathbf{s}_1 - \mathbf{s}_2)^{\mathrm{T}}\mathbf{C}^{-1}\mathbf{r}|H_1\right\}$$

$$= E\left\{\left[(\mathbf{s}_1 - \mathbf{s}_2)^{\mathrm{T}}\mathbf{C}^{-1}(\mathbf{r} - \mathbf{s}_1)\right]\left[(\mathbf{s}_1 - \mathbf{s}_2)^{\mathrm{T}}\mathbf{C}^{-1}(\mathbf{r} - \mathbf{s}_1)\right]^{\mathrm{T}}|H_1\right\}$$

$$= (\mathbf{s}_1 - \mathbf{s}_2)^{\mathrm{T}}\mathbf{C}^{-1}E\{(\mathbf{r} - \mathbf{s}_1)(\mathbf{r} - \mathbf{s}_1)^{\mathrm{T}}|H_1\}\mathbf{C}^{-1}(\mathbf{s}_1 - \mathbf{s}_2)$$

$$= (\mathbf{s}_1 - \mathbf{s}_2)^{\mathrm{T}}\mathbf{C}^{-1}\mathbf{C}\mathbf{C}^{-1}(\mathbf{s}_1 - \mathbf{s}_2) = d_{1,2}^2 \tag{4.21}$$

where $d_{1,2}$ denotes the Mahalanobis distance between the PDFs for the two hypotheses. From Eqs. (4.20) and (4.21), we see that the probability of error is the probability that a Gaussian PDF with mean $(-d_{1,2}^2/2)$ and variance $d_{1,2}^2$ takes on positive values, i.e.,

$$P_\mathrm{e} = \frac{1}{\sqrt{2\pi d_{1,2}^2}}\int_0^\infty \exp\left[-\frac{\left(l + \frac{d_{1,2}^2}{2}\right)^2}{2d_{1,2}^2}\right]\mathrm{d}l \tag{4.22}$$

$$= \frac{1}{\sqrt{\pi}}\int_{\sqrt{d_{1,2}^2/8}}^\infty \exp[-p^2]\,\mathrm{d}p = \frac{1}{2}\mathrm{erfc}\left(\sqrt{d_{1,2}^2/8}\right)$$

Thus, the probability of error P_e is a monotonically decreasing function of $d_{1,2}$, the Mahalanobis distance between the two classes.

In this section, we have deliberately focused on the simple case where the PDFs are Gaussian, the covariance matrices for all cases are identical and all classes are equally likely. Under these assumptions, the minimum error decision boundaries are linear, making it easy to evaluate the resulting probability of error. However, in practice, these assumptions may not hold. Even for a 2-class problem using Gaussian PDFs, minimum error decision boundaries will be nonlinear (in fact, quadratic) if the covariance matrices are not equal. In such cases, it will not be easy to evaluate the error probabilities since the decision statistic is not a linear function of the observation vector. However, using central limit theorem arguments, we may still be able to approximate the decision statistic as Gaussian and evaluate approximate error probabilities. Another practical problem is that we rarely know the PDFs with certainty as they are estimated from measurements, which themselves may contain errors. Thus model mismatch (i.e., the difference between assumed PDFs and the actual PDFs) is a real concern, but such deep topics are beyond the scope of this book, which is focused on CPR. The interested reader can find detailed treatments elsewhere [22, 29].

4.3 Estimation theory

Detection deals with the issue of selecting one of a finite number of hypotheses, and cannot handle issues such as locating a particular target in a given scene, or determining the error probability of a classification scheme. In such cases, the parameter of interest (e.g., error rate or target coordinates) take on either a continuum of values or just too many possible values to treat the problem as one of a finite number of hypotheses. It is more convenient to treat it as estimating an unknown value and this section is aimed at providing a brief look at basic estimation techniques. Good estimation methods take advantage of the a-priori information as well as available data.

In this section, we will look at the problem of estimating a parameter θ using the data set $X = \{\mathbf{x}_1, \mathbf{x}_2, \ldots, \mathbf{x}_N\}$ consisting of N independent observation vectors. Since the particular set of N vectors that constitute the observation set is a randomly drawn sample, the resulting estimate θ is also random and thus has an associated mean, variance and, in fact, a PDF, i.e., θ is really a random variable. Ideally, we want these PDFs to be delta functions centered at the correct value of the parameter. If this can happen, then the estimator will always yield the same, correct estimate of the parameter. This is not practical and instead we must settle for the following attributes for good estimates.

Bias For a good estimate, its PDF should have a mean that equals the true value of the parameter. This attribute of an estimate makes it *unbiased* and can be stated mathematically as follows:

$$E\{\theta(\mathbf{X})\} = \theta \qquad (4.23)$$

The difference between the mean of the parameter estimate and the true parameter is known as the *bias*.

Variance Just forcing the estimate to be unbiased is not good enough since an estimate can be unbiased even if it is never close to the true value. For example, an estimate that takes values of $+1$ and -1 with equal probability has a mean of 0 and is unbiased if the true parameter value is 0. However, the estimate never equals the correct parameter value. This is achieved by demanding that the variance of the estimate be small. An estimate is said to be *consistent* provided that

$$\lim_{N\to\infty} \text{var}\{\theta(\mathbf{X})\} = 0 \qquad (4.24)$$

Efficiency An estimate is said to be *efficient* if it uses the available N data vectors to provide the smallest possible variance in the estimate. In general, it is difficult to test whether an estimate is efficient or not. However, for maximum likelihood estimation methods introduced in the next sub-section, there exists a theorem that helps us answer the efficiency question.

We will consider the case of a single parameter in this section. However, results presented can be extended to the case of more than one parameter. For more details and more advanced treatment of this topic, readers are advised to consult some of the excellent references [13, 30].

4.3.1 Maximum likelihood estimation

Maximum likelihood (ML) estimation is based on the idea that the parameter to choose is the one that maximizes the likelihood of observing the data set. More precisely, θ_{ML} maximizes $f(\mathbf{X}|\theta) = f(\mathbf{x}_1, \mathbf{x}_2, \ldots, \mathbf{x}_N|\theta) = \prod_{i=1}^{N} f(\mathbf{x}_i|\theta)$ with respect to θ, where we use the assumption that the observation vectors are statistically independent. Since the PDF is non-negative, we can maximize the logarithm of the PDF if it is more convenient. Thus, for smooth PDFs, we can use the following equivalent condition for the ML estimate:

$$\sum_{i=1}^{N} \frac{\partial \ln f(\mathbf{x}_i|\theta)}{\partial \theta}\bigg|_{\theta=\theta_{ML}} = 0 \qquad (4.25)$$

ML estimate of the mean of a Gaussian To illustrate the use of Eq. (4.25), we will consider the problem of estimating the mean m of a univariate Gaussian PDF with known variance σ^2 from N independent samples $\{x_1, x_2, \ldots, x_N\}$ drawn from that PDF. Applying Eq. (4.25) leads to the following ML estimate for the mean:

$$\sum_{i=1}^{N} \frac{\partial}{\partial m} \left[-\ln\left(\sigma\sqrt{2\pi}\right) - \frac{1}{2\sigma^2}(x_i - m)^2 \right]\Bigg|_{m=m_{\mathrm{ML}}} = 0$$

$$\Rightarrow \sum_{i=1}^{N}(x_i - m_{\mathrm{ML}}) = 0 \Rightarrow m_{\mathrm{ML}} = \frac{1}{N}\sum_{i=1}^{N} x_i \tag{4.26}$$

Satisfyingly, the ML estimate for the unknown mean is the sample average of the N observations. However, we must remember that such a nice result is the consequence of the assumption that the samples are independently drawn from a Gaussian PDF with known variance.

Sample mean properties Is the ML estimate in Eq. (4.26) a good estimate? As stated before, we should determine the bias and the variance of the estimate. Firstly, we will show that this ML estimate is unbiased:

$$E\{m_{\mathrm{ML}}\} = E\left\{\frac{1}{N}\sum_{i=1}^{N} x_i\right\} = \frac{1}{N}\sum_{i=1}^{N} E\{x_i\}$$

$$= \frac{1}{N}\sum_{i=1}^{N} m = m \Rightarrow m_{\mathrm{ML}} \text{ is unbiased} \tag{4.27}$$

Let us now determine the variance of the ML estimate:

$$\mathrm{var}\{m_{\mathrm{ML}}\} = E\left\{(m_{\mathrm{ML}} - m)^2\right\} = E\left\{\left(\left[\frac{1}{N}\sum_{i=1}^{N} x_i\right] - m\right)^2\right\}$$

$$= E\left\{\left(\frac{1}{N}\sum_{i=1}^{N}(x_i - m)\right)^2\right\} = \frac{1}{N^2}\sum_{i=1}^{N}\sum_{j=1}^{N} E\{(x_i - m)(x_j - m)\} \tag{4.28}$$

$$= \frac{1}{N^2}\sum_{i=1}^{N} E\left\{(x_i - m)^2\right\} = \frac{N\sigma^2}{N^2} = \frac{\sigma^2}{N}$$

where we use the fact that the observations are statistically independent to remove the cross terms in the double summation. From Eq. (4.28), we see that the variance goes to zero as N goes to infinity, indicating that the ML estimate is consistent.

Cramer–Rao bound We have shown that the sample average is an unbiased and consistent estimator of the mean of a Gaussian PDF. Is that estimate efficient? A bound known as the Cramer–Rao bound [31] helps address this issue. For unbiased estimators, it can be shown that the variance has the following lower bound:

$$\text{var}\{\theta(X)\} \geq \left[E\left\{ \left(\frac{\partial \ln f(X|\theta)}{\partial \theta} \right)^2 \right\} \right]^{-1}$$

$$= \left[-E\left\{ \left(\frac{\partial^2 \ln f(X|\theta)}{\partial \theta^2} \right) \right\} \right]^{-1} = \text{CRLB}$$

(4.29)

where CRLB stands for Cramer–Rao lower bound. Since the sample average is an unbiased estimate, we know from Eq. (4.29) that its variance cannot be smaller than the CRLB determined below:

$$\text{CRLB} = \left[-E\left\{ \left(\frac{\partial^2 \ln f(X|m)}{\partial m^2} \right) \right\} \right]^{-1}$$

$$= \left[-E\left\{ \left(\frac{\partial^2}{\partial m^2} \left[-\frac{N}{2} \ln(2\pi\sigma^2) - \frac{1}{2\sigma^2} \sum_{i=1}^{N} (x_i - m)^2 \right] \right) \right\} \right]^{-1}$$

(4.30)

$$= \left[-E\left\{ \left(\left[-\frac{2N}{2\sigma^2} \right] \right) \right\} \right]^{-1} = \frac{\sigma^2}{N}$$

Thus, we see that the variance of the sample average estimate (see Eq. (4.28)) equals the CRLB indicating that the estimate under study is efficient, i.e., no other estimate can yield smaller variance for the given observation set. We can show that if an estimate reaches the CRLB, then it must be an ML estimate [29]. This is one of the main reasons for the popularity of the ML estimate.

4.3.2 Other estimators

While ML estimation is perhaps the most popular estimation method, it has one significant drawback. It models the unknown parameter as a deterministic constant, and thus ML estimates cannot take any advantage of any a-priori knowledge we may have about the parameters. Suppose we know a-priori that the unknown mean of a Gaussian PDF is known to lie in the interval [−2, 2]. How do we incorporate this information? There is nothing to prevent an ML estimate of 3.5, for example. In this section, we will present two estimators that model the unknown parameter as random and allow such a-priori information to be included.

MAP estimation   In MAP estimation, we choose the estimate to maximize the a-posteriori PDF, $f(\theta|\mathbf{X}) = f(\mathbf{X}|\theta)f(\theta)/f(\mathbf{X})$. Thus the ML and MAP estimates differ only through the a-priori PDF $f(\theta)$ of the parameter. If this a-priori PDF is relatively constant over the range of interest, then there is no significant difference between the ML and the MAP estimates. On the other hand, the inclusion of the a-priori PDF in the maximization enables the MAP estimate to take advantage of that information.

To appreciate the role that the a-priori information can play, let us revisit the problem of estimating the mean of a Gaussian PDF (with known variance) using N independent observations. Suppose our a-priori information is that the unknown mean itself can be modeled by another Gaussian PDF with mean m_0 and variance σ_0^2. Using this information, one can show (we will omit the rather tedious derivation and refer the interested readers to references [1, 2]) that the MAP estimate for the mean is as follows:

$$m_{\text{MAP}} = \frac{\sigma^2/N}{(\sigma^2/N) + \sigma_0^2} m_0 + \frac{\sigma_0^2}{(\sigma^2/N) + \sigma_0^2} m_N \quad \text{where } m_N = \frac{1}{N} \sum_{i=1}^{N} x_i \quad (4.31)$$

We see from Eq. (4.31) that the MAP estimate is a weighted average of the a-priori mean m_0 and sample mean m_N obtained from the N observations. The relative emphasis between the two means depends on the a-priori variance σ_0^2 as well as the variance σ_0^2/N of the mean estimate from the N samples. If σ_0^2 is low, then the MAP estimate weighs the a-priori mean m_0 more heavily. On the other hand, if σ^2/N is low (because the number of observations N is large or the observation noise variance σ^2 is small or both), the MAP estimate depends more strongly on the sample mean m_N.

Minimum mean squared error estimation   Another estimation method that uses the a-priori information is the minimum mean squared error (MMSE) estimator. In MMSE estimation, we choose the estimate to minimize the average squared error between the estimate and the true value, i.e., we minimize $E\left\{ \left(\hat{\theta} - \theta \right)^2 \right\}$, where we use the hat to distinguish the estimate from the true parameter. Setting the derivative of this mean squared error to zero results in the MMSE estimate as shown below:

$$\frac{\partial}{\partial \hat{\theta}} E\left\{\left(\hat{\theta} - \theta\right)^2\right\} = \frac{\partial}{\partial \hat{\theta}} \left[\int \left(\hat{\theta} - \theta\right)^2 f(\theta | \mathbf{X}) d\theta\right] = 2 \int \left(\hat{\theta} - \theta\right) f(\theta | \mathbf{X}) d\theta = 0$$

$$\Rightarrow \hat{\theta} \int f(\theta | \mathbf{X}) d\theta = \int \theta f(\theta | \mathbf{X}) d\theta \Rightarrow \hat{\theta}_{\text{MMSE}} = \int \theta f(\theta | \mathbf{X}) d\theta = E\{\theta | \mathbf{X}\}$$

$$(4.32)$$

Thus, the MMSE estimator is nothing but the conditional mean of the parameter, whereas the MAP estimate is the mode of the conditional PDF as the MAP estimate maximizes the conditional PDF. For some PDFs (e.g., Gaussian PDFs), the mean and the mode occur at the same place and thus the MAP and the MMSE estimates are identical.

4.3.3 Error rate estimation

When a classifier such as a correlation-based recognition system is designed we would, of course, want to know how well it works. It would be great if we could derive analytical expressions for the error probabilities, since such theoretical expressions allow us to figure out which system parameters would improve the classification performance. However, deriving theoretical error probability expressions is often impossible and we may have to resort to Monte Carlo methods.

We would test the designed classifier using N statistically independent data vectors whose correct classes are known. The classifier will classify each of these data vectors either correctly or incorrectly. Let ε denote the unknown probability of error of this classifier. Then the probability of observing M errors in data vectors can be modeled by the Binomial distribution, i.e.,

$$\Pr\{M \text{ errors in } N \text{ vectors} | \varepsilon\} = \binom{M}{N} \varepsilon^M (1 - \varepsilon)^{(N-M)}, \quad M = 0, 1, \ldots, N$$

$$(4.33)$$

To obtain the ML estimate of ε, we set the derivative of the logarithm of the likelihood in Eq. (4.33) to zero and obtain the ML estimate as follows:

$$\frac{\partial}{\partial \varepsilon} [M \ln \varepsilon + (N - M) \ln(1 - \varepsilon)] = \frac{M}{\varepsilon} - \frac{N - M}{1 - \varepsilon} = 0 \Rightarrow \varepsilon_{\text{ML}} = \frac{M}{N} \quad (4.34)$$

Thus, the ML estimate of the unknown error probability ε is the ratio of M (the number of errors) to N (the number of data vectors). For the binomial distribution in Eq. (4.33), the mean is $N\varepsilon$ and the variance is

$N\varepsilon(1 - \varepsilon)$. As a result, the mean and variance of the error rate estimate $\varepsilon_{\mathrm{ML}}$ are as follows:

$$E\{\varepsilon_{\mathrm{ML}}\} = \frac{E\{M\}}{N} = \frac{N\varepsilon}{N} = \varepsilon$$

$$\mathrm{var}\{\varepsilon_{\mathrm{ML}}\} = \frac{\mathrm{var}\{M\}}{N^2} = \frac{N\varepsilon(1 - \varepsilon)}{N^2} = \frac{\varepsilon(1 - \varepsilon)}{N}$$

(4.35)

We see from Eq. (4.35) that $\varepsilon_{\mathrm{ML}}$ is unbiased and consistent. Using this mean and variance, we can derive confidence bounds on error estimates. Using the CRLB, we can show that $\varepsilon_{\mathrm{ML}}$ is also efficient.

4.4 Chapter summary

The goal of this chapter was to provide a brief review of tools and techniques needed to deal with detection of signals and estimation of parameters. Following is a brief summary of the main observations.

- For the additive Gaussian noise model, minimum error probability is achieved by using the maximum a-posteriori (MAP) detection scheme. For the case of two equally likely hypotheses, the resulting minimal probability of error can be seen to be a monotonic function of the Mahalanobis distance between the two PDFs. If all of the noise components are uncorrelated and are of equal variance, the Mahalanobis distance simplifies to the Euclidean distance. Otherwise, the Mahalanobis distance uses smaller weights for the more noisy components.
- For the case of multiple signals corrupted by additive, zero-mean Gaussian noise, a minimum probability of error is achieved by assigning the received vector to the hypothesis that is closest, in the Mahalanobis distance sense. Assuming that all signals are equally likely, this leads to a linear classifier with decision boundaries that bisect the lines connecting the various signals. If the noise components are uncorrelated and are of equal variance, these bisectors are also orthogonal to the line connecting the two class means.
- For non-Gaussian PDFs, the decision boundaries can be complicated and estimation of error probabilities can be difficult. Even for Gaussian PDFs, the decision boundaries are not linear if the covariance matrices for all classes are not equal. In such cases, one may still be able to use the results based on Gaussian PDFs, if central limit theorem type arguments can be used to approximate the detection statistic as a Gaussian random variable.
- Good estimates should be unbiased and consistent. Unbiased estimates require that the mean of the estimate should equal the true parameter value. Consistent estimates exhibit variance that approaches zero as the number of observations goes to infinity.

- Maximum likelihood estimation chooses the estimate that results in the highest likelihood for the observations. The ML estimate is attractive in that it has the potential to be efficient, i.e., to provide the smallest variance for the given data. The Cramer–Rao lower bound is useful in assessing the efficiency of estimators.
- Maximum likelihood estimates model the unknown parameter as deterministic and thus cannot take advantage of any a-priori information one may have about the parameter being estimated. Maximum a-posteriori and minimum mean squared error estimators treat the unknown parameter as an RV and thus can include the a-priori information.

5

Correlation filter basics

The basic concept of correlation is illustrated in Figure 1.4 with the help of a simple character recognition example. In this figure, black pixels take on a value of 1 and white pixels take on a value of 0. Suppose we are trying to locate all occurrences of the reference or target image (C in this example) in the test image (also called the input scene). One way to achieve this is to cross-correlate the target image with the input scene. The target image is placed in the upper left corner of the input scene and pixel-wise multiplication is carried out between the two arrays; all of the values in the resultant product array are summed to produce one correlation output value. This process is repeated by shifting the target image by various shifts to the right and down, thus producing a two-dimensional (2-D) output array called the correlation output. Ideally, this correlation output would have two large values corresponding to the two "C" letters in the input scene and zeros for other letters. Thus, large cross-correlation values indicate the presence and location of the character we are looking for. However, this will not always be achievable because some other letters may have high cross-correlation. For example, letter "C" and letter "O" have large cross-correlation. One of the goals of this book is to develop methods that preserve large cross-correlation with desired targets, while suppressing cross-correlation with undesired images (sometimes called the clutter), and reducing sensitivity to noise and distortions such as rotations, scale changes, etc. The goal of this chapter is to provide the basic ideas underlying the use of correlation as a pattern recognition tool.

Correlation can be thought of as the output from a *matched filter* (a linear, shift-invariant (LSI) filter whose impulse response is the reflected version of the reference signal or image), and it can be shown to be "optimal" for detecting known signals corrupted by additive white noise. We will start this chapter by establishing this notion of optimality of correlation. Correlation can be implemented using either optical or digital processing, and this chapter will also discuss some basic correlator implementation methods.

A major milestone in the development of correlation for pattern recognition was the pioneering work by VanderLugt [5] who represented complex-valued matched filters using holograms and thus implemented correlation operation using coherent optical processors [32]. In particular, that work made possible the use of optical correlators to detect and locate reference images in observed scenes. While optical correlators will be considered in more detail in Chapter 8, we will introduce the VanderLugt correlator briefly in this chapter mainly to motivate the development of the many variants of the classical matched filter.

We will also consider digital methods for computing the correlations. While digital implementation of correlations using discrete Fourier transforms (DFTs) appears rather straightforward, some important issues arise. (The fast Fourier transform (FFT) is literally just an efficient algorithm for computing the DFT, but we will often use FFT and DFT synonymously.) Firstly the use of DFTs results in circular rather than linear correlations and care must be taken to make the circular correlation match the desired linear correlation. Another issue is that if a direct (i.e., in time or space domain) digital correlation must be carried out, it is more hardware-efficient (hence faster) if we can reduce the number of bits used to represent the signals or images. We will consider the consequences of using binarized (2 levels, or 1 bit per pixel) or other quantized images (e.g., 4 levels, or 2 bits per pixel) for correlation.

We will use 1-D CT signal notation throughout making only occasional use of 2-D notation. All results presented using 1-D notation have obvious extensions to 2-D unless specifically indicated otherwise.

Section 5.1 introduces the notion of a matched filter and shows how the matched filter maximizes the output signal-to-noise ratio (SNR). Implementations of correlators are discussed in Section 5.2 and performance metrics to evaluate correlation outputs are presented in Section 5.3. Generalizations of the matched filter are discussed in Section 5.4, and Section 5.5 presents our model for the optical correlation process, including how noise affects the statistics of the measured correlation. This section also unifies several of the filter-optimizing algorithms under the minimum Euclidean distance optimal filter (MEDOF) scheme. Finally, Section 5.6 deals with non-overlapping noise which arises when the object obscures the background. More advanced correlation filter concepts are discussed in Chapter 6.

5.1 Matched filter

The popularity of correlation methods for pattern recognition owes much to the role that matched filters play in detecting signals in received radar returns corrupted by additive noise. This section will be devoted to reviewing the basic theory of matched filters. Consider the example where a known signal is transmitted and the received signal is examined to answer three questions:

1. Is there a target in the path of the transmitted energy?
2. If there is a target, what is its range from the transmitter?
3. If there is a target, what is its velocity?

In this section, we will focus on the first question, detecting the presence of a target. Once a target is detected, we can use the matched filter output to estimate the relative time shift between the transmitted and the received signals. Provided the speed of signal propagation is constant and known, this time delay yields the range of the target. Even without directional antennas, by using at least three transmitted signals whose transmission locations are known, we can estimate the range of the target to three known positions and thus triangulate the position of the target. If the target moves with a velocity component towards a receiver, it will introduce a Doppler shift in the received signal thus causing a frequency shift. We can estimate this frequency shift and hence the target velocity.

5.1.1 *Known signal in additive noise*

Let $s(t)$ denote the transmitted signal and $r(t)$ denote the received signal containing effects such as attenuation, time delay, Doppler shift, and noise. In this simple model, we will consider the effects of additive noise only. As mentioned earlier, time delay and frequency shifts can be estimated from the received signal. Attenuation causes a decrease in the SNR, which degrades the detection performance. However, attenuation does not change the optimality of the maximal SNR filter we derive in this section.

For this additive noise model, the detection problem simplifies to that of choosing between the following two hypotheses:

$$H_0 : r(t) = n(t)$$
$$H_1 : r(t) = s(t) + n(t)$$

(5.1)

where $n(t)$ denotes the noise. This noise is modeled as a wide sense stationary (WSS) random process with zero mean and power spectral density (PSD) $P_n(f)$. Note that we have not yet assumed anything about the noise probability density function (PDF). Our task is to select between the two hypotheses using $r(t)$ and our knowledge of $s(t)$ and $P_n(f)$.

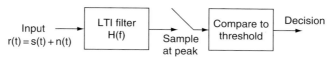

Figure 5.1 Linear filter for the binary signal detection problem

5.1.2 Maximal SNR filter

The basic approach used for this binary signal detection problem is the linear filter paradigm shown in Figure 5.1. The received signal $r(t)$ is passed through an LSI system with impulse response $h(t)$ (or equivalently, frequency response $H(f)$, the FT of $h(t)$). The output signal $y(t)$ is searched for its maximal value $y_{\max}$ and this maximum is compared to a pre-selected threshold T. If $y_{\max}$ equals or exceeds T, then the received signal is declared to contain the transmitted signal (i.e., H_1 is selected), whereas if $y_{\max}$ is less than T, then the received signal is declared to contain only noise (i.e., H_0 is selected). In fact, the position of this maximal value yields the relative time shift between the transmitted and received signals and thus the target range. If the threshold T is low, then the probability of a miss will be small (few H_1 cases will be missed), but the probability of a false alarm (case of H_0 being mis-detected as H_1) will be large. If T is large, the converse will occur.

Signal-to-noise ratio In this approach, the most important step is the design of the filter $H(f)$. A good filter should make the average $y_{\max}$ large (under hypothesis H_1) and make the average noise-induced variance as small as possible. Thus, it is desirable that the filter $H(f)$ maximize the SNR defined as follows:

$$\mathrm{SNR} = \frac{|E\{y_{\max}|H_1\}|^2}{\mathrm{var}\{y_{\max}\}} \tag{5.2}$$

where $E\{\cdot\}$ denotes expectation and "var" denotes the variance. Since the variance arises strictly as a result of noise in this additive noise model, and since the noise process has the same characteristics under both hypotheses, output noise variance is the same for both hypotheses.

Optical and digital correlators have different expressions for SNR, owing to the processors' different properties. The digital processor's output can be exactly linear with the input; however, the optical processor's cannot. The sensed output in optical correlators is the electromagnetic intensity – the squared magnitude of the field. Different expressions for SNR result. Having paid this polite nod, we will concentrate on the strictly linear form in this chapter and clarify the optical situation in Section 5.5 and Chapter 8.

Since the mean of the noise is assumed to be zero, $E\{y_{\max}|H_1\}$ is the maximal value of the filter output when $s(t)$ is the input signal. For the purposes of determining the optimal $H(f)$, we can assume without loss of generality that the output $y(t)$ has its maximal value at the origin. If a given filter's output peak happens to occur somewhere else, it can be brought to the origin by simply multiplying $H(f)$ by an appropriate linear-phase function in frequency. This multiplication of $H(f)$ by a phase function will not affect the filter's noise response since the output noise PSD and the variance it induces are affected only by the magnitude of the filter frequency response, and not its phase. Thus, the numerator of Eq. (5.2) can be simplified as follows:

$$|E\{y_{\max}|H_1\}|^2 = |E\{y(0)|H_1\}|^2$$
$$= \left|\left\{\int s(t)h(-t)\mathrm{d}t\right\}\right|^2 = \left|\left\{\int S(f)H(f)\mathrm{d}f\right\}\right|^2 \quad (5.3)$$

where we assume that the signal $s(t)$ and the impulse response $h(t)$ are real. To express the denominator of Eq. (5.2) similarly in terms of known quantities, we note that the variance depends only on the noise and is thus independent of the signal $s(t)$. Since the input noise $n(t)$ is WSS with PSD $P_n(f)$, the output noise from this LSI system is also WSS with PSD $P_n(f)|H(f)|^2$. Since the variance of a zero-mean random process equals the total area under its PSD, we can express the denominator of Eq. (5.2) as follows:

$$\mathrm{var}\{y_{\max}\} = \int P_n(f)|H(f)|^2\mathrm{d}f \quad (5.4)$$

Using Eqs. (5.3) and (5.4) in Eq. (5.2), we obtain the following expression for SNR in terms of $S(f)$, the FT of the known transmitted signal $s(t)$, $P_n(f)$, the PSD of the additive input noise $n(t)$ and $H(f)$, filter frequency response:

$$\mathrm{SNR} = \frac{|\int S(f)H(f)\mathrm{d}f|^2}{\int P_n(f)|H(f)|^2\mathrm{d}f} \quad (5.5)$$

Before determining the $H(f)$ that maximizes the SNR, a few remarks based on Eq. (5.5) are in order. Multiplying the filter $H(f)$ by a complex constant α does not affect the SNR since it scales both the numerator and the denominator of Eq. (5.5) identically. Also, the phase of the filter affects the numerator of this SNR expression, but not its denominator. Finally, if there are any frequency regions where the noise PSD is zero and the signal FT is not zero, we can hypothetically achieve infinite SNR simply by setting the filter magnitude to be non-zero in only those frequency regions. In practice, this does not occur and we need a more realistic filter to achieve high SNR.

Signal-to-noise ratio maximization Our goal is to find the filter $H(f)$ that maximizes the SNR in Eq. (5.5). To obtain this filter in the digital instance, we use the Cauchy–Schwarz inequality (discussed in Chapter 2) rewritten below in terms of two arbitrary functions $A(f)$ and $B(f)$.

$$\left|\int A(f)B(f)\mathrm{d}f\right|^2 \leq \left[\int |A(f)|^2\mathrm{d}f\right]\left[\int |B(f)|^2\mathrm{d}f\right] \qquad (5.6)$$

with equality if and only if $A(f) = \beta B^*(f)$, where β is a complex constant. We can apply Eq. (5.6) to the numerator of Eq. (5.5) to obtain the following upper bound on the SNR:

$$\mathrm{SNR} = \frac{|\int S(f)H(f)\mathrm{d}f|^2}{\int P_n(f)|H(f)|^2\mathrm{d}f} = \frac{\left|\int \left[\dfrac{S(f)}{\sqrt{P_n(f)}}\right]\left[H(f)\sqrt{P_n(f)}\right]\mathrm{d}f\right|^2}{\int P_n(f)|H(f)|^2\mathrm{d}f}$$

$$\leq \frac{\left[\int \dfrac{|S(f)|^2}{P_n(f)}\mathrm{d}f\right] \cdot \left[\int P_n(f)|H(f)|^2\mathrm{d}f\right]}{\int P_n(f)|H(f)|^2\mathrm{d}f} = \int \frac{|S(f)|^2}{P_n(f)}\mathrm{d}f = \mathrm{SNR}_{\max}$$

$$(5.7)$$

where we are allowed to take the square root of $P_n(f)$ since it is real and non-negative. Equation (5.7) shows that the SNR achievable with any filter must be less than or equal to $\mathrm{SNR}_{\max}$, which depends only on $S(f)$, the signal FT and $P_n(f)$, the noise PSD, and not on the filter $H(f)$. The Cauchy–Schwarz inequality in Eq. (5.6) also tells us when the equality holds. Using the equality condition, we see that the maximal SNR is obtained if and only if

$$\left[\frac{S(f)}{\sqrt{P_n(f)}}\right] = \beta\left[H(f)\sqrt{P_n(f)}\right]^* \Rightarrow H(f) = \alpha\frac{S^*(f)}{P_n(f)} \qquad (5.8)$$

where α is any complex constant. Thus the filter in Eq. (5.8) is optimal in the sense that it maximizes the SNR.

We will call a filter *phase-canceling* if its phase and that of $S(f)$, the signal FT, sum to a constant. Thus, the maximal-SNR filter is phase-canceling. It is satisfying to see that the optimal filter has a frequency response magnitude that is proportional to the ratio of the signal FT magnitude to the noise PSD.

At those frequencies where the known signal is weak compared to the noise, the optimal filter has low gain and thus the received signal is attenuated. At those frequencies where the signal is strong compared to the noise, the filter gain is high and the received signal is amplified. From Eq. (5.7), we see that the maximal SNR can be increased by amplifying the signal and/or reducing the noise level.

Signal-to-noise ratio maximization using vectors We used the Cauchy–Schwarz inequality to derive the maximal-SNR filter in the CT domain. We will show that the same result can be obtained in the DT domain using matrix/vector results presented in Chapter 2.

Towards that end, let us denote the sampled version of the desired filter by column vector $\mathbf{h}$; i.e., $\mathbf{h} = [H(-N\Delta f) \ldots H(0) \ldots H(N\Delta f)]^{\mathrm{T}}$ where we sample the CT filter frequency response $H(f)$ at uniform intervals of Δf and where we truncate the discretized frequency response to $(2N+1)$ samples centered at zero frequency. Similarly $\mathbf{s}$ is a column vector whose $(2N+1)$ elements are the samples of $S(f)$, the signal FT. The noise PSD $P_n(f)$ is sampled at uniform intervals of Δf, and the resulting $(2N+1)$ samples are placed along the diagonal of a $(2N+1)$ by $(2N+1)$ diagonal matrix $\mathbf{P}$. Assuming that the sampling interval Δf is sufficiently small, the SNR in Eq. (5.5) can be approximated as follows:

$$\mathrm{SNR} = \frac{\left|\int S(f)H(f)\mathrm{d}f\right|^2}{\int P_n(f)|H(f)|^2\mathrm{d}f} \cong \frac{\left|\Delta f \sum_{k=-N}^{N} S(k\Delta f)H(k\Delta f)\right|^2}{\Delta f \sum_{k=-N}^{N} P_n(k\Delta f)|H(k\Delta f)|^2}$$

$$= \Delta f \frac{\left|\mathbf{s}^{\mathrm{T}}\mathbf{h}\right|^2}{\mathbf{h}^+\mathbf{P}\mathbf{h}} = \Delta f \frac{\mathbf{h}^+\mathbf{s}^*\mathbf{s}^{\mathrm{T}}\mathbf{h}}{\mathbf{h}^+\mathbf{P}\mathbf{h}} \tag{5.9}$$

where superscript $+$ denotes the conjugate transpose. Once again, multiplying vector $\mathbf{h}$ by a complex scalar does not affect the SNR. To find the filter vector $\mathbf{h}$ that maximizes the SNR, we set the gradient of SNR with respect to $\mathbf{h}$ to zero as follows. (A similar gradient technique will be developed for optimal optical filters.)

$$\nabla_{\mathbf{h}^+}(\mathrm{SNR}) = \Delta f \, \nabla_{\mathbf{h}^+}\left(\frac{\mathbf{h}^+\mathbf{s}^*\mathbf{s}^{\mathrm{T}}\mathbf{h}}{\mathbf{h}^+\mathbf{P}\mathbf{h}}\right)$$

$$= \Delta f \frac{\left(\mathbf{h}^+\mathbf{P}\mathbf{h}\right)\left(\mathbf{s}^*\mathbf{s}^{\mathrm{T}}\mathbf{h}\right) - \left(\mathbf{h}^+\mathbf{s}^*\mathbf{s}^{\mathrm{T}}\mathbf{h}\right)\mathbf{P}\mathbf{h}}{\left(\mathbf{h}^+\mathbf{P}\mathbf{h}\right)^2} = \mathbf{0} \tag{5.10}$$

$$\Rightarrow \mathbf{P}\mathbf{h} = \frac{\left(\mathbf{h}^+\mathbf{P}\mathbf{h}\right)\left(\mathbf{s}^{\mathrm{T}}\mathbf{h}\right)}{\left(\mathbf{h}^+\mathbf{s}^*\mathbf{s}^{\mathrm{T}}\mathbf{h}\right)}\mathbf{s}^* \Rightarrow \mathbf{h} = \alpha\mathbf{P}^{-1}\mathbf{s}^*$$

where

$$\alpha = \frac{(\mathbf{h}^+\mathbf{Ph})(\mathbf{s}^T\mathbf{h})}{(\mathbf{h}^+\mathbf{s}^*\mathbf{s}^T\mathbf{h})}$$

We see that $\mathbf{h} = \alpha\mathbf{P}^{-1}\mathbf{s}^*$ yields the maximal SNR, and in Eq. (5.9) we further observe that α does not affect the SNR, so in fact is arbitrary. Since $\mathbf{P}$ is a diagonal matrix, this is the same as the sampled version of the maximal SNR filter derived in Eq. (5.8).

5.1.3 White noise case

An important special case occurs when the input noise is white noise. The PSD for white noise is a constant; i.e., $P_n(f) = N_0/2$, where the denominator 2 is included to indicate that the PSD is a two-sided spectrum. For this special case, the maximal-SNR filter and the resulting maximal SNR simplify as follows:

$$H(f) = \alpha S^*(f) \tag{5.11}$$

and

$$\text{SNR}_{\max} = \int \frac{|S(f)|^2}{N_0/2}\,df = \frac{\int |s(t)|^2 dt}{N_0/2} = \frac{E_s}{N_0/2}$$

where E_s denotes the energy in the transmitted signal.

Matched filter The maximal-SNR filter in Eq. (5.11) is known as the *matched filter* (MF) since $H(f)$ is proportional to $S^*(f)$ or equivalently $h(t)$ is proportional to $s(-t)$. Thus, for the white noise case, the filter that maximizes the output SNR has an impulse response that is the reflected version of the transmitted signal. For time-domain signals, this time reversal may appear to be impractical in that $h(t)$ may be non-zero for negative arguments and thus the filter may be non-causal and thus unrealizable. If the signal $s(t)$ is of finite length or can be approximated as of some finite length (as all practical signals can be), one can overcome the non-causality problem by using $h(t) = s(t - T)$ where T represents a sufficiently long delay. For spatial signals, such as images, the impulse response's being non-zero for negative arguments is not an issue since the causality-type concept is not relevant for spatial systems; i.e., there is no fundamental problem in using image pixels to the left or to the right of (or above or below) the current pixels. Some sources refer to the maximal-SNR filter in Eq. (5.8) as the MF even when the noise is not white, but we will reserve the phrase "matched filter" strictly for the filter in Eq. (5.11).

The SNR of the MF is seen to equal $E_s/(N_0/2)$ and thus we can increase the output SNR by either increasing the energy of the transmitted signal or by decreasing the input noise level. Another important observation is that the output SNR of this matched filter is a function of only E_s and $(N_0/2)$, and does not depend on the *shape* of signal. Thus all transmitted signals with the same energy and same noise level result in the same matched filter output SNR, independently of their shape.

Cross-correlation When the MF is used, the output is the cross-correlation of the received signal with the known signal as shown below.

$$y(t) = \text{IFT}\{R(f)H(f)\} = \text{IFT}\{R(f)S^*(f)\}$$
$$= r(t) \otimes s(t) = \int r(p)s(p-t)\mathrm{d}p \qquad (5.12)$$

where IFT is the inverse Fourier transform and $\otimes$ indicates the cross-correlation operation. Thus cross-correlation provides the maximal-output SNR when the input noise is additive and white.

Suppose the received signal contains the reference signal and no noise. Then the matched filter output is the correlation of $s(t)$ with itself, i.e., the auto-correlation of $s(t)$. We have shown in Chapter 3 that the auto-correlation function (ACF) always peaks at the origin. If the received signal $r(t)$ is $s(t-t_0)$, a shifted version of $s(t)$, then the MF output peaks at t_0 (because of the shift-invariance of the matched filter) allowing us to estimate the time delay between the transmitted and the received signal.

5.1.4 Colored noise

The previous section established the result that the cross-correlator is theoretically optimal (in the sense of maximizing the output SNR) when the input noise is additive and white. But if the noise is colored (i.e., non-white), we will show that the maximal-SNR filter can be viewed as a matched filter operating on the pre-whitened signal. The maximal-SNR filter in Eq. (5.8) can be expressed as a cascade of two filters $H_{\text{pre}}(f)$ and $H_{\text{MF}}(f)$; i.e.,

$$H(f) = \alpha \frac{S^*(f)}{P_n(f)} = H_{\text{pre}}(f) \cdot H_{\text{MF}}(f) \qquad (5.13)$$

where

$$H_{\text{pre}}(f) = \frac{1}{\sqrt{P_n(f)}} \quad \text{and}$$

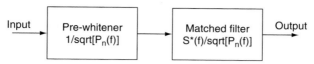

Figure 5.2 Interpretation of the maximal-SNR filter for colored noise as the cascade of a pre-whitening filter and a matched filter matched to the output of the pre-whitener

$$H_{\mathrm{MF}}(f) = \alpha \frac{S^*(f)}{\sqrt{P_n(f)}}$$

Figure 5.2 represents this cascade of two filters. When the received signal $r(t)$ is passed through $H_{\mathrm{pre}}(f)$, the filter output is given by $y'(t) = s'(t) + n'(t)$, where $S'(f) = S(f)H_{\mathrm{pre}}(f) = S(f)/\sqrt{P_n(f)}$, and where the output noise PSD is $P_{n'}(f) = P_n(f) \cdot |H_{\mathrm{pre}}(f)|^2 = P_n(f)/P_n(f) = 1$. Thus, the noise coming out of the first filter is white and that is why the first filter is called the *pre-whitener*. The input to the second filter is the pre-whitened signal $s'(t)$ corrupted by white noise. The second filter in the cascade is matched to $s'(t)$ and thus $H_{\mathrm{MF}}(f) = \alpha S'^*(f) = \alpha S^*(f)/\sqrt{P_n(f)}$, where α is an arbitrary constant.

5.2 Correlation implementation

As shown in Section 5.1.3, cross-correlation provides the maximal-output SNR when the input noise is additive and white. In this section, we will consider the implementation of this cross-correlation. Optical implementations will be discussed in more detail in Chapter 7. We will also use 2-D signals in this section to emphasize that the MF concept that we introduced using 1-D signals is applicable to detecting 2-D targets in images.

We mention that the quality of a correlation is considerably dependent on the set of values (called the *domain*) from which the filter may be drawn. Performing digital computation using the huge dynamic range of perhaps 64-bit complex filter values is, essentially, to have a continuum of complex filter values. For reasons of hardware speed, size, or electrical power draw, though, one might perform computations using fewer bits. In an optical correlator the filter domain is physically restricted to a curvilinear or discrete subset of the complex unit disk, because the spatial light modulators (SLMs) used to represent the filter values can accommodate only a subset of complex values. The fastest SLMs (e.g., ferroelectric liquid crystal [33]) might be restricted to binary values such as $\{-1,1\}$ or $\{0,1\}$. A real-positive SLM (e.g., a film transparency) might be restricted to the $[0,1]$ continuum. A nematic liquid crystal SLM [34] will have a continuum locus that is curvilinear in the unit disk. We seek to optimize a filter's performance in view of its domain, a point to be elaborated in Chapter 8.

Consider the problem of detecting the presence of, and locating the position of, a known target image $s(x, y)$ in an observed scene $r(x, y) = s(x, y) + n(x, y)$, where $n(x, y)$ denotes the additive noise. Based on what we have proved already, it is easy to see that we can achieve maximal SNR by cross-correlating $r(x, y)$ with the known target image $s(x, y)$ to produce the correlation output $c(x, y)$. This correlation output is then searched for its peak and if this peak exceeds a pre-chosen threshold, then it is declared that the scene contains the desired target at the location given by the peak coordinates. Because of the shift-invariance of the correlation operation, this correlation peak will shift by the same amount as the shift of the target image in the observed scene.

In fact, there is no need to limit our attention to only one target image in the observed scene. If $r(x, y) = \sum_{i=1}^{N} s(x - x_i, y - y_i) + n(x, y)$, then the cross-correlation of $r(x, y)$ with $s(x, y)$ will result in the output $\sum_{i=1}^{N} c(x - x_i, y - y_i) + [n(x, y) \otimes s(x, y)]$; i.e., the correlation output is a noisy version of the sum of N correlations centered at the coordinates of the targets. As long as these correlation outputs do not overlap significantly, their peaks can be easily located to provide the coordinates of the multiple targets. This is illustrated in Figure 5.3. This figure contains a target image (the word "PROFESSOR") in part (a) and a larger input image (containing multiple occurrences of the target) in part (b). Part (c) shows a "helicopter view" of the cross-correlation output. It can be seen that the three bright peaks in this output correspond to the three locations of the word "PROFESSOR" in the input image. We also show in part (d) the same correlation output except using an isometric view where the three correlation peaks are more visible. Note that the positions of the three peaks correspond to the three locations where the word "PROFESSOR" occurs in the input scene. In the rest of this section, we will look at the basic methods of implementing this cross-correlation operation.

5.2.1 *VanderLugt correlator*

An efficient way to compute the cross-correlation of the scene with the target image is in the frequency domain. From Chapter 3, we know that the correlation output can be obtained as follows:

$$c(x, y) = r(x, y) \otimes s(x, y) = \text{IFT}\{R(u, v) \cdot S^*(u, v)\} \tag{5.14}$$

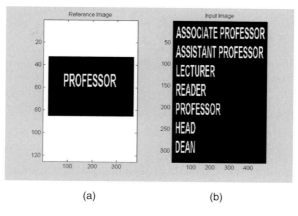

(a)   (b)

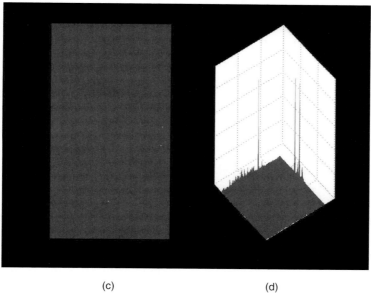

(c)   (d)

Figure 5.3 Illustration of the ability of cross-correlation operation to detect and locate multiple occurrences of a target image in an observed scene, (a) the target image, (b) a larger input image containing multiple target occurrences, (c) top–down view of the cross-correlation output of the target in part (a) with the input in part (b), and (d) an isometric view of the cross-correlation output in part (c)

Eq. (5.14) provides the basis for efficiently computing the cross-correlation of two images.

Optical FT We can form the 2-D FT of an image using an optical system shown schematically in Figure 5.4. The input image $r(x, y)$ is represented by an

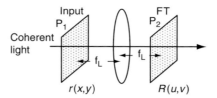

Figure 5.4 Schematic of optical two-dimensional Fourier transformation

illuminated transparency (e.g., a film negative) placed in plane P_1. The illumination is a uniform plane wave of coherent light (e.g., the single-wavelength light from a laser spatially filtered through a pinhole). The modulation of the coherent light by the transparency, coupled with the propagation of light, leads to the formation of a diffraction pattern in the back focal plane of the spherical lens. This diffraction pattern is proportional to the 2-D FT of the image encoded on the transparency in plane P_1.

In fact, by placing a spherical lens of focal length f_L at distance f_L from plane P_1, we can obtain the following complex amplitude light distribution in plane P_2:

$$R(x_2, y_2) = \eta \iint r(x_1, y_1) e^{-j2\pi(x_1 x_2 + y_1 y_2)/\lambda f_L} dx_1 dy_1$$
$$= \eta R(u, v)|u = x_2/\lambda f_L, v = y_2/\lambda f_L \tag{5.15}$$

where λ is the wavelength of the coherent illumination, η is a complex constant and where (x_1, y_1) denotes the coordinates in plane P_1 and (x_2, y_2) denotes the coordinates in plane P_2. It is easy to see that the amplitude of the light in plane P_2 is indeed a scaled version of the 2-D FT $R(u, v)$ of the image in plane P_1. Since the 2-D FT is available essentially instantaneously after the input image is placed, we can claim that the optical system carries out the 2-D FT "at the speed of light." Of course, extracting this FT from plane P_2 requires the use of photo-detectors or other sensing systems that usually reduce the process speed significantly, if the FT is all that we are interested in. Another matter is that the FT is generally complex and its phase information is lost when detected.

Serial correlator VanderLugt [5] showed how to take advantage of the optical FTs in computing cross-correlations. The basic idea is presented schematically in Figure 5.5. The observed scene $r(x, y)$ is placed in plane P_1 and is illuminated by a uniform plane wave of coherent light. With proper placement of the lens L_1, the complex amplitude of the light incident on plane P_2 is a scaled version of the 2-D FT of the image $R(u, v)$. In VanderLugt's embodiment, plane P_2

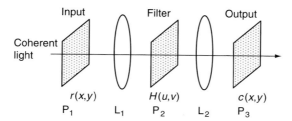

Figure 5.5 Schematic of serial optical correlator

contains a fully complex holographic transparency that represents filter $H(u, v) = S^*(u, v)$, the complex conjugate of the 2-D FT of the target image $s(x, y)$. Thus the light leaving plane P_2 is the product of $R(u, v)$ and $S^*(u, v)$. With the second lens L_2 placed at a distance of f_L from plane P_2, we can obtain the 2-D FT of this product. But we have deliberately reversed the direction of the (x_3, y_3) axes so that the second lens system yields an inverse FT in plane P_3. Thus, the complex amplitude distribution of the light in plane P_3 is given by IFT $\{R(u, v)S^*(u, v)\}$, which is the desired cross-correlation as can be seen from Eq. (5.14). Ordinarily we are uninterested in the phase of the cross-correlation (e.g. see Eq. (5.2)), which is a good thing since the phase is lost during photodetection. This optical architecture is known by several names including frequency plane correlator, VanderLugt correlator, serial correlator and 4-f correlator (to emphasize that the distance from the input plane to the correlator plane is four times the focal length).

Some important distinctions between the processing described so far in this chapter and the real-world optical implementation are that: (1) we must deal with the complex nature of light modulators, so much of the real-valued arithmetic used so far is not strictly appropriate; (2) the optical implementation noisily detects the squared magnitude, with ramifications to the noise models; and (3) the necessity of physical implementation restricts the set of values that $H(f)$ can actually take. Finally, whereas α is usually arbitrary in digital processing, it has an influence in the optical processing and is usually an optimizing tool.

The optical correlator in Figure 5.5 is a parallel processor in that the entire observed scene is compared against the target image in a single step. If the input scene contains multiple occurrences of the target, they are all identified in parallel rather than in a one-after-another manner. This ability to handle multiple targets in a test scene is true for digital correlation also.

In spite of its parallelism and potential for high speed, the serial correlator suffers from a few drawbacks. Most important is that the transparency in plane P_2 represents $S^*(u, v)$, ideally a complex function with individually controllable magnitude and phase variation. VanderLugt's original work used holographic methods to record this complex transmittance via interferometric techniques. For practical recognition systems which require that the observed scene be compared to multiple reference images in rapid succession, we need to use filter plane devices whose transmittances can be electronically controlled and quickly changed. Such devices with controllable transmittances are known as SLMs. A more detailed discussion of SLMs will appear in Chapter 7, but it is sufficient at this time to realize that current SLMs cannot accommodate arbitrary complex filter functions. This SLM limitation leads to the need for designing correlation filters with constraints such as phase-only (i.e., the magnitude can be only 1), binary (only two values), etc.

There are other difficulties in using the optical serial correlator. The observed scene must be displayed in plane P_1 with enough dynamic range such that the target details are not lost. Since SLMs used to represent the scene may have limited gray-scale capability (6 to 8 bits), it is important to understand the consequence of reduced dynamic range representations. In linear correlation theory, a correlation output can be negative. However, when the optical correlation output is detected by a photodetector array in plane P_3, we lose all the polarity information and obtain only the squared correlation magnitude. Finally, although the correlation output may be formed at the speed of light, we are usually limited by the speed of the modulators and of available detector arrays that convert this light amplitude distribution to electronic signals that can be further processed. As we will see later in this chapter, these practical difficulties have led to the development of a variety of correlation filter variants.

Other optical architectures to compute cross-correlations exist. These include the joint transform correlator [35, 36] and acousto-optic (AO) correlator [37, 38].

5.2.2 Digital correlation

The target image and the observed scene are usually in the form of discrete arrays and are denoted by $s[m, n]$ and $r[m, n]$, respectively. These image arrays are used to address SLM pixels in an optical correlator, but can also be used in a digital computer to compute the cross-correlation in one of at least two different ways. The first method is a direct computation. The second is based on the use of DFTs.

Direct correlation computation We can determine the cross-correlation of an N_r by N_r digitized input scene $r[m, n]$ with an N_s by N_s digitized target image

$s[m, n]$ by a direct summation. Since the target image is usually smaller than the observed scene, we can express the correlation output $c[m, n]$ as follows:

$$c[m, n] = r[m, n] \otimes s[m, n] = \sum_{k=1}^{N_s} \sum_{l=1}^{N_s} s[k, l] r[k - m, l - n] \qquad (5.16)$$

In Eq. (5.16), $s[k, l]$ and $r[k, l]$ are zero whenever k and l take on values outside the support for the images. The resulting correlation output is of size N_c by N_c, where $N_c = (N_s + N_r - 1)$. Thus, computing the complete cross-correlation according to Eq. (5.16) requires $(N_c N_s)^2$ multiplications. Each of these multiplications usually involves either floating point representations or multi-bit fixed point representations, and the complexity of these multiplications can be reduced by using fewer bits to represent these values. In fact, an extreme example of using a limited number of bits is using just one bit ($+1$ for positive values and -1 for negative values). The resulting binarized correlations can be computed efficiently and we will consider them in more detail in Section 5.4.5.

Sometimes, we may not need to determine the entire correlation plane. If it is known a priori that the target image center is located in an N_t by N_t area, then we need only $(N_t N_s)^2$ multiplications.

Digital correlation via FFT In general, the FFT gives a more efficient method for computing cross-correlations than the direct method. As discussed before, we can obtain the desired correlation by performing the following three steps:

- Compute $S[k, l]$ and $R[k, l]$, the $N \times N$ 2-D DFTs of the target image $s[m, n]$ and the observed scene $r[m, n]$, respectively. Obviously N must be larger than both N_s and N_r and thus requires zero padding; i.e., both images must be padded with zeros to make them of size N by N. If we need to compute the entire correlation plane in one step, then N must be at least as large as $N_c = (N_s + N_r - 1)$.
- Multiply $S[k, l]$ by $R^*[k, l]$ to obtain $C[k, l]$.
- Perform an $N \times N$ inverse 2-D DFT of $C[k, l]$ to obtain the cross-correlation output $c[m, n]$.

Thus, computing the cross-correlation involves three 2-D DFTs each of size and N^2 multiplications needed for the array multiplication in the second step above. Since the 2-D FFT of size $N \times N$ requires $2N^2 \log_2(N)$ operations, the total number of multiplications needed for this method is $N^2(1 + 6 \log_2(N))$.

The computational complexity can be reduced by taking advantage of the symmetries present. If the images are real, then their FTs are conjugate symmetric, thus reducing the number of points necessary to be computed by a factor of two. We can either use DFT routines designed to work with real images or we can compose a complex image from two real images (one as the real part and the other as the imaginary part). Also, we may want to test multiple scenes for the same target in which case there is no need to re-compute the DFT of the target image every time. Similarly, if we are checking a scene for different targets, there is no need to re-compute the DFT of the scene.

Circular correlation As discussed in Chapter 3, using DFTs results in circular correlation rather than linear correlation. As long as the DFT size is large enough, the resulting circular correlation will equal the linear correlation. This is achieved by padding the two images with enough zeros so that the zero-padded arrays are at least of the size of the linear correlation. This may be inefficient if we end up padding with too many zeros. Instead, we can use overlap–save and overlap–add methods as presented in Chapter 3 and avoid excessive zero-padding. Use of these overlap–add and overlap–save techniques requires careful identification of which circular correlation outputs equal linear correlations and which do not.

Circular correlations are aliased versions of linear correlations; i.e., circular correlation is the sum of linear correlation shifted by integer multiples of N in both axes when $N \times N$ DFTs are used. The correlation with itself of an image containing energy at high spatial frequencies results in a sharply peaked correlation output. As a result, the circular correlation of an image with itself is an aliased version of sharply peaked linear correlation functions, leading to perhaps only a small amount of aliasing. However, cross-correlations are not generally so sharply peaked and can result in more aliasing. Thus, circular correlation may not be a problem in computing auto-correlations, whereas it may lead to inaccuracies when carrying out cross-correlations.

Correlation using overlap–save method As indicated in Chapter 3, it may be computationally better to partition the larger image into many smaller images and carry out smaller sized correlations and assemble them, rather than carrying out the full correlation. Let us illustrate this with the help of an example.

Suppose we want to locate a 32×32 target image in a 256×256 scene; i.e., $N_s = 32$ and $N_r = 256$. Thus, $N_c = (N_s + N_r - 1) = 256 + 32 - 1 = 287$. Direct computation of the full 287 by 287 cross-correlation requires $((287) \cdot (32))^2 = 84.3$ million operations.

If we want to compute the full 287 by 287 correlation using FFT methods, we can use an FFT of size 512 by 512 since it represents the nearest power-of-2 FFT size larger than 287. Since $287 = 7 \times 41$, a 287-point DFT can be obtained in terms of 41 DFTs each of size 7 points. Such a DFT algorithm is known as

Table 5.1. *Computational complexities of different digital correlation approaches to correlate a 32 by 32 target image with a 256 by 256 input scene*

Direct correlation	Correlation using 512 by 512 FFTs	Correlation using 128 by 128 FFTs	Correlation using 64 by 64 FFTs
84.3 million	14.4 million	4.5 million	6.6 million

a mixed-radix FFT. While we may be able to use mixed-radix FFTs of a smaller size, they are not as efficient as power-of-2 FFTs. For the 512 by 512 FFT, the computational complexity of the correlation operation is $N^2(1 + 6 \log_2(N)) = (512)^2(1 + 6 \log_2(512)) = 14.4$ million operations.

We can also use the overlap–save method to obtain the complete correlation. Suppose that we decide to use DFTs of size 64 by 64. Since the target image is of size 32 by 32, we can use scene segments of size 32 by 32 in order for the resulting 64 by 64 circular correlations to equal the linear correlations. Thus we must partition the observed 256 by 256 scene into 64 segments each of size 32 by 32. This means that we need to compute 64 partial cross-correlations that can be assembled to provide the complete cross-correlation. Each partial cross-correlation involves three DFTs of size 64 by 64 and one array multiplication of size 64 by 64. However, we need to compute the target image's 64 by 64 DFT only once. As a result, we need a total of 129 DFTs of size 64 by 64, and 64 multiplications of 64 by 64 arrays, leading to a total computational complexity given by $(129)(64)^2(\log_2(64)^2) + 64(64)^2 = 6.6$ million operations.

If we decide to employ 128 by 128 DFTs instead of 64 by 64 DFTs, then we need to segment the 256 by 256 input scene into nine segments each of size 96 by 96. This means that we need a total of 19 DFTs of size 128 by 128 along with nine array multiplications of size 128 by 128, leading to a total complexity of $(19)(128)^2(\log_2(128)^2) + 9(128)^2 = 4.5$ million operations. We summarize the complexities of these different approaches in Table 5.1.

At least two observations can be made from Table 5.1. Firstly all three DFT-based methods are superior to the direct method, with the best one requiring about 20 times fewer operations. In fact, unless one of the images to be correlated is very small (e.g., of size 3 by 3 or 5 by 5), it does not make computational sense to use the direct correlation method. The second point is that there is an optimum DFT size to be used. Of the three DFT sizes considered, 128 by 128 DFTs result in the smallest number of operations. This is of course a function of the sizes of the two images and needs to be carefully investigated for each application.

5.3 Correlation performance measures

As shown in Section 5.1, the matched filter yields the maximal SNR when the input noise is additive. SNR is an important metric for designing correlation filters in that it quantifies the filter's noise sensitivity. But that is not the only performance criterion of interest. We would also want the correlation peak to be sharp so that it is easily discriminated from the background. In a later section, we will also consider more advanced pattern recognition performance metrics such as the area under the receiver operating curve. In this section, we will quantify metrics useful for characterizing the correlation outputs. We will switch back to the 1-D CT notation to keep the mathematics reader-friendly.

5.3.1 *Signal-to-noise ratio*

We have already defined the output SNR in Eq. (5.2) as the ratio of the square of average correlation peak to its variance. While maximizing the SNR is attractive because of its tractability, what we are really interested in is minimizing the probability of error. For the special case where the noise is additive and Gaussian, we can relate the SNR to the probability of error, but for non-Gaussian noise, it is not so straightforward.

Gaussian noise In deriving the maximal SNR filter in Section 5.1.2, we needed only the noise PSD. Thus, the maximal-SNR filter expression in Eq. (5.8) is valid for non-Gaussian noise also. The mean and variance of $c(0)$ for the two hypotheses are as follows:

$$E\{c(0)|H_0\} = E\left\{\int n(t)h(-t)\mathrm{d}t\right\} = 0$$

$$E\{c(0)|H_1\} = E\left\{\int [s(t) + n(t)]h(-t)\mathrm{d}t\right\}$$

$$= \int s(t)h(-t)\mathrm{d}t = \int S(f)H(f)\mathrm{d}f \tag{5.17}$$

$$\mathrm{var}\{c(0)|H_0\} = \mathrm{var}\{c(0)|H_1\} = \int P_n(f)|H(f)|^2\mathrm{d}f$$

The mean and variance expressions in Eq. (5.17) are valid for any WSS noise process, even if it is not Gaussian. But for now let us assume that noise $n(t)$ is zero-mean Gaussian with PSD $P_n(f)$. Without loss of generality we may assume that $r(t)$ contains the un-shifted signal $s(t)$ and the noise, and thus the filter output achieves its maximal value at the origin. Let us denote this output value by $c(0)$. Since the input noise to the linear filter is Gaussian, $c(0)$ is also a Gaussian RV. We have shown in Chapter 4 that the probability of error for a two-class Gaussian problem is a monotonically decreasing function of the Mahalanobis distance d.

$$
d^2 = \frac{|E\{c(0)|H_1\} - E\{c(0)|H_0\}|^2}{\mathrm{var}\{c(0)\}}
$$

$$
= \frac{\left|\int S(f)H(f)\mathrm{d}f\right|^2}{\int P_n(f)|H(f)|^2\mathrm{d}f} = \mathrm{SNR} \tag{5.18}
$$

where we see that the squared Mahalanobis distance d^2 is equal to the SNR. Thus, for the case of Gaussian noise, maximizing the SNR is equivalent to minimizing the probability of error.

Non-Gaussian noise Even for non-Gaussian noise, the relationship in Eq. (5.18) is valid. However, when the input noise is non-Gaussian, we do not necessarily have a monotonic relationship between the Mahalanobis distance and the probability of error. Thus maximizing SNR does not necessarily minimize the probability of error. Unfortunately, it is often impractical to obtain closed-form expressions for the filters that minimize the probability of error in the presence of non-Gaussian noise. As a result, we settle for optimizing quadratic metrics (e.g., SNR).

5.3.2 Peak sharpness measures

Many correlation filter designs are aimed at producing sharp correlation peaks. Sharp and large peaks easily stand out from the background and they also typically exhibit good discrimination properties. Maximizing SNR does not usually produce sharp correlation peaks. In fact, the magnitude frequency response of the MF is the same as that of the target image and is thus usually of a low-pass nature since most reference images are of low-pass type. As a result, MF output correlation peaks tend to be broad. However, sometimes the image may be dominated by high frequency content and some of our generalizations (based on the low-pass assumption about images) will not hold.

Peak-to-sidelobe ratio To characterize the sharpness of a correlation peak, several measures have been introduced. One such measure is the *peak-to-sidelobe ratio* (PSR), defined in different ways. According to one definition, the PSR is the ratio of the correlation peak to the maximal value outside

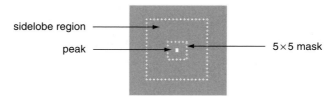

Figure 5.6 Estimation of peak-to-sidelobe ratio (PSR)

a central region surrounding the peak. A second definition, illustrated in Figure 5.6, is that it is the ratio of the correlation peak to the standard deviation of correlation values in a region that is centered on the peak, but excluding a small region around the peak. As shown in Figure 5.6, correlation output is first searched to locate its maximal value. Then a small region centered at the peak (in Figure 5.6, a 5×5 region) is excluded for sidelobe determination since it may include part of the peak. The mean and standard deviation in a square annular region (in Figure 5.6, a 20×20 region centered at the peak) are computed and PSR is determined as follows:

$$\text{PSR} = \frac{(\text{peak} - \text{mean})}{\text{std}} \tag{5.19}$$

Other definitions are possible, but these are essentially similar. These definitions result in large values for sharply peaked correlations and small values for broad correlations.

PSR is often used to characterize how well a region in the observed scene matches the target image; i.e., when the correlation output contains multiple peaks, they are sorted according to their PSR values. Thus PSR is valuable for correlation-based classification. However, PSR does not lend itself to closed-form mathematical analysis, as it involves nonlinear steps such as selecting the peak value and excluding a small region centered at a correlation peak. As a result, PSR is not very convenient as an optimization criterion, even though it very well describes the quality of a correlation.

Peak-to-correlation energy A more tractable measure of the peak sharpness is the peak-to-correlation energy (PCE), defined below:

$$\text{PCE} = \frac{|c(0)|^2}{\int |c(x)|^2 \text{d}x} = \frac{\left| \int S(f)H(f)\text{d}f \right|^2}{\int |S(f)H(f)|^2 \text{d}f} \tag{5.20}$$

where we express the PCE in both space and frequency domains. Note that the PCE expression completely ignores noise effects. PCE is deliberately formulated that way since the SNR measure is already available to characterize the noise sensitivity of the correlation filter. The role of the PCE is to characterize the peak sharpness resulting from a correlation filter. Of course, noise will affect the peak location and that effect can be captured by another metric known as the peak location error (PLE) [39]. We will not discuss PLE in this book because, for most images of interest to us, this can be pretty small and can mostly be ignored.

One can define a modified PCE by not including the $|c(0)|^2$ value at the origin in the denominator of Eq. (5.20). For CT signals, this amounts to eliminating one point in the denominator integral and thus will not make any difference. For DT signals, the resulting modified PCE is monotonically related to the PCE and thus maximizing one is equivalent to maximizing the other. We prefer the simpler form of Eq. (5.20).

It is easy to see that the PCE is maximal when the correlation output is a delta function, and that it is zero or close to zero when the output is a constant. The more energy there is in the peak compared to the rest of the correlation plane, the higher its PCE value. Since maximal PCE is obtained when the correlation output is a delta function, the filter that maximizes PCE is the inverse filter; i.e., $H(f) = 1/S(f)$. Most images of interest have more energy at lower frequencies than at higher frequencies. As a consequence, the corresponding inverse filter suppresses low frequencies more than the high frequencies. While the inverse filter leads to maximal PCE, its "other-pass" emphasis results in unacceptably high noise characteristics.

5.3.3 Optimal tradeoff correlation filters

Maximizing SNR leads to the MF whose spectral magnitude shape is that of the whitened target image, whereas maximizing the PCE results in the inverse filter whose spectral magnitude is the reciprocal of the target image's spectral magnitude. Thus, these two measures are usually conflicting and we cannot design a single filter that maximizes both SNR and PCE. However, we can optimally trade off between these two metrics using a multi-criteria optimization method.

Multi-criteria optimization The basic idea underlying the optimal tradeoff approach is to optimize one of the metrics (e.g., SNR) while holding the other (e.g., PCE) constant at some value. The SNR term in Eq. (5.5) and PCE in Eq. (5.20) have a common numerator, but differing denominators. Using this and using the quadratic nature of the two denominators, Réfrégier [40] showed that maximizing the SNR while holding the PCE constant is equivalent to the problem of maximizing the following figure of merit (FOM):

$$\text{FOM} = \frac{\left|\int S(f)H(f)\mathrm{d}f\right|^2}{\int |H(f)|^2\left[\beta|S(f)|^2+\sqrt{1-\beta^2}P_n(f)\right]\mathrm{d}f} \tag{5.21}$$

where β is a parameter that varies between 0 and 1 and characterizes the relative emphasis on the PCE compared to the SNR. For $\beta=0$, the FOM is the same as the SNR whereas FOM equals PCE for $\beta=1$. For other β values, we get an optimal compromise between the two, in the sense that maximizing the FOM will lead to an optimal tradeoff (OT) correlation filter which yields SNR and PCE values both of which cannot be *simultaneously* exceeded by any other correlation filter. The best β value depends on the distribution of values for the noise spectrum and the signal spectrum and is usually found through systematic search. In many applications, $\beta=0$ gives poor results whereas a very small non-zero value of β leads to improved results.

Optimal tradeoff filter The FOM in Eq. (5.21) is similar to the SNR in Eq. (5.5) except that $P_n(f)$ is replaced by $\left[\beta|S(f)|^2+\sqrt{1-\beta^2}P_n(f)\right]$ which is also real and non-negative. Thus, we can use the Cauchy–Schwarz inequality to find the filter that maximizes the FOM just as we maximized the SNR. Just as in Eq. (5.8), the maximal-FOM filter turns out to be as follows:

$$H_{\text{FOM}}^{\beta}(f) = \gamma\frac{S^*(f)}{\left[\beta|S(f)|^2+\sqrt{1-\beta^2}P_n(f)\right]} \tag{5.22}$$

where γ is any arbitrary complex constant and where β controls the tradeoff between the SNR and the PCE maximization. Note that $\beta=0$ leads to the MF and $\beta=1$ results in the inverse filter (IF). For other β values, we get an OT filter.

SNR versus PCE tradeoff Let us illustrate this SNR versus PCE tradeoff using the image shown in Figure 5.7(a). We show the spectral magnitudes of the MF, the IF and the OT filter for $\beta=10^{-10}$ in parts (b), (c), and (d), respectively. This extremely small value of β results from the fact that the noise spectral density is assumed to be constant at 1, whereas the signal spectrum takes on a very wide range of values. The best value of β depends on the range of signal values and noise values, and is difficult to predict. In practice, several β values are tried and the one yielding the best results on data (with known ground truth) is used. We have assumed $P_n(f)=1$ in this example. The MF magnitude spectrum is

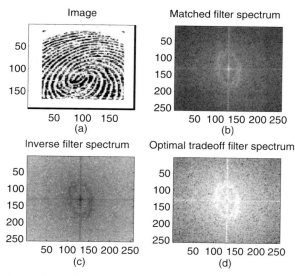

Figure 5.7 Illustration of the spectral magnitudes of different filters, (a) the reference image, (b) matched filter magnitude spectrum, (c) inverse filter magnitude spectrum, (d) magnitude spectrum of the optimal tradeoff filter for $\beta = 10^{-10}$. Whiter regions denote larger filter magnitudes

strong at low frequencies whereas the spectrum of the IF is dominant at high frequencies. The OT emphasizes intermediate frequencies depending on the tradeoff parameter β that is selected. The best way to choose β appears to be by trial and error as it depends on the image size and image values.

There are some features and practical matters that must be explained here. The Cartesian rays in Figure 5.7 result from the pixelation of the image and they are usually removed (i.e., set to zero) from a filter. The DC value of the filter (i.e., filter value at the (0,0) frequency) is suppressed, since the average illumination is not useful information for classification. Also note that the ringish structure in the image produces a ringish structure in the filter.

For the maximal-FOM filter in Eq. (5.22), we can see that the resulting SNR and PCE are functions of β, noise PSD $P_n(f)$, and signal PSD $|S(f)|^2$. For illustration purposes, we assume that the input noise is white leading to a noise PSD of 1 at all frequencies. We also use as the signal PSD the squared magnitude of the 2-D DFT of the image in Figure 5.7(a). Using these, we computed the SNR and PCE for various values of β and show in Figure 5.8 SNR versus PCE with β as a parameter. Note that the SNR decreases as the PCE increases. The ends of this curve refer to the metrics achieved by the MF and the IF; we will refer to this as the OT curve. We also show on this plot the SNR and PCE values of the phase-only filter (POF) and the binary phase-only

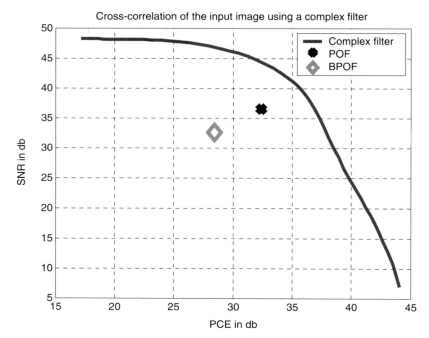

Figure 5.8 SNR versus PCE for the OT filter of the image in Figure 5.7(a)

filter (BPOF), to be discussed in Sections 5.4.1 and 5.4.2, respectively. Note that the POF and the BPOF provide SNR and PCE pairs that are below the OT curve. Thus, OT filters outperform other correlation filters.

Sub-pixel shift estimation Sometimes we are interested in estimating the peak position to sub-pixel accuracy. Such sub-pixel accuracy may be needed even in CT correlators such as optical correlators, since the continuous correlation output is sensed using a discrete array such as a charge-coupled detector (CCD) array. One way to determine sub-pixel correlation peak location is to employ a larger size DFT than the image size; i.e., by padding with zeros. Such a zero-padding requires the use of larger FFTs and we can achieve sub-pixel accuracy by fitting a model to the correlation output near the correlation peak. A simple non-parametric method is to model the correlation function $c(\tau)$ near the peak by a second-order polynomial, i.e.:

$$c(\tau) = a + b\tau + d\tau^2 \tag{5.23}$$

where a, b, and d are unknown constants. We will once again assume without any loss of generality that the peak is nominally at the origin and the sampling interval is 1. Let us denote the correlation peak value by $c[0]$ and its two neighboring values by $c[-1]$ and $c[1]$. From these three known correlation values, constants a, b, and d can be determined.

$$c[-1] = a - b + d \quad c[0] = a \quad c[1] = a + b + d$$

$$\Rightarrow a = c[0] \quad b = \frac{c[1] - c[-1]}{2} \quad d = \frac{c[1] + c[-1]}{2} - c[0] \tag{5.24}$$

Setting to zero the derivative of $c(\tau)$ in Eq. (5.23) with respect to τ yields an expression for the sub-pixel peak shift as follows:

$$\frac{\partial}{\partial \tau} \left[a + b\tau + d\tau^2 \right] \Big|_{\tau = \tau_p} = 0$$

$$\Rightarrow b + 2d\tau_p = 0 \tag{5.25}$$

$$\Rightarrow \tau_p = -\frac{b}{2d} = \frac{(c[1] - c[-1])}{2(2c[0] - c[1] - c[-1])}$$

The sub-pixel peak shift estimator in Eq. (5.25) is attractive in that it needs only three correlation values centered at the peak. One can use higher-order polynomials by using more correlation samples, but this is usually not recommended for at least two reasons. Firstly as we include more samples around the peak, the local structure (which is important for estimating sub-pixel shifts) is given less emphasis. Also, setting the derivatives of higher-order polynomials to zero can result in multiple candidate peak positions that will need to be compared using other methods. In general, the simple sub-pixel shift estimator in Eq. (5.25) is recommended.

5.4 Correlation filter variants

We have discussed in Section 5.2 the basic theory underlying some optical and digital implementations of correlators. These implementations impose certain requirements on the correlation filter. As already shown, the MF is complex and thus cannot be easily represented by available SLMs used in optical correlators. We will need correlation filters whose values fall within the set of values that can be accommodated by an SLM. Similarly, high-speed digital hardware might be restricted to heavily quantized inputs and filters. Such deviations from the MF will result in performance degradation. The objective of this section is to present some of these alternative filter designs. We will assume white noise unless specified otherwise. We will present some special cases in this section and then present the more general MEDOF formulation [41] in the next section.

5.4.1 Phase-only filters

Consider the serial correlator shown in Figure 5.5. The spatial filter in plane P_2 encodes the matched filter, namely $H_{MF}(u, v) = S^*(u, v)$ where $S(u, v)$ is the 2-D FT of the target image $s(x, y)$. Although the magnitude of $S(u, v)$ can take on any value, the spatial filter $H(u, v)$ placed in the frequency plane of the serial correlator cannot have a magnitude greater than 1 (since a transparency can only attenuate the incident light). Thus, strictly speaking, the filter used in plane P_2 is given as follows:

$$H_{MF}(u, v) = \frac{S^*(u, v)}{S_{max}} \tag{5.26}$$

where S_{max} denotes the maximal magnitude of the 2-D FT of the target image. For most images, the magnitude spectrum exhibits a large dynamic range, resulting in the MF magnitude's being close to zero at many spatial frequencies. This means that the filter blocks the incident light at these frequencies with low magnitude and thus much of the incident light will not make it to the output plane. It may be desirable to get as much light into the correlation output plane as possible when detector noise is present.

Phase-only filter (POF) definition Mainly to improve the light throughput efficiency and because more advanced optical filtering concepts were not then available, Horner and Gianino [42] proposed the following POF:

$$H_{POF}(u, v) = \frac{S^*(u, v)}{|S(u, v)|} = e^{-j\theta(u, v)} \tag{5.27}$$

where $\theta(u, v)$ denotes the phase of $S(u, v)$. From Eq. (5.27), we can see that the POF has a magnitude of 1 at all spatial frequencies thus passing all the light through.

   Another motivation for the use of the phase-only filter is that it is considered that the phase of the 2-D FT of an image appears to contain more information about that image than the FT magnitude [43]. Phase "carries more information" because it is a minimum Euclidean distance (MED) mapping, whereas magnitude is not.

   We illustrate this with the help of Figure 5.9. In this figure, the top row contains two original images, namely kitchen appliances and a cash register. Below each of these original images are shown the magnitudes of the inverse DFTs of the combinations of magnitude of the DFT of one of the images and phase of the DFT of the other. The lower rows combine random phase or

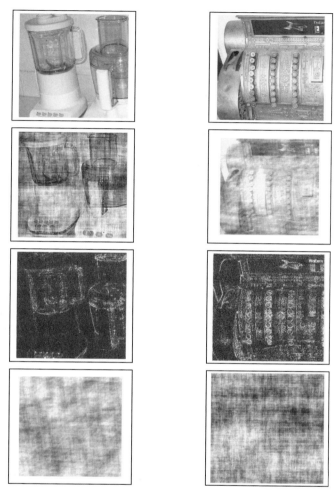

Figure 5.9 The top row contains two original images: some kitchen appliances and a cash register. The second row has phase of the appliances and magnitude of the register on the left, phase of register, magnitude of the appliances on the right. The third row has phase from the first row, but pseudo-random magnitude. The fourth row has magnitudes from the first row, but pseudo-random phase. Phase dominates magnitude in their relative effects for representing image information

random magnitude with the phase and magnitude of the original images. It is clear that the reconstructions look more like the image whose DFT phase is used for reconstruction.

MED mappings We now introduce MED mappings. The abbreviation MED stands for minimum Euclidean distance (it should be "minimal" but we will

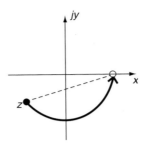

Figure 5.10 In this example, complex value $z = x + jy$ is converted to the real positive value $m = \sqrt{x^2 + y^2}$. Complex reconstruction error is the dashed line, but the magnitude reconstruction error is zero. This is not an MED mapping

stick with the published terminology). Many of the filters we discuss are reduced-information filters, in that a computed filter is altered in some fashion (binarizing, suppressing magnitude variation, etc.), which reduces the information in the result. The question is how to carry out the alteration in a way that minimizes the adverse effect. We are most often working with complex objects – images, image transforms, filters, and correlations – and the pervasive question is how to represent the original item on the reduced domain. The images printed in this book are an example; in the previous figure the result of crossing phase and magnitude spectra is assuredly complex, but the images' ink density on the paper is decidedly real and positive. How does one convert a complex direct-domain quantity into a viewable real positive quantity? Our solution was to print images whose halftoned ink density is directly related to the magnitude of the crossed-spectra result – i.e., we suppressed the phase of the computed result. This method is motivated by our visual experience – that the phase of light we view is most often unimportant. Figure 5.10 indicates how we reduced the complex image to a real one for display.

In comparison with visual *images*, for the optimal representation of *filters* we need to stay more engaged in the complex nature of the values being converted to a reduced domain. The principle of MED mapping is significant for filter representation, as shall be explained further in Chapter 8. But for now, we need to know only that the closest member of the representing domain – the one having minimum Euclidean distance from the computed value – is the one to be selected. This principle supports the optimization of the reduced-dimension correlation filters. For now we will look at just the POF and binary filters in view of the MED principle. Consider the two filter domains in Figure 5.11. The first is a set of binary values (A and B) and the second is the unit circle $\exp(j\theta)$, $\theta \in [0, 2\pi]$. A complex value is mapped to the

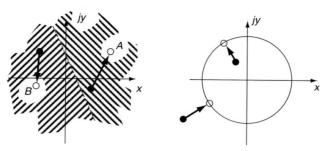

Figure 5.11 On the left, the line between patterns is the perpendicular bisector of the line segment between the values *A* and *B*. Minimum Euclidean distance mappings bring all values in a half-plane to the binary value within it. On the right, MED mappings bring values radially onto the unit circle so as to maintain phase. Note that unless *A* and *B* are disposed such that their perpendicular bisector passes through the origin, multiplying a value by a real scalar factor can take it across the bisector and thus change its MED binarization. This factor would not change the phase-only representation of a point, however

closer of the binary values, which amounts to the binary value on the same side of the perpendicular bisector of the line segment between A and B.

Maximizing the correlation peak intensity It seems reasonable that the POF uses the phase of the MF. We will prove now that the POF in Eq. (5.27) yields the maximal correlation peak intensity among all unit-magnitude filters. Let $\psi(u, v)$ denote the phase of a unit-magnitude filter. The resulting correlation output is given as follows:

$$c(0,0) = |c(0,0)|e^{j\lambda} = \iint |S(u,v)|e^{j[\theta(u,v)+\psi(u,v)]}dudv$$

$$\Rightarrow |c(0,0)| = \iint |S(u,v)|e^{j[\theta(u,v)+\psi(u,v)-\lambda]}dudv \leq \iint |S(u,v)|dudv$$

(5.28)

where the equality is satisfied when $\theta(u, v) + \psi(u, v)$ equals a constant. Thus correlation peak intensity is maximized when $\psi(u, v) = -\theta(u, v) + \lambda$, where λ can be any constant. Thus, the POF in Eq. (5.27) yields the maximal correlation peak intensity.

Correlation peak position The POF has the same phase as the MF and is thus phase-canceling. As a result, the POF will lead to a correlation peak at the origin when the input is the same as the target image and there is no noise. Since the POF is a shift-invariant filter, this correlation peak will shift by the same amount as the target shifts in the observed scene. A word of caution is in

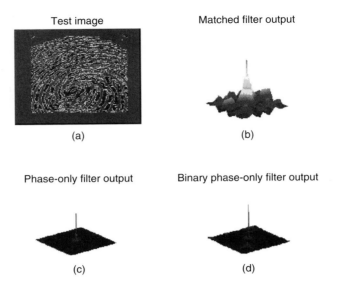

Figure 5.12 Correlation outputs for several correlations with one noiseless image: (a) target image, (b) correlation output using matched filter (MF), (c) correlation output using phase-only filter (POF), and (d) correlation output using binary phase-only filter (BPOF)

order here. This feature of the MF, the POF, and all other phase-canceling correlation filters – to guarantee correlation peak at the origin – is only for input scenes with no noise. As soon as noise is introduced in the input, the location of the correlation peak becomes random.

Peak sharpness Another perhaps unintended benefit of the POF is that, for most images which have more energy at low frequencies, the resulting correlation peaks tend to be sharper than those provided by the matched filter. On the other hand, if an image is dominated by high frequencies, the POF (since its magnitude is equal to 1 at all frequencies) does not amplify high frequencies as much as an MF and thus leads to correlation peaks that are not as sharp as those produced by an MF. We show in Figure 5.12(a) a target image and in (b) and (c) the correlation outputs resulting from the MF and the POF, respectively. We also show in part (d) of that figure the correlation output resulting from the use of a BPOF, to be introduced later. Note that the correlation output is more sharply peaked for POF compared to the MF. Such sharp peaks may be easier to locate if the noise is low.

The sharp peak produced by a POF is not a coincidence. The POF in Eq. (5.27) can be thought of as the cascade of the MF with a filter whose frequency response is $1/|S(u, v)|$. For most images of interest, $|S(u, v)|$ is of low-pass nature, i.e., it is large for low frequencies and small for high frequencies. As a consequence, $1/|S(u, v)|$ is small for low spatial frequencies and large for high frequencies. Such enhancement of high frequencies is equivalent to increasing the bandwidth and hence narrowing the width of the correlation peak. Had our reference image been of a high-pass nature rather than of a low-pass type, its POF would have led to broader peaks than the MF.

POF noise sensitivity While POF can yield higher light throughput and sharper correlation peaks compared to the MF, it obviously yields lower output SNR since MF was expressly designed to maximize the output SNR. How bad is the SNR loss for POFs? We can see from Eq. (5.27) that POF has a magnitude of 1 at all frequencies; i.e., it is an all-pass filter. Since the noise response of a filter depends on only the magnitude of its frequency response, the POF (and in fact any all-pass filter) has no noise suppression ability. Thus, the output noise from a POF will be white if the input noise is white. In practice, we limit the frequency support of any correlation filter (including the POF) to the bandwidth of the signal to avoid the out-of-band noise. Thus, if the target image energy is reasonably contained in the frequency range $|u| \leq B_u/2$ and $|v| \leq B_v/2$, then the output noise variance from the POF can be approximated as $N_0 B_u B_v/2$, where the input white noise PSD is assumed to be $N_0/2$.

POF with a region of support One way to improve the noise tolerance of a POF is to set some of the filter frequencies to zero. Strictly speaking, the resultant filter is not a unit-magnitude filter. Let R denote the set of frequencies for which the filter magnitude is non-zero. We can select R to maximize the SNR. Let us denote this optimal region of support (ROS) as R^*. For notational simplicity, we will switch back to 1-D notation. From Eq. (5.18), the resultant SNR is given as follows:

$$\text{SNR}(R) = \frac{\left|\int_R S(f)H(f)\,df\right|^2}{\int_R P_n(f)|H(f)|^2\,df} = \frac{\left|\int_R |S(f)|\,df\right|^2}{\int_R P_n(f)\,df} \cong \Delta f \frac{\left[\sum_{k \in R} S_k\right]^2}{\sum_{k \in R} P_k} \quad (5.29)$$

where S_k and P_k are samples of $|S(f)|$ and $P_n(f)$ sampled at uniform intervals of Δf. Our goal is to find the region R^* that maximizes the SNR in Eq. (5.29). To determine this region, let us re-index the frequencies such that the ratio S_k/P_k is sorted as follows:

$$\frac{S_1}{P_1} \geq \frac{S_2}{P_2} \geq \cdots \geq \frac{S_N}{P_N} \geq 0 \quad (5.30)$$

where N is the total number of frequency samples. We can prove that R^* includes all frequencies with an index lower than n if it includes frequency n from the sorted list in Eq. (5.30) [44]. This means that we don't have to try the 2^N different ways of partitioning the set of N frequencies. Instead, we have to try only N different partitionings (namely, each partition of interest includes the first M frequencies indexed according to (5.30), but M can vary from 1 to N).

To determine M, we compute the SNR for M values from 1 to N and select the M yielding the highest SNR. We assume that the input noise is additive and white. As a result, the sorting in Eq. (5.30) refers to sorting just the FT magnitudes of the target image. The SNR as a function of M is given as follows:

$$\text{SNR}(M) = \frac{2\Delta f}{N_0 M}\left(\sum_{k=1}^{M} S_k\right)^2 \tag{5.31}$$

For many images, maximal SNR is obtained when M is a small fraction of the total number of frequencies. When $M = N$ (i.e., all frequencies are used), none of the frequencies is excluded and thus we have the POF. We can get higher SNR by not including all frequencies.

The POF with the optimal ROS is of a low-pass nature and thus the resulting correlation peak is not as sharp as one would get with the conventional POF. The advantage of using the optimal ROS becomes clear when the target image is corrupted by additive white noise. The optimal-ROS POF produces a correlation output that is not as sensitive to input noise as the POF. In the absence of noise, POF yields a sharp correlation peak. However, the all-pass nature of the conventional POF allows all of the input noise through, thus making the detection of the target almost impossible in the presence of noise.

Optimal tradeoff POFs Our discussion regarding the region of support selection for POFs focused entirely on the SNR metric since noise sensitivity is one of the main problems with the use of the POF. However, an ROS can also be used to maximize the PCE. The PCE and SNR definitions (see Eqs. (5.18) and (5.20)) differ only in that PCE uses $|S(f)|^2$, whereas the SNR uses $P_n(f)$. Thus the optimal ROS selection algorithm discussed for maximizing the SNR can be used except that we should sort the sequence $|S_k|/|S_k|^2 = 1/|S_k|$ in descending order; i.e., the frequency indexes should be sorted so that $|S_k|$ form an ascending sequence. The maximal-PCE ROS will thus include the M frequencies where the signal is the weakest. We can compute the PCE as a function of M and select the M and hence the ROS that results in the highest PCE. It is interesting

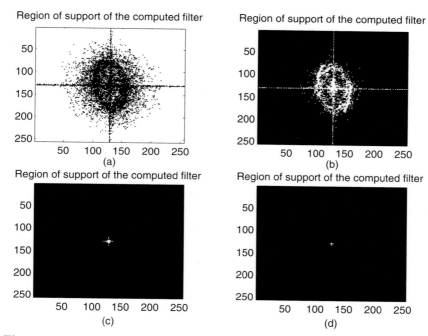

Figure 5.13 Optimal regions of support for the POF of the target image in Figure 5.7(a) for (a) $\beta = 1$, (b), $\beta = 0.1$, (c) $\beta = 0.01$, and (d) $\beta = 0$. White areas denote the frequencies where the filter is non-zero

to note that the conventional POF does not necessarily maximize PCE since PCE-maximizing ROS may not include all frequencies.

It is disappointing to see that the SNR-maximizing ROS includes the strongest signal frequencies (assuming white noise) whereas the PCE-maximizing ROS includes the weakest signal frequencies. Obviously, this suggests the need for an optimal tradeoff between these two metrics. As discussed in Section 5.3.3, we can optimally tradeoff between SNR and PCE by using the FOM introduced in Eq. (5.21). Again, the SNR and the FOM differ only in that the FOM uses $\left[\beta|S(f)|^2 + \sqrt{1 - \beta^2}P_n(f)\right]$, with tradeoff parameter β, whereas the SNR uses $P_n(f)$, the noise PSD. For a given β, the maximal-FOM region of support must include the first M frequencies from the following sorted list:

$$\frac{S_1}{\beta S_1^2 + \sqrt{1 - \beta^2}P_1} \geq \frac{S_2}{\beta S_2^2 + \sqrt{1 - \beta^2}P_2} \geq \cdots \geq \frac{S_N}{\beta S_N^2 + \sqrt{1 - \beta^2}P_N} \geq 0$$

(5.32)

It is illustrative to see in Figure 5.13 the dependence of the optimal ROS on the tradeoff parameter β.

5.4.2 *Binary phase-only filters*

The POF appears to yield acceptable performance even though it completely ignores the magnitude information. Can we get adequate performance even if we use only a few bits of the phase? This is a question of practical significance since some SLMs can provide reasonably high frame rates, but can accommodate only two phase levels (namely 0 and π). Phase-only filters that are constrained to take on only two phases are termed binary phase-only filters.

Binary phase-only filters (BPOFs) can be obtained from the MF in a variety of ways, such as binarizing the real part of the MF, binarizing the imaginary part of the MF or binarizing the sum of the real part and the imaginary part (this sum is related to the Hartley transform of the image). [45] These different binarizations can be unified into a single method with the help of the threshold line angle (TLA) [46] and insight from Figure 5.11. In Figure 5.11, suppose A and B are of equal magnitude. Then their perpendicular bisector passes through the origin.

Threshold line angle A BPOF can be defined as follows using the TLA, θ, that can vary between $-\pi/2$ and $\pi/2$:

$$H_{\mathrm{BPOF}}(f) = \mathrm{sgn}\big[\mathrm{Re}\{e^{-j\theta}S(f)\}\big] = \begin{cases} +1 & \text{if } -\pi/2 \le (\theta(f) - \theta) < \pi/2 \\ -1 & \text{otherwise} \end{cases}$$

$$\tag{5.33}$$

where

$$\mathrm{sgn}(x) = \begin{cases} +1 & \text{if } x \ge 0 \\ -1 & \text{if } x < 0 \end{cases}$$

For $\theta = 0$, the BPOF is obtained by binarizing the real part of the FT of the image. If $\theta = \pi/2$, then the binarization refers to the imaginary part. Using $\theta = \pi/4$ results in a BPOF that binarizes the Hartley transform of the image. Of course, we can use other TLA values.

The advantage of using BPOFs is that the filters require a lot less storage (one bit per frequency pixel), less computation (no need for multiplications in frequency domain, only additions and subtractions), and they can be represented on some high frame-rate binary SLMs.

BPOF disadvantages Since the BPOF is also an all-pass filter, it suffers from the same disadvantage as POFs in that there is no noise suppression and thus the output SNR can be very low. Another drawback is that, unlike POF, BPOF is not phase-canceling since the BPOFs have only two phase levels (0 and π). Thus, even if there were no noise and even if the input scene is an un-shifted target image, there is no guarantee that the correlation peak will be at the origin. What is more problematic is that the location of the BPOF correlation peak is a function of the target image and can vary from one target to another. This causes errors in estimating the target coordinates from the correlation peak position. In practice, this correlation peak shift is small in most cases and thus may not be a concern.

SNR of ternary filters The only way to improve the noise behavior of a BPOF is to set some frequencies to zero value. Strictly speaking, a BPOF should have a magnitude of 1 at all frequencies, and filters accommodating -1, 0, and $+1$ values should be called three-level or *ternary* filters. Allowing the filter to take on a zero value helps in improving its noise suppression abilities. It has been shown that a magneto-optic SLM [47] can be operated such that we can effectively get a zero transmittance level in addition to its -1 and $+1$ levels.

A key design question is the selection of the frequencies that should be set to zero. Let R denote the ROS for the ternary filter; i.e., for all frequencies in R the ternary filter is either $+1$ or -1, whereas for all frequencies outside R, the filter is zero. From Eq. (5.18), the SNR for the ternary filter is given as follows:

$$
\mathrm{SNR} = \frac{\left| \int S(f)H(f)\mathrm{d}f \right|^2}{\int P_n(f)|H(f)|^2\mathrm{d}f} = \frac{\left| \int_R S(f)H(f)\mathrm{d}f \right|^2}{\int_R P_n(f)\mathrm{d}f}
$$

$$
= \frac{\left| \int_{R^+} S(f)\mathrm{d}f - \int_{R^-} S(f)\mathrm{d}f \right|^2}{\int_R P_n(f)\mathrm{d}f}
$$

(5.34)

where we have further partitioned R into R^+ (the set of frequencies for which the filter is $+1$), and R^- (the set of frequencies for which the filter is -1). The denominator of the SNR expression in Eq. (5.34) depends on the area or the number of pixels in R, and not on how it is partitioned into R^+ and R^-. Thus, we should partition a given R into R^+ and R^- to maximize the numerator of Eq. (5.34).

Maximal peak intensity BPOF Let $c(0) = |c(0)|e^{j\theta}$ be the correlation output at the origin due to the use of a BPOF with a region of support R containing R^+ and R^-. Then $|c(0)|$ can be rewritten as follows:

$$
|c(0)| = c(0)e^{-j\theta} = \left[\int_{R^+} S(f)e^{-j\theta}\mathrm{d}f - \int_{R^-} S(f)e^{-j\theta}\mathrm{d}f \right]
$$

$$
= \left[\int_{R^+} \mathrm{Re}\{S(f)e^{-j\theta}\}\mathrm{d}f - \int_{R^-} \mathrm{Re}\{S(f)e^{-j\theta}\}\mathrm{d}f \right]
$$

(5.35)

where we utilize the fact that $|c(0)|$ is real and, as a result, the imaginary parts of the integrals in Eq. (5.35) must add up to zero. A closer look at Eq. (5.35) indicates that $|c(0)|$ will be as large as it can be if we assign all frequencies in R that yield non-negative $\text{Re}\{S(f)e^{-j\theta}\}$ to R^+, and to R^- all those frequencies that lead to negative $\text{Re}\{S(f)e^{-j\theta}\}$. Of course, angle θ of the correlation output depends on the filter and is thus unknown. But this result establishes that the correlation peak of a BPOF is maximized by the following filter design:

$$H_{\text{BPOF}}(f) = \text{sgn}\left[\text{Re}\{S(f)e^{-j\theta}\}\right] \tag{5.36}$$

where θ is an unknown parameter to be determined. The unknown angle θ is usually determined by trying all angles from $-\pi/2$ to $\pi/2$ and selecting the one that results in the maximal $|c(0)|$. By substituting the filter expression in Eq. (5.36) into the correlation peak intensity expression in Eq. (5.35), we obtain the following expression for maximal peak intensity in terms of the unknown angle θ:

$$
\begin{aligned}
|c(0)|_{\max} &= \left[\int_{R^+}\text{Re}\{S(f)e^{-j\theta}\}\mathrm{d}f - \int_{R^-}\text{Re}\{S(f)e^{-j\theta}\}\mathrm{d}f\right] \\
&= \int_R \left|\text{Re}\{S(f)e^{-j\theta}\}\right|\mathrm{d}f
\end{aligned}
\tag{5.37}
$$

For a ROS, the best TLA can be found by determining the integral in Eq. (5.37) for various TLA values, and choosing the θ that results in the maximal peak intensity.

Optimum ROS for ternary filters Our next task is to choose the ROS to maximize the SNR. Using Eq. (5.37) in Eq. (5.18), we can get the SNR as a function of R.

$$
\begin{aligned}
\text{SNR}(R,\theta) &= \frac{\left|\int_R S(f)H(f)\mathrm{d}f\right|^2}{\int_R P_n(f)\mathrm{d}f} \\
&= \frac{\left[\int_R \left|\text{Re}\{S(f)e^{-j\theta}\}\right|\mathrm{d}f\right]^2}{\int_R P_n(f)\mathrm{d}f} \cong \Delta f\frac{\left[\sum_{k\in R}S_k^\theta\right]^2}{\sum_{k\in R}P_k}
\end{aligned}
\tag{5.38}
$$

where $S_k^\theta = \left|\text{Re}\{S(k\Delta f)e^{-j\theta}\}\right|$, $P_k = P(k\Delta f)$ and Δf is the sampling interval as before. Because of the similarity with the SNR expression in Eq. (5.29) for the POF with an ROS, we know that the ROS that maximizes the SNR in Eq. (5.38) must include the first M frequencies in the following sorted list:

$$\frac{S_1^\theta}{P_1} \geq \frac{S_2^\theta}{P_2} \geq \cdots \geq \frac{S_N^\theta}{P_N} \geq 0 \tag{5.39}$$

Thus, the design of a maximal-SNR ternary filter involves the following steps:

- For a given TLA θ, determine Re $\{S(f)e^{-j\theta}\}$ and sample it at intervals of Δf to obtain N samples $S_1, S_2, \ldots, S_N$, where we have omitted the superscript θ for notational simplicity. Let $P_1, P_2, \ldots, P_N$ denote the corresponding samples of the noise PSD.
- Arrange the ratio S_k/P_k in the descending order as in Eq. (5.39).
- Determine the SNR$(M) = \Delta f \left[\sum_{k=1}^{M} S_k\right]^2 / \left[\sum_{k=1}^{M} P_k\right]$ as a function of M for M values from 1 to N.
- Select the M that maximizes the SNR and note corresponding M and SNR as $M_{max}(\theta)$ and SNR$_{max}(\theta)$, respectively.
- Repeat the above four steps for various θ values from $-\pi/2$ to $\pi/2$ and select the θ_{max} that maximizes SNR. Then the maximal-SNR ternary filter is given as follows:

$$H_{ternary-SNR}(f) = \begin{cases} \text{sgn}\left[S(f)e^{-j\theta_{max}}\right] & \text{for } f \in R_{max} \\ 0 & \text{otherwise} \end{cases} \quad (5.40)$$

where R_{max} is the set of M_{max} frequencies selected for the threshold line angle θ_{max}. This algorithm can be easily adapted for determining the ROS that maximizes the PCE or the FOM for a given tradeoff parameter β.

BPOF extensions We have focused so far on BPOF and ternary filters. This is mainly because high frame-rate SLMs can usually accommodate only two or three levels. However, some SLMs may be capable of accommodating more phase levels. An interesting question is whether we need more phase levels.

A quad phase-only filter (QPOF) [48] has unit magnitude and four possible phase values, namely 0, $\pi/2$, π, and $3\pi/2$. In its simplest version, QPOF is obtained from the MF by quantizing the MF phase to four regions each covering the $\pi/2$ phase and centered at 0, $\pi/2$, π, and $3\pi/2$. The noise behavior of the QPOF is the same as that of the POF and BPOF since all three filters are all-pass filters. However, the average correlation peak achievable using the QPOF is usually higher than that due to the BPOF resulting in improved SNR.

To improve the noise properties of the QPOF, the complex ternary matched filter (CTMF) is suggested [49], for which the real part and the imaginary part of the filter take on values -1, 0, and $+1$ separately. While this leads to a filter taking on nine different complex filter values (namely, 0, $-1, +1, -j, +j, -1-j, -1+j, 1-j,$ and $1+j$), CTMF can be implemented optically using an interferometric setup. Numerical experiments with the CTMF for a real image appear to yield SNR values that are within 2 dB of the theoretical maximal SNR values.

We still need to answer the question of how many phase levels are needed. Let us consider a POF obtained by quantizing the phase of the matched filter to L uniform levels in the range 0 to 2π. The noise behavior of these filters is independent of L since they all have unit magnitude. The correlation peak depends on L as follows:

$$
\begin{aligned}
c_{\mathrm{L}}(0) &= \int S(f)H(f)\mathrm{d}f \\
&= \int |S(f)|e^{j[\theta(f)-\theta_q(f)]}\mathrm{d}f = \int |S(f)|e^{j\varepsilon(f)}\mathrm{d}f
\end{aligned}
\tag{5.41}
$$

where $\theta(f)$ is the phase of the target signal FT and $-\theta_q(f)$ is the phase of the filter which is obtained by quantizing $-\theta(f)$ to L uniform levels. The difference $\varepsilon(f)=[\theta(f)-\theta_q(f)]$ is the error due to quantizing the phase to L levels, and this phase error can be modeled as uniformly distributed in the interval $[-\pi/L, \pi/L]$. For this simple model, we can compute the average correlation peak as follows [50]:

$$
\begin{aligned}
E\{c_{\mathrm{L}}(0)\} &= \int |S(f)|E\left\{e^{j\varepsilon(f)}\right\}\mathrm{d}f = \int |S(f)|\left[\frac{L}{2\pi}\int_{-\pi/L}^{\pi/L}e^{j\theta}\mathrm{d}\theta\right]\mathrm{d}f \\
&= \operatorname{sinc}\left(\frac{1}{L}\right)\int |S(f)|\mathrm{d}f = c(0)\operatorname{sinc}\left(\frac{1}{L}\right)
\end{aligned}
\tag{5.42}
$$

where $c(0)$ is the correlation output due to an unquantized POF. As L increases, $\operatorname{sinc}(1/L)$ approaches 1 indicating that using more phases, as expected, leads to higher average correlation output values. Thus, the average correlation peak is $\operatorname{sinc}(1/L)$ of the peak from the unquantized POF. This correlation peak variation as a function of L is shown in Figure 5.14. Note that using only four phase levels results in average peak reduction by less than 1 dB, explaining why the QPOF performs as well as it does.

5.4.3 Saturated filters

The SNR expression in Eq. (5.18) is unaffected when we multiply the filter frequency response $H(f)$ by any complex constant α. This behavior is unrealistic for an optical correlator primarily for two reasons. Firstly optical filters cannot amplify light and thus the filter magnitude must be less than or equal to 1. The second and perhaps more important reason is that the detector elements in the correlator output plane introduce noise which affects the recognizability of a correlation peak. The stronger the detector noise, the stronger the correlation peak needs to be for it to be discernible. As we will

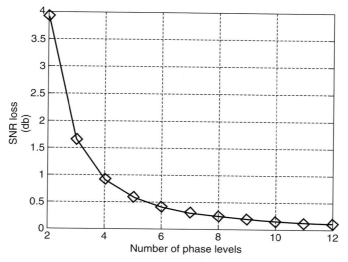

Figure 5.14 SNR loss as a function of L, the number of phase quantization levels

show in this section, the inclusion of detector noise has an impact on the optimal filter form and design strategy. Both aspects are related as seen by the definition of the SNR_d that includes detector noise.

SNR with detector noise In the presence of detector noise, we can define the following SNR_d where the numerator is still the magnitude squared of the correlation peak, but the denominator now consists of two variance terms: one due to detector noise and the other due to input noise that passes through the correlation filter. We make the reasonable assumption that the input noise and detector noise are statistically independent.

$$SNR_d = \frac{\left| \int S(f)H(f)df \right|^2}{\sigma_d^2 + \int P_n(f)|H(f)|^2 df}$$

$$\cong \frac{\left| \sum_k S_k H_k \right|^2}{\left(\frac{\sigma_d^2}{\Delta f^2} \right) + \sum_k \left(\frac{P_k}{\Delta f} \right) |H_k|^2} \tag{5.43}$$

When the detector noise level σ_d^2 is small, SNR_d approaches the conventional SNR. On the other hand when σ_d^2 is large, the denominator of SNR_d can be approximated as a constant independent of the filter. In such a case, maximizing SNR_d is equivalent to maximizing the numerator of Eq. (5.43) which is the correlation peak intensity. Note that multiplying the filter by a complex constant changes SNR_d, whereas such a constant gain does not affect the SNR without detector noise. Thus, gain constants are important in maximizing SNR_d. In the rest of this discussion, we will assume that $\Delta f = 1$ for notational simplicity.

Maximizing SNR$_d$ Let us now determine the filter H_k that maximizes SNR_d in Eq. (5.43). Note that the phase of the filter affects only the numerator of the SNR_d, and thus can be chosen to maximize the correlation peak intensity as was done in the POF case. Thus the filter phase is negative in relation to the phase of the FT of the target image. It is more challenging to determine the magnitude M_k of the filter. Let A_k denote the magnitude of the signal FT. For the optimal filter phase selected, SNR_d can be written as follows:

$$\text{SNR}_d \cong \frac{\left[\sum_k A_k M_k\right]^2}{\sigma_d^2 + \sum_k P_k M_k^2} \qquad (5.44)$$

Our goal is to select the filter magnitudes M_k to maximize the SNR_d in Eq. (5.44). Since the filter will be implemented optically, the filter magnitudes must be less than or equal to 1. Similarly, we will assume that the filter magnitudes must be greater than or equal to ρ, the minimal allowed filter magnitude; i.e., $0 \le \rho \le M_k \le 1$.

Maximizing the SNR_d in Eq. (5.44) involves setting to zero its derivatives with respect to M_k, provided that the resulting M_k values are within the allowed range, namely $\rho \le M_k \le 1$. If that derivative does not become zero in that interval, then M_k must take on one of the boundary values (ρ and 1 in this case). Setting the derivative of SNR_d with respect to M_m to zero results in the following condition:

$$\frac{\partial(\text{SNR}_d)}{\partial M_m} \cong \frac{\partial}{\partial M_m}\left\{\frac{\left[\sum_k A_k M_k\right]^2}{\sigma_d^2 + \sum_k P_k M_k^2}\right\}$$

$$= \frac{2A_m\left(\sigma_d^2 + \sum_k P_k M_k^2\right)\left(\sum_k A_k M_k\right) - 2M_m P_m\left[\sum_k A_k M_k\right]^2}{\left(\sigma_d^2 + \sum_k P_k M_k^2\right)^2} = 0$$

$$\Rightarrow 2\left(P_m\left[\sum_k A_k M_k\right]^2\right)\left(G\frac{A_m}{P_m} - M_m\right) = 0$$

$$\Rightarrow M_m = G\frac{A_m}{P_m} \quad \text{where } G = \frac{\left(\sigma_d^2 + \sum_k P_k M_k^2\right)}{\left(\sum_k A_k M_k\right)}$$

(5.45)

where G is the gain parameter that is independent of frequency k. We can choose $M_m = (GA_m/P_m)$ provided that (GA_m/P_m) is between ρ and 1. If (GA_m/P_m) is larger than 1, then the derivative $\partial(\text{SNR}_d)/\partial M_m$ in Eq. (5.45) is positive for all allowed filter magnitude values (since M_m cannot be larger than 1), and thus maximal SNR_d is obtained by choosing M_m to be the largest allowed value (i.e., 1). Similarly, if (GA_m/P_m) is smaller than ρ, the derivative $\partial(\text{SNR}_d)/\partial M_m$ in Eq. (5.45) is negative for all allowed filter magnitude values (since M_m must not be smaller than ρ), and thus maximal SNR_d is obtained by choosing M_m to be the smallest allowed value (i.e., ρ). Thus SNR_d is maximized by the following filter:

$$M_m = \text{sgn}_\rho^1 \left[G \frac{A_m}{P_m} \right] \quad \text{where } \text{sgn}_\rho^1[x] = \begin{cases} \rho & \text{if } x < \rho \\ x & \text{if } \rho \le x \le 1 \\ 1 & \text{if } x > 1 \end{cases} \quad (5.46)$$

The filter magnitudes in Eq. (5.46) coupled with the earlier assertion that filter phase is the negative of the phase of the target image FT produce the maximal-SNR correlation filter. We call these filters *saturated filters* [51] since this filter is identical to the matched filter $H_k = (GA_k/P_k)$, except that the filter magnitude saturates at the boundary values of 1 and ρ. The gain G in Eq. (5.46) is not at all arbitrary. It satisfies conditions that are found to be verified by the filter when SNR is optimized. You have to know G to compute the filter, and you have to know the filter to get G. The search method is simply breaking the impasse. But when all is said and done, the values of G and the filter must have the stated compatibility.

Detector noise effect As the detector noise level increases, the gain G in Eq. (5.45) increases. For larger gains, more filter frequencies will be saturated to value 1 thus making the correlation filter 100% transparent (i.e., filter magnitude is 1) at more frequencies. This will lead to more light throughput and hence larger correlation peak value, which is needed to combat the increased detector noise. In the extreme case of infinite detector noise, all frequencies will saturate to 1 resulting in a phase-only filter. If there is no detector noise, then the SNR is independent of G. This independence of SNR from the filter gain allows us to compress the filter values to within the allowed dynamic range as long as the matched filter is not zero at any frequency. When the unsaturated filter takes the zero value at a particular frequency, it will be saturated to the value ρ in the saturated filter.

We show in Figure 5.15 the magnitude of the maximal-SNR_d filter for the target image in Figure 5.7(a) for differing levels of detector noise. The SLM is assumed to allow filter magnitudes in the range 0.001 to 1. For small detector noise levels, the filter preferentially passes those frequencies of higher

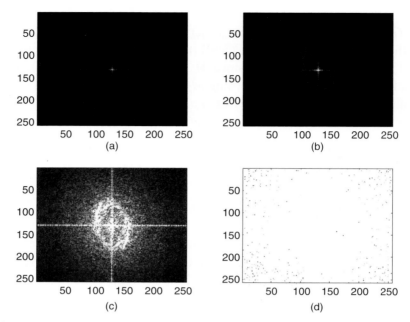

Figure 5.15 Magnitude of the maximal-SNR$_d$ filter for the target image in Figure 5.7(a) for differing levels of detector noise: (a) $\sigma_d^2 = 0$, (b) $\sigma_d^2 = 1$, (c) $\sigma_d^2 = 100$, and (d) $\sigma_d^2 = 10\,000$. The filter is allowed to take on values in the range 0.001 to 1.0. White denotes large magnitudes and black denotes low magnitudes

magnitude, just like the MF, whereas for high levels of detector noise, the filter becomes all-pass like a phase-only filter.

5.4.4 *Constrained filters*

Many SLMs can accommodate more than two phase levels and at least a few in magnitude. Phase-only filters and BPOFs are unit-magnitude filters that cannot take advantage of the SLM ability to represent multiple magnitudes. Ternary filters and POFs with a region of support can take advantage of the zero magnitude level, but not the intermediate filter magnitudes. Thus, it would be better if the filter were designed keeping in mind the filter SLM constraints. Since the more popular liquid crystal device (LCD) SLMs can be made to provide a limited set of complex values, the correlation filter must be designed to take on only this limited set of complex values. What is even more challenging is that the set of operating points can vary from one SLM to another even if all of the SLMs are supposed to be same. In this sub-section,

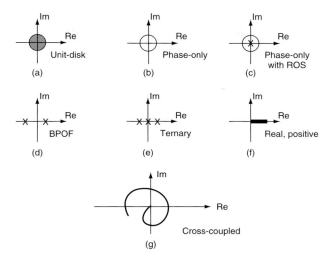

Figure 5.16 Different SLM operating regimes: (a) unit-disk, (b) phase-only, (c) phase-only with a region of support (ROS), (d) binary phase-only, (e) ternary, (f) real, positive, and (g) cross-coupled

we will consider the design of correlation filters when the SLM is cross-coupled, i.e., its magnitude and phase values are not independently controllable.

SLM operating curve The transmittance of an electrically addressed SLM pixel is usually controlled by the drive voltage or current applied to it. For popular LCD SLMs, the drive voltage causes a change in the polarization of the transmitted light. By placing polarization analyzers in different orientations, we can obtain different effective transmittances from the same pixel. Thus, the same SLM can be made to yield different operating curves by changing the way it is set up in relation to other polarization-sensitive elements in the optical path. An SLM operating curve specifies the achieved complex transmittance (or reflectance if it is a reflecting SLM) as a function of the applied drive voltage.

We show in Figure 5.16 some example operating curves or regimes. This figure shows complex planes with the solid curves and crosses indicating achievable transmittances from an SLM pixel. Figure 5.16(a) shows the unit-disk operating region where any phase is allowed and any magnitude less than or equal to 1 is allowed. This is the most accommodating operating regime from any passive SLM in that it covers all the points in the complex plane that have magnitudes less than or equal to 1. Speaking historically, VanderLugt envisioned this operating region and that $\sigma_d^2 = 0$ in the SNR equation. He would have, in essence, varied G by adjusting the exposure in making his holographic filters, even though he would not have had the analytical development at hand then. The POF can accommodate only magnitudes of 1 and thus the operating curve in Figure 5.16(b) is a unit circle in the complex plane. When a phase-only SLM accommodates a zero transmittance option as in Figure 5.16(c), we can obtain a POF with an ROS. The BPOF can accommodate only two transmittance values (namely -1 and $+1$) as shown in Figure 5.16(d). When we introduce a zero transmittance capability, we get the ternary filter in Figure 5.16(e) and obtain some noise suppression ability. The real, positive constraint requires the filter to take on only real and positive values. This constraint is a reasonable model when a photographic transparency is employed in the filter plane. Finally, Figure 5.16(g) shows a more realistic SLM operating curve where changing the drive voltage leads to changes in both the magnitude and the phase in a coupled but characterizable and predictable manner. The locus of the points in the complex plane as a function of the drive variable leads to an operating curve.

Maximizing the SNR_d Let Ω denote the set of allowed complex transmittances of the SLM being considered. Our objective is to maximize the SNR_d in Eq. (5.44) while allowing the filter pixels H_k to take on values only from Ω. We can see from Eq. (5.44) that the denominator of the SNR_d expression is independent of the filter phase θ_k, whereas its numerator is a function of both the filter phase θ_k and filter magnitude M_k.

Finding the cross-coupled filter H_k that maximizes SNR_d involves setting to zero the derivatives of the SNR_d expression with respect to filter phase θ_k and filter magnitude M_k. If the filter value H_k that yields zero derivatives is one of the values in the operating region, that filter value is used. On the other hand, if H_k is not within the operating region, then we need to find an operating point for that filter pixel that maximizes SNR_d, but does not set the derivative to zero.

Instead of presenting the detailed optimal cross-coupled filter design, we will first provide an intuitive explanation of the filter design and then summarize the filter design algorithm.

For the annulus operating region, we have derived the maximal-SNR_d filter in Eq. (5.46), which indicated that the saturated filter is the same as the fully complex optimal filter whenever that filter value is in the annular operating regime. On the other hand, when the fully complex optimal filter value is outside the annular region, the saturated filter takes on the same phase as the complex filter, but a saturated magnitude (namely 1 or ρ, depending on which is closer to the magnitude of the complex filter). We have also shown in [51] that the saturated filter values are the MED points in the operating regime from the fully complex optimal filter. This MED property of the optimal correlation filters was shown by Juday [41] for any arbitrary operating curve including the cross-coupled curve in Figure 5.16(g). Thus, we can find the maximal-SNR$_d$ filter that takes on values from Ω by first maximizing the SNR$_d$ with no SLM constraints, and then finding the MED version of this filter within the given operating curve.

However, the SNR of a fully complex filter is unaffected by multiplying the complex filter by a complex scalar (with gain G and angle λ) when the detector noise is absent. We have stated earlier that gain parameter G needs to be searched to determine the optimal G value to be used. For circularly symmetric operating regions such as an annulus operating region, all possible phase values are accommodated. As a consequence, we do not have to search over the angle parameter λ in the optimal saturated filter design for such circularly symmetric operating regions. On the other hand, cross-coupled operating curves such as those in Figure 5.16(g) are not circularly symmetric and thus we have to search over both parameters, namely gain G and angle λ. Based on the above discussion, we can summarize the design of the MEDOF as below.

MEDOF design algorithm Let $S_k = A_k e^{j\theta_k}$ denote the samples of the FT of the target image and let P_k denote the samples of the input additive noise PSD. Let σ_d^2 denote the variance of the detector noise. The filter design goal is to determine a correlation filter H_k that maximizes the SNR$_d$ expression in Eq. (5.44) while taking values from the operating set Ω. We will assume without any loss of generality that $\Delta f = 1$.

- Compute the fully complex optimal filter Z_k.

$$Z_k = \left[\frac{A_k}{P_k} e^{-j\theta_k} \right] \tag{5.47}$$

- Initialize the parameters gain $G = 1$ and angle $\lambda = 0$. For the chosen G and λ, determine the minimum Euclidean distance version H_k of Z_k i.e.,

$$H_k(G, \lambda) = \arg\min_{H_k \in \Omega} \left| H_k - G e^{j\lambda} Z_k \right| \quad k = 1, 2, \ldots, N \tag{5.48}$$

where N is the number of filter pixels.
- Compute the SNR$_d$ resulting from the use of filter H_k in Eq. (5.48) using the following SNR$_d$ expression:

$$\mathrm{SNR}_d(G, \lambda) \cong \frac{\left[\sum_k S_k H_k(G, \lambda) \right]^2}{\sigma_d^2 + \sum_k P_k |H_k(G, \lambda)|^2} \tag{5.49}$$

where the SNR_d is shown explicitly to depend on G and λ.

- Repeat the second and third steps for various G and λ values. Select the parameter pair (G^*, λ^*) that results in the maximal SNR_d. The optimal cross-coupled filter is $H_k(G^*, \lambda^*)$.

Optimal gain and angle selection   The gain and angle parameters can be selected in at least three different ways. The first is the most straightforward method of trying all parameter values exhaustively. We can limit the range of gain values by considering the minimal and maximal values of the ratio A_k/M_k. Filter angle values must be discretized in the range 0 to 2π. The second method is an iterative one. We start by assuming the gain G to be 1 and using that to design an MED cross-coupled filter. This computed filter can in turn be used to determine the next G value using an expression similar to that in Eq. (5.45). This process can be iterated until G does not change significantly. Unfortunately, there is no proof establishing that this iterative search will converge to G^*. Filter angle values once again must be discretized in the range 0 to 2π. The third method is based on expressing the Euclidean distance between H_k and Z_k for different G and λ values in terms of a 2-D cross-correlation between log-polar transformed versions of two arrays computed from the fully complex filter and from the operating curve. In this log-polar version, gain and angle appear as translations which can be identified as the coordinates of the minimum in the cross-correlation surface. Thus, optimal parameters can be obtained in a direct manner.

5.4.5 *Binarized correlations*

The correlation filter variants discussed so far (e.g., POF, BPOF, cross-coupled filter, etc.) all pertain to constraints imposed in the frequency domain. Sometimes, DFT hardware may not be available and direct digital computation of correlation may be too computationally intensive to be feasible. In such cases, we may have to resort to quantizing the signals to a few bits and correlating those quantized signals. An example of this is binarizing the images prior to their direct correlation.

In optical serial correlators, an SLM is used to represent the input scene. Although the frame-rate requirements for the input SLM are more lenient compared to that of the filter SLM, many input SLMs can represent only a few bits with acceptable accuracy. Thus, it makes sense to quantize the input scene even in optical implementations. One example of such input image quantization is the binarization of the input image.

Modeling the target image as Gaussian In this section, we will discuss the implications of binarizing (i.e., quantizing to one bit) the signals prior to cross-correlation. We will use 1-D DT signal notation for notational simplicity. As quantizing is a nonlinear operation, available theory is somewhat limited and we will end up making some assumptions in order to derive some results. For example, we will assume that the input image $r[k]$ is the unshifted target image $s[k]$ with no noise. As we will be modeling the target image itself as a random process, the correlation output will be random even in the absence of any noise. We will assume that the signal can be modeled as a sample realization from a stationary Gaussian random process with zero-mean and auto-correlation function (ACF) $R^s[k]$.

We will investigate what happens to the correlation output as either one or both signals are binarized. We will use the peak-to-sidelobe ratio (PSR) defined below as a measure of the quality of the correlation output.

$$\mathrm{PSR} = \frac{|E\{c[0]\}|^2}{\mathrm{var}\{c[l]\}}\Bigg|_{l \gg 0}. \tag{5.50}$$

where the denominator in Eq. (5.50) refers to the variance of the correlation output away from its peak assumed to be nominally at the origin. We will observe the effect of image binarization on the PSR values.

Gray-scale correlation The gray-scale correlation of $r[k] = s[k]$ with the target image $s[k]$ leads to the following correlation output.

$$c_{\mathrm{g,g}}[l] = \sum_{k=1}^{N} r[k]s[k-l] = \sum_{k=1}^{N} s[k]s[k-l] \tag{5.51}$$

where the subscript g,g is used to indicate that the observed gray-scale scene is correlated with the gray-scale target image. Let us now determine the statistics of this correlation output at the origin.

$$E\{c_{\mathrm{g,g}}[0]\} = \sum_{k=1}^{N} E\{s[k]s[k]\} = \sum_{k=1}^{N} E\left\{[s[k]]^2\right\}$$

$$= NE\left\{[s[k]]^2\right\} = NR^s[0] \tag{5.52}$$

For large l, the expectation of correlation $c_{\mathrm{g,g}}[l]$ can be approximated by zero since most signals have auto-correlation functions that decay to zero for large lags. Thus the variance of $c_{\mathrm{g,g}}[l]$ for large l can be expressed as follows:

$$\text{var}\{c_{g,g}[l]\} = \text{var}\left\{\sum_{k=1}^{N} s[k]s[k-l]\right\} = E\left\{\left[\sum_{k=1}^{N} s[k]s[k-l]\right]^2\right\}$$

$$= E\left\{\sum_{k=1}^{N}\sum_{m=1}^{N} s[k]s[k-l]s[m]s[m-l]\right\} \qquad (5.53)$$

$$= \sum_{k=1}^{N}\sum_{m=1}^{N} E\{s[k]s[k-l]s[m]s[m-l]\}$$

The variance determination in Eq. (5.53) requires a fourth-order moment. Since we assume that $s[k]$ can be modeled as a Gaussian process, we can express the fourth moment in terms of its second-order statistics as follows:

$$\text{var}\{c_{g,g}[l]\} = \sum_{k=1}^{N}\sum_{m=1}^{N} E\{s[k]s[k-l]s[m]s[m-l]\}$$

$$= \sum_{k=1}^{N}\sum_{m=1}^{N}\left[\begin{array}{l} E\{s[k]s[k-l]\}E\{s[m]s[m-l]\} \\ +E\{s[k]s[m]\}E\{s[k-l]s[m-l]\} \\ +E\{s[k]s[m-l]\}E\{s[m]s[k-l]\} \end{array}\right]$$

$$= \sum_{k=1}^{N}\sum_{m=1}^{N}\left\{(R^s[l])^2+(R^s[k-m])^2+R^s[k-m+l]R^s[k-m-l]\right\}$$

$$\cong \sum_{k=1}^{N}\sum_{m=1}^{N}\left\{(R^s[k-m])^2\right\} = N\sum_{n=-N}^{N}\left(1-\frac{|n|}{N}\right)(R^s[n])^2$$

$$(5.54)$$

where we use the assumption that $R^s[l]$ is zero for large l. Using Eqs. (5.52) and (5.54), we can express the PSR of the gray-scale correlation as follows:

$$\text{PSR}_{g,g} = \frac{|E\{c_{g,g}[0]\}|^2}{\text{var}\{c_{g,g}[l]\}} = \frac{(NR^s[0])^2}{N\sum_{n=-N}^{N}\left(1-\frac{|n|}{N}\right)(R^s[n])^2}$$

$$(5.55)$$

$$= N\left[\sum_{n=-N}^{N}\left(1-\frac{|n|}{N}\right)\left(\frac{R^s[n]}{R^s[0]}\right)^2\right]^{-1}$$

A few observations regarding the PSR in Eq. (5.55) are in order. Increasing the signal length N appears to increase the PSR. The second observation is more subtle. Suppose the normalized signal ACF can be approximated as having significant non-zero values for $L \ll N$. Then the summation in Eq. (5.55) can be approximated by L thus yielding an approximate PSR of N/L. Since the ACF and the PSD form an FT pair, $1/L$ is approximately the signal bandwidth. Thus, the gray-scale correlation yields a PSR that is equal to N times the signal bandwidth or the *space-bandwidth product*; i.e., the product of the signal extent with its bandwidth.

To illustrate the dependence of the PSR of the gray-scale correlation on signal length and signal bandwidth, let us consider the following exponential ACF for the signal.

$$R^s[k] = R^s[0]\mathrm{e}^{-a|k|} \tag{5.56}$$

where a is a measure of the signal bandwidth. We substitute the ACF in Eq. (5.56) into the PSR expression in Eq. (5.55) to obtain the PSR as a function of N and a. We show in Figure 5.17 the PSR of the gray-scale correlation as a function of N and a. Note that the PSR increases with increased signal length and increased signal bandwidth (i.e., larger a).

Binarizing one image Suppose we binarize one of the two images and let $s_\mathrm{b}[k]$ denote the binarized target image.

$$s_\mathrm{b}[k] = \mathrm{sgn}[s[k]] = \begin{cases} +1 & \text{if } s[k] \geq 0 \\ -1 & \text{if } s[k] < 0 \end{cases} \tag{5.57}$$

Let us once again assume that the input image $r[k]$ is the unshifted target image with no noise. The correlation of $r[k]$ with this binarized target image $s_\mathrm{b}[k]$ leads to the following correlation output:

$$c_{\mathrm{g,b}}[l] = \sum_{k=1}^{N} r[k]s_\mathrm{b}[k-l] = \sum_{k=1}^{N} s[k]s_\mathrm{b}[k-l]$$

$$= \sum_{k \in R_l^+} s[k] - \sum_{k \notin R_l^+} s[k] \tag{5.58}$$

where the subscript g,b indicates that a gray-scale scene is correlated with a binarized target image. The region R_l^+ denotes the set of k values for which $s_\mathrm{b}[k-l]$ is $+1$. Equation (5.58) shows the main advantage of this method. The direct cross-correlation $c_{\mathrm{g,b}}[l]$ can be obtained without any multiplications, only additions and subtractions.

We will now look at the statistics of $c_{\mathrm{g,b}}[0]$. Using the same assumptions as before, the average correlation output at the origin can be seen to be as follows:

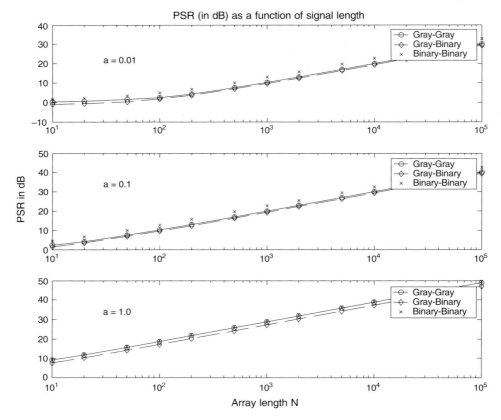

Figure 5.17 PSR as a function of signal length N for different types of correlation methods assuming an exponential ACF for the signal: (a) $a = 0.01$, low-bandwidth, (b) $a = 0.1$, medium-bandwidth, and (c) $a = 1.0$, high-bandwidth

$$E\{c_{\mathrm{g,b}}[0]\} = \sum_{k=1}^{N} E\{s[k]s_{\mathrm{b}}[k]\} = \sum_{k=1}^{N} E\{|s[k]|\} = NE\{|s[k]|\} \qquad (5.59)$$

Since $s[k]$ is a Gaussian RV with zero mean and variance $R^s[0]$, we can use the average of its absolute value to determine the average of the correlation output.

$$
\begin{aligned}
E\{c_{\mathrm{g,b}}(0)\} &= NE\{|s[k]|\} \\
&= N \frac{1}{\sqrt{2\pi R^s[0]}} \int_{-\infty}^{\infty} |s| \exp\left[-\frac{s^2}{2R^s[0]}\right] \mathrm{d}s \\
&= N \sqrt{\frac{2R^s[0]}{\pi}}
\end{aligned}
\qquad (5.60)
$$

For large *l*, the expectation of $c_{g,b}[l]$ is zero because the signal ACF decays to zero quickly. Using this, the variance $c_{g,b}[l]$ for large *l* can be approximated as follows:

$$\text{var}\{c_{g,b}[l]\} \cong \sum_{k=1}^{N} \sum_{m=1}^{N} E\{s[k]s_b[k-l]s[m]s_b[m-l]\}$$

$$\cong \sum_{k=1}^{N} E\{s^2[k]\} + \sum_{k=1,k\neq m}^{N} \sum_{m=1}^{N} E\{s[k]s[m]\} E\{s_b[k-l]s_b[m-l]\}$$

$$= NR^s[0] + N \sum_{n=-N,n\neq 0}^{N} \left(1 - \frac{|n|}{N}\right) (R^s[n]) (R_b^s[n]) \qquad (5.61)$$

where $R_b^s[n] = E\{s_b[k]s_b[k-n]\}$ can be related to $R^s[n]$ as follows because of the Gaussian assumption:

$$R_b^s[n] = E\{s_b[k]s_b[k-n]\}$$

$$= \Pr\{s[k]s[k-n] \geq 0\} - \Pr\{s[k]s[k-n] < 0\} \qquad (5.62)$$

$$= \frac{2}{\pi} \sin^{-1}\left(\frac{R^s[n]}{R^s[0]}\right)$$

where we used the zero-mean Gaussian nature of $s[k]$ and $s[k-n]$ to evaluate the required probabilities in deriving Eq. (5.62). Note from Eq. (5.62) that $R_b^s[0]$ equals 1. Using Eqs. (5.62), (5.61), and (5.60), we obtain the following expression for the PSR of the correlation of a gray-scale image with its binarized version:

$$\text{PSR}_{g,b} = \frac{|E\{c_{g,b}[0]\}|^2}{\text{var}\{c_{g,b}[l]\}}$$

$$= \frac{N \frac{2R^s[0]}{\pi}}{\frac{2}{\pi} \sum_{n=-N}^{N} \left(1 - \frac{|n|}{N}\right) (R^s[n]) \sin^{-1}\left(\frac{R^s[n]}{R^s[0]}\right)} \qquad (5.63)$$

$$= \frac{N}{\sum_{n=-N}^{N} \left(1 - \frac{|n|}{N}\right) \left(\frac{R^s[n]}{R^s[0]}\right) \sin^{-1}\left(\frac{R^s[n]}{R^s[0]}\right)}$$

We can compute the $\text{PSR}_{g,b}$ in Eq. (5.63) once we know the normalized signal ACF. Once again, we use the exponential ACF in Eq. (5.56) in the PSR expression in Eq. (5.63) and sketch the results in Figure 5.17. Note that the PSR obtained from binarizing one of the two images is not significantly different from the gray-scale correlation case.

Binarizing both images Suppose we binarize both images. Let us once again assume that the input image $r[k]$ is the unshifted target image with no noise. The correlation of $r_b[k]$ (i.e., the binarized version of the input scene) with the binarized target image $s_b[k]$ leads to the following correlation output:

$$c_{b,b}[l] = \sum_{k=1}^{N} r_b[k]s_b[k-l] = \sum_{k=1}^{N} s_b[k]s_b[k-l]$$

$$= N^+ - N^- = N^+ - (N - N^+) = 2N^+ - N \tag{5.64}$$

where the subscript b,b indicates that both images are binarized prior to correlation, and where $N^+(N^-)$ indicates the number of positions in which $s[k]$ and $s[k-l]$ have the same (opposite) polarity. Equation (5.64) shows that the binarized correlation $c_{b,b}[l]$ can be obtained without any multiplications. In fact, all we need to do is to find N^+, the number of positions in which signs of $s[k]$ and $s[k-l]$ agree. Using the same assumptions, the average correlation output at the origin can be seen to be as follows:

$$E\{c_{b,b}[0]\} = \sum_{k=1}^{N} E\{s_b[k]s_b[k]\} = N \tag{5.65}$$

Before we determine the variance of the correlation output at l, a position away from the peak, we should point out that the expectation of $c_{b,b}[l]$ is approximately zero. This is because the signal ACF decays to zero quickly for large l making the Gaussian RVs $s[k]$ and $s[k-l]$ uncorrelated and hence statistically independent. Hence their binarized versions $s_b[k]$ and $s_b[k-l]$ are also statistically independent. Using this, the variance $c_{b,b}[l]$ for large l can be approximated as follows:

$$\text{var}\{c_{b,b}[l]\} \cong \sum_{k=1}^{N} \sum_{m=1}^{N} E\{s_b[k]s_b[k-l]s_b[m]s_b[m-l]\}$$

$$\cong \sum_{k=1}^{N} (1) + \sum_{k=1,k\neq m}^{N} \sum_{m=1}^{N} E\{s_b[k]s_b[m]\}E\{s_b[k-l]s_b[m-l]\}$$

$$= N + N \sum_{n=-N,n\neq 0}^{N} \left(1 - \frac{|n|}{N}\right)(R_b^s[n])^2 = N \sum_{n=-N}^{N} \left(1 - \frac{|n|}{N}\right)(R_b^s[n])^2 \tag{5.66}$$

Using Eqs. (5.65) and (5.66), we obtain the following expression for the PSR of the auto-correlation of a binarized image:

$$
\text{PSR}_{b,b} = \frac{\left| E\{c_{b,b}[0]\} \right|^2}{\text{var}\{c_{b,b}[l]\}} = \frac{N^2}{N \sum\limits_{n=-N}^{N} \left(1 - \frac{|n|}{N}\right) \left(R_b^s[n]\right)^2}
$$

$$
= \frac{N}{\frac{4}{\pi^2} \sum\limits_{n=-N}^{N} \left(1 - \frac{|n|}{N}\right) \left(\sin^{-1}\left(\frac{R^s[n]}{R^s[0]}\right)\right)^2}
$$

(5.67)

We can compute the $\text{PSR}_{b,b}$ in Eq. (5.67) once we know the signal ACF. Let us consider the simple case where $R^s[l]$ can be approximated as a constant over a support of length L. Strictly speaking, the signal ACF cannot be a rect[·] function as the corresponding PSD will be a sinc[·] function which takes on negative values and that is unacceptable in a PSD. With this approximation, $\text{PSR}_{b,b}$ simplifies to $(N\pi^2/4L)$, which is higher than the PSR of the gray-scale correlation by a factor of $(\pi^2/4)$, a PSR gain of about 4 dB.

Once again, we use the exponential ACF in Eq. (5.56) in the PSR expression in Eq. (5.67) and sketch the results in Figure 5.17. Note that the PSR obtained from binarizing both images is slightly higher than the PSR from the gray-scale correlation.

It is somewhat surprising that the binary correlation is not only easier to implement, but can also yield better PSRs. The increase in PSR can be understood from the fact that binarizing in the time or space domain is a nonlinear point mapping which can be thought of as a high-order polynomial mapping in the input. In the frequency domain, this transformation leads to auto-convolutions of several orders which increase the effective bandwidth of the signal. This increased bandwidth leads to increased space-bandwidth product and higher PSR. However, we should keep in mind that this increase in bandwidth is artificial in that it cannot generate new information.

Numerical examples We show in Figure 5.18 various correlation outputs for the target image from Figure 5.7(a). Figures 5.18(a), (b), and (c) show the correlation outputs obtained with gray-scale correlation, one image binarized correlation, and both images binarized correlation, respectively when the images contain no noise. Clearly, binarizing the images does not appear to degrade correlation peaks significantly. Parts (d), (e), and (f) show the results when the input contains a large amount of additive noise. Note that the gray-scale correlation peak in part (d) is more easily discernible than those in parts (e) and (f). Thus, binarized correlation may be a good computational

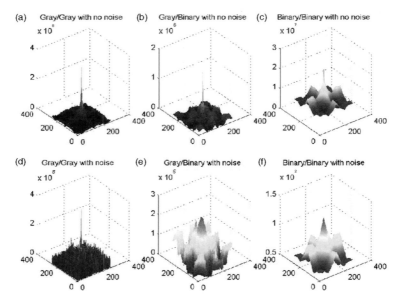

Figure 5.18 Illustration of the effect of binarizing the image (in Figure 5.7(a)) prior to correlation: (a) correlation of the gray-scale image with itself, (b) correlation of the gray-scale image with its binarized version, (c) correlation of the binarized version with itself, (d) correlation of the gray-scale image with noisy version of itself, (e) correlation of the noisy gray-scale image with binarized template, (f) correlation of the binarized version of noisy image with binarized template

convenience in high SNR situations, but gray-scale correlation is warranted in low SNR cases.

5.5 Minimum Euclidean distance optimal filter

Let us examine an optical implementation of correlation – its mathematical model, its filtering limitations, and its optimization of correlation pattern recognition. We are borrowing some results not fully developed until Chapter 8, but these results fit well with the present chapter's dealing with limited-domain filters.

The minimum Euclidean distance optimal filter (MEDOF) approach to optimal optical filtering has several distinctions from the theory presented so far in this chapter. Firstly, we recognize that the input objects are encoded on a wavefront and are complex thereby, not real-valued as is the custom in digital processing. Secondly, we model the effect of input noise and clutter as being an RV (it turns out to be a circular Gaussian), that is the sum of many complex numbers having uniform phase distribution. That complex RV is added to the correlation electromagnetic field, and the result is intensity detected, whereas in digital processing we can use distinctions that are not visible optically (the difference between $+1$ and -1, for example). Thirdly, statistics of the correlation *intensity* are used, rather than statistics of the magnitude of the complex correlation *field*. Finally, the optimization is done explicitly with the restricted filter domain in mind. That is, although it might seem that a casually selected fully complex filter is projected onto the restricted filter domain, in fact the fully complex filter to be projected is well determined, although a few parameters must be searched in order to find it (and the consequent limited-domain filter). The filter optimization statement is that no permissible change in the filter can move the metric being optimized further to an extreme.

The input object is, say, a voltage image $v(i, j)$ that is applied to an SLM whose (generally complex) encoding of v is $e(v)$, so the input object appears in the correlator as $s(i, j) = e[v(i, j)]$. It is diffractively Fourier transformed for its arrival at the filter plane as $S(f)$, where it is multiplied by the filter $H(f)$. Another Fourier transform, evaluated at the center of the correlation plane (for the known-to-be-centered appearance of the reference object), produces the central electromagnetic field $B \exp(j\beta) = \sum_k H(f_k)S(f_k)$. The noise-induced variance in the value of B has been presented earlier as being $\sigma_{\text{mag}}^2 = \sum_k |H_k|^2 P_{nk}$, where P_n is the noise power spectral density from noise and clutter objects in the input scene, and k indexes spatial frequency. The resulting intensity variance is $\sigma_{\text{int}}^2 = \left(\sigma_{\text{mag}}^2\right)^2 + 2B^2\sigma_{\text{mag}}^2$.

Optimization of several metrics has been analytically accomplished for optical correlation as implemented on arbitrary subsets of the complex unit disk. Those metrics include the simple intensity, the SNR, the area under the receiver operating characteristic curve, the Bayes' error, and others [52]. It will be shown in Chapter 8 that the optimization boils down to the following algorithm (for a single reference object). Although the metrics being optimized are different in the digital portions of this chapter, optimizing those digital metrics is accomplished by the same algorithm (the optimizing constants Γ and γ will be different, not only among the statistical pattern recognition metrics based on optical intensity, but also among the digital versions of the statistical pattern recognition metrics and the simpler digital metrics introduced in this chapter).

1. For the current values of the search parameters Γ and γ, form the ideal filter value

$$H_k^I = \Gamma \exp(j\gamma)\frac{S_k^*}{P_{nk}} \tag{5.68}$$

which may be regarded as equivalent to $\exp[j(\gamma - \text{Arg } S_k)]$ if P_{nk} is zero for the frequency in question.

2. Choose for H_k the realizable filter value closest, by Euclidean measure in the complex plane, to H_k^I. For those frequencies for which P_{nk} is zero, we select the realizable value with the largest projection in the complex direction $\exp[j(\gamma - \text{Arg } S_k)]$.
3. Evaluate the metric, which is now a function of only the independent parameters Γ and γ. Adjust Γ and γ to drive the metric to an extreme.

A critical step is finding the closest realizable value (by Euclidean measure in the complex plane) to the computed ideal filter value H_k^I, and hence the name MEDOF. Optimization of the digital filters as discussed in Section 5.4 can be understood according to the MEDOF principle. Consider the phase-only, binary, ternary, and saturated filters:

Phase-only filter

$$H_k = \exp[j(\gamma - \text{Arg } S_k)] \tag{5.69}$$

Binary (± 1)

$$H_k = \begin{cases} 1, & \cos(\gamma - \text{Arg } S_k) > 0 \\ \text{coin–flip}, & \cos(\gamma - \text{Arg } S_k) = 0 \\ -1, & \cos(\gamma - \text{Arg } S_k) < 0 \end{cases} \tag{5.70}$$

Ternary ($\pm 1, 0$)

$$H_k = \begin{cases} -1, & \text{Re}\left(H_k^{I<}\right) < -0.5 \\ \text{coin–flip } \{-1, 0\}, & \text{Re}\left(H_k^{I<}\right) = -0.5 \\ 0, & \left|\text{Re}\left(H_k^{I<}\right)\right| < 0.5 \\ \text{coin–flip } \{0, +1\}, & \text{Re}\left(H_k^{I<}\right) = 0.5 \\ +1, & \text{Re}\left(H_k^{I<}\right) > 0.5 \end{cases} \tag{5.71}$$

Saturated (in the annulus $\rho \leq |H| \leq 1$)

$$H_k = \begin{cases} \exp[j(\gamma - \text{Arg } S_k)], & \Gamma \dfrac{A_k}{P_{nk}} > 1 \\ H_k^I, & \rho \leq |H_k^I| \leq 1 \\ \rho \exp[j(\gamma - \text{Arg } S_k)], & \Gamma \dfrac{A_k}{P_{nk}} < \rho \end{cases} \tag{5.72}$$

5.6 Non-overlapping noise

Our analysis so far has assumed the standard model of additive noise. When an observed scene is dominated by sensor noise and/or propagation uncertainties, the additive noise model is reasonable. But there are occasions when additive noise may not be an appropriate model. For example, if the target image is placed in an unknown random background, the scene statistics would be different in the target region compared with outside that target region. Recently, optimal detectors have been introduced to cope with this case of *non-overlapping* noise. In this section, we will provide a relatively brief introduction to this topic using 1-D signals. Readers interested in more details should consult some of the references [53–55].

5.6.1 Effect of constant background

Let us consider the problem of designing and using a correlation filter for the target signal $s(t)$. The additive noise model considered so far assumes that the received signal $r(t)$ contains the target signal $s(t)$ corrupted at all locations by additive noise $n(t)$. However, the received signal may contain a background component in addition to noise. Let us model this background as a constant value B_r. Then the received signal can be modeled as follows:

$$r(t) = s(t) + n(t) + B_r(1 - w_s(t)) \tag{5.73}$$

where $w_s(t)$ is the target signal support function defined as follows:

$$w_s(t) = \begin{cases} 1 & \text{if } s(t) \neq 0 \\ 0 & \text{if } s(t) = 0 \end{cases} \tag{5.74}$$

The role of the window $w_s(t)$ is to separate the target from the background, but the window definition in Eq. (5.74) is deficient in that this window function will be zero even if some interior pixels of a target are legitimately zero. The correlation output of this received signal with the target signal is made up of several components as shown below.

$$\begin{aligned} r(t) \otimes s(t) &= [s(t) + n(t) + B_r(1 - w_s(t))] \otimes s(t) \\ &= s(t) \otimes s(t) + n(t) \otimes s(t) + B_r A_s - B_r w_s(t) \otimes s(t) \end{aligned} \tag{5.75}$$

where A_s is the constant that results from correlating a constant signal of 1 with the target signal $s(t)$. In addition to the desired auto-correlation function of the target image with itself and the undesired cross-correlation of the noise with the target image, we see two other terms in Eq. (5.75): a constant bias and another that is the cross-correlation of the support function $w_s(t)$ with the target signal $s(t)$. Unless the background B_r in the received signal is zero, these two terms can affect the correlation deleteriously.

One method that was proposed to deal with this background issue is the deliberate introduction of a constant background B_s in the target image prior to forming the correlation filter. The resulting correlation output then contains even more terms; i.e.,

$$[r(t)] \otimes [s(t) + B_{\mathrm{s}}(1 - w_{\mathrm{s}}(t))]$$

$$= [s(t) + n(t) + B_{\mathrm{r}}(1 - w_{\mathrm{s}}(t))] \otimes [s(t) + B_{\mathrm{s}}(1 - w_{\mathrm{s}}(t))] \qquad (5.76)$$

$$= s(t) \otimes s(t) + \text{noise} + \text{bias}$$

where we simply pool all of the random terms together as "noise" and all of the constant terms as "bias." While there is no guarantee that we can diminish this background problem by just introducing a deliberate bias in the target signal, simulation results [53] have indicated that using a non-zero background in the target image prior to designing the correlation filter can result in better correlation output.

For optical correlators, input image background may not be a problem. When matched spatial filters are constructed by exposing a holographic plate, the high intensity at the origin in the frequency plane results in a low-frequency blocking spatial filter whose effect is to remove the average value from the input scene. When the input scene contains a small target in a background, the background effect is diminished significantly by this DC block. Even when SLMs are used instead of holographic plates, we can introduce deliberate DC blocks to reduce the background effects.

5.6.2 *Non-overlapping noise*

Often, the background cannot be modeled as a constant and we must use a random process to describe the background variations. We can use a model similar to that in Eq. (5.73) except that the constant background B_{r} is replaced by $b(t)$, a background noise term. The background noise and the target are *non-overlapping* or *disjointed* in that background noise is present only outside the target region.

The classical matched filter is designed to maximize the output SNR in the presence of additive input noise and is thus not equipped to handle non-overlapping scene noise. The relatively poor correlation peaks due to the classical matched filter provide the motivation for designing optimal detection strategies to handle non-overlapping noise. However, we must emphasize that the classical MF was designed to overcome additive noise and thus was not designed to deal with non-overlapping noise. The desirability of the correlation peak's being "visible" has more to do with PSR than SNR. The PSR measures the sharpness of a correlation peak compared to its background. On the other hand, SNR measures how large the average of a correlation peak is compared to its standard deviation as different random noise realizations appear in the input scene. The standard deviation is arguably shift-invariant and thus applies to all locations, not just at the center.

In fact, in the absence of additive noise, we can avoid the input non-overlapping noise altogether if we can use a target *segmentation mask* that allows only those pixels in the scene belonging to a target silhouette to pass through. The result of such a masking contains only the target image and no noise. In such a case, there is no variability at the output. In the absence of input noise, many different correlation filter schemes including the inverse filter can be applied to obtain sharp correlation peaks. In general, the position of the target in the input scene is unknown and we may not know exactly where to place the segmentation mask in the input scene. This difficulty can be overcome by placing the segmentation mask at every possible position in the input scene and selecting the position that results in the "best" correlation peak. However, such a method loses the simplicity of correlation operation.

5.6.3 Optimal detection strategy for non-overlapping noise

If the scene contains both non-overlapping noise and additive noise, then the MF may not provide the best detection strategy. Although this field is still evolving, recent research [56] in the design of an optimal detection strategy for this situation is worth mentioning. Our review will be brief. This method is based on multiple hypotheses decision theory.

Received signal model The received signal can be modeled as consisting of three terms: the target signal at location j, background or non-overlapping noise $n_b[k]$, and noise $n[k]$ added to the target. Since the target location j is unknown, we can define one hypothesis H_j for each unknown location and attempt to select the most likely hypothesis. Hypothesis H_j can be characterized by the following model:

$$\text{Hypothesis } H_j : r[k] = w[k-j]s[k-j] + n_b[k]\{1 - w[k-j]\} + n[k]$$
$$= w[k-j]\{s[k-j] + n[k]\} + n_j[k] = r^{\text{in}}[k] + r^{\text{out}}[k]$$

where

$$r^{\text{in}}[k] = w[k-j]\{s[k-j] + n[k]\}$$

and

$$r^{\text{out}}[k] = n_j[k] = \{n[k] + n_b[k]\}\{1 - w[k-j]\} \tag{5.77}$$

where $w[k-j]$ is the support function for the target located at j. We will assume that the received signal is available at N samples or pixels of which the target occupies M samples. Thus, $r^{\text{in}}[k]$ denotes the M samples from inside the target window and $r^{\text{out}}[k]$ denotes the $(N-M)$ samples from outside the window. Note that $r^{\text{in}}[k]$ contains the target signal and additive noise whereas $r^{\text{out}}[k]$ contains only additive noise and background noise. We will also use I_j to denote the set of k values for which $s[k-j]$ is non-zero.

We will assume that the background noise $n_b[k]$ and noise $n[k]$ are statistically independent, white Gaussian stationary processes with variance σ_b^2 and σ_n^2, respectively. Additive noise is assumed to be of zero mean whereas the background noise is assumed to have mean m_b. We will also assume all hypotheses are equally likely, i.e., the target can be located with equal probability at any of the candidate positions.

Maximal likelihood detection Since all hypotheses are assumed to be equally likely, maximal a-posteriori (MAP) detection that yields the minimal probability of error is the same as the maximal likelihood (ML) detection. Maximal likelihood detection requires assigning the received signal $r[k]$ to hypothesis H_j that yields the maximal probability $f(r[k]|H_j)$ among all hypotheses. Since the background noise $n_b[k]$ and additive noise $n[k]$ are assumed to be Gaussian, the received signal $r[k]$ is also Gaussian.

From Eq. (5.77), we can see that the randomness in $r[k]$ is different inside the target window $w[k-j]$ from outside it. Inside this window, we have only additive noise $n[k]$ whereas we have both $n[k]$ and $n_b[k]$ outside the window. We have already assumed that the background noise and additive noise are statistically independent, white Gaussians. Thus the noise inside the target window is statistically independent of the noise outside the target window and we can express the log-likelihood functions as follows:

$$\ln\left\{f\left(r[k]|H_j\right)\right\} = \ln\left\{f\left(r^{\text{in}}[k]|H_j\right)\right\} + \ln\left\{f\left(r^{\text{out}}[k]|H_j\right)\right\}$$

$$= A - \frac{1}{2\sigma_n^2}\sum_{k\in I_j}\left(r[k] - s[k-j]\right)^2 - \frac{1}{2\left(\sigma_n^2 + \sigma_b^2\right)}\sum_{k\notin I_j}\left(r[k] - m_b\right)^2$$

$$\tag{5.78}$$

where A is a constant independent of the hypothesis. Maximal likelihood detection is achieved by selecting the hypothesis j that results in the maximal log-likelihood. Let us now consider two special cases.

Low background noise Let us consider the special case where the background noise variance is sufficiently small to be approximated by zero. Then maximizing the log-likelihood in Eq. (5.78) can be seen to be equivalent to minimizing the following "distance" measure between received signal $r[k]$ and a template $t[k-j]$ centered at $k=j$.

$$d_j^2 \triangleq \sum_{k \in I_j} (r[k] - s[k-j])^2 + \sum_{k \notin I_j} (r[k] - m_b)^2$$

$$= \sum_k (r[k] - t[k-j])^2 \tag{5.79}$$

where

$$t[k] = \begin{cases} s[k] & \text{if } w[k] = 1 \\ m_b & \text{if } w[k] = 0 \end{cases}$$

Minimizing the squared distance d_j^2 in Eq. (5.79) with respect to j is equivalent to maximizing the cross-correlation between the received signal $r[k]$ and the template $t[k]$ as shown below.

$$d_j^2 = \sum_k (r[k] - t[k-j])^2 = E_r + E_t - 2 \sum_k r[k] t[k-j] \tag{5.80}$$

where

$$E_r = \sum_k (r[k])^2 \text{ and } E_t = \sum_k (t[k-j])^2$$

where the template energy E_t is unaffected by shifts in the target signal, as long as the shifts do not take the target outside the observation window of N samples.

Note that cross-correlation once again proves to be the optimal solution. Eq. (5.79) shows that the optimal template should include the mean of the background. This is consistent with the experimental observations that using a carefully selected non-zero background in correlation filter design usually leads to better performance than using zero background. Some investigations [53] have shown that this background level should equal the average value of the target.

Low additive noise Let us now consider the special case where the additive noise variance is negligible compared to the background noise variance. From Eq. (5.78), we see that maximizing the log-likelihood in this case is equivalent to minimizing the following distance measure η_j.

$$\eta_j^2 \triangleq \sum_{k \in I_j} (r[k] - s[k-j])^2 = E_r^j + E_s - 2 \sum_k r[k] s[k-j] \tag{5.81}$$

where

$$E_r^j = \sum_k r^2[k]w[k-j]$$

and

$$E_s = \sum_k s^2[k-j]w[k-j] = \sum_k s^2[k-j]$$

where E_s is independent of target shift j as long as the target is not shifted out of the observation window. In contrast, the received signal running energy E_r^j computed over the target window centered at j can change with j. Thus minimizing the distance measure η_j is equivalent to maximizing the correlation of $r[k]$ with $s[k]$, only if the running energy E_r^j of the received signal is relatively constant with shift j. Also, the template needed for this correlation is zero outside the target support and is independent of the constant background level.

This methodology has been extended to other cases including colored noise, multiplicative noise, combinations of these noises, and target orientation estimation. Those topics are beyond the scope of this book and references are provided for the interested reader.

5.7 Chapter summary

The aim of this chapter was to provide an overview of the use of correlation in detecting and locating a signal in noise. It is desirable that the correlation be tolerant to input noise and provide sharp correlation peaks that are easily discernible. We have defined performance measures that quantify these desirable attributes. Correlation can be implemented using optical or digital methods and these implementations impose their own requirements. The optical correlator peculiarities can result in special correlation filter designs such as phase-only filters, binary phase-only filters and cross-coupled filters. For digital implementation, quantizing the input signals to one bit results in binarized correlations that are easy from a hardware perspective. We also briefly looked at the case of non-overlapping noise. The major points are as follows:

● An important metric for quantifying the noise tolerance of a digital correlation filter is the signal-to-noise ratio (SNR), defined as the ratio of the square of the average correlation peak to its variance. If the underlying noise is Gaussian, SNR is monotonically related to the probability of error and maximizing the SNR is equivalent to minimizing the probability of error. For non-Gaussian noise, there may be no such monotonic relationship, but SNR is still a useful metric for quantifying a correlation filter's noise tolerance.

- The SNR is maximized by the filter $H(f) = \alpha S^*(f)/P_n(f)$, where α is an arbitrary complex constant, $S^*(f)$ is the conjugated FT of the known reference signal, and $P_n(f)$ is the power spectral density (PSD) of the additive noise. For the case of white noise, the maximal-SNR filter is the matched filter (MF) given digitally as $H(f) = S^*(f)$, but optically must be dimensionally converted from S's units of [V/m] to the dimensionless complex transmittance of the filter SLM. The MF magnitude is the same as the magnitude of the reference signal Fourier transform (FT), whereas the MF phase is the negative of the phase of the signal FT. Use of the MF results in an output signal that is the cross-correlation of the input signal with the reference signal.

- The MF is phase-canceling in that its phase and the phase of the FT of the signal sum to a constant. We can show that the correlation peak is guaranteed to be at the origin when the input contains the unshifted target signal with no noise, and when the filter is phase-canceling. Some correlation filters such as binary phase-only filters (BPOFs) are not phase-canceling and cannot guarantee that the peak will occur at the origin when the input contains the unshifted target signal with no noise.

- Correlation can be implemented optically or digitally. Digital implementations can be either direct or the usually more efficient FFT based methods. The DFT based methods perform circular correlations rather than linear correlations and care must be taken to ensure that we obtain the desired linear correlations.

- Matched filters of most images are usually of low-pass nature with a large dynamic range and thus absorb much of the light incident on the frequency plane of a VanderLugt optical correlator. At the expense of increasing the deleterious effects of noise, we can achieve higher light throughput by using a phase-only filter (POF) that has the same phase as the MF, but a unit magnitude. A side benefit of using the POF is that it usually leads to sharper correlation peaks for low-pass images. The use of a POF is also somewhat justified by the empirical result that the phase of the FT of an image appears to contain more information than its magnitude. However, the POF is an all-pass filter and thus possesses no noise suppression capabilities.

- The peak sharpness of a correlation output can be quantified using peak-to-correlation energy (PCE), defined as the ratio of correlation peak intensity to the total noiseless input image energy that is passed through to the correlation plane. The PCE is maximized by the inverse filter (IF), but the IF is of little practical value since it amplifies high frequencies, and hence the input noise, to unacceptably high levels.

- Correlation peak position is affected by input noise and peak location error (PLE) quantifies the variance in the correlation peak position. Using the Cauchy–Schwarz inequality, we can prove that the filter that maximizes the SNR also minimizes the PLE. An interesting point is that the PLE is affected more positively (i.e., it becomes smaller) by increasing the SNR at higher frequencies. The peak position in each direction can be estimated to sub-pixel accuracy using a second-order polynomial fit for the three pixels centered on a correlation peak.

- A rigorous way of trading off SNR versus PCE is through the design of optimal tradeoff (OT) correlation filters. The OT filters are obtained by maximizing the figure of merit (FOM), which is a ratio with the same numerator as that of the SNR and the PCE. The denominator, however, is a weighted sum of the denominators in PCE and SNR. The weighting parameter β can be chosen to emphasize one measure relative to the other. Maximizing FOM results in an OT filter that provides SNR and PCE values that cannot be simultaneously exceeded by any other correlation filter.

- The poor noise tolerance of a POF can be improved by allowing certain filter frequencies to take on zero values thus providing some noise suppression capability. The set of frequencies for which the filter has a non-zero magnitude is known as its region of support (ROS). The ROS that maximizes the output SNR can be determined by sorting S_k/P_k, the ratio of signal FT magnitude samples to noise PSD samples. This algorithm to select maximal SNR can also be used to select the ROS to maximize PCE or to maximize FOM.

- Some SLMs can accommodate only two phases and, for such SLMs, binary phase-only filters (BPOFs) may be convenient. The BPOFs, like all other all-pass filters, have no noise suppression capability. The BPOF that maximizes the correlation peak intensity can be proven to be of the form that binarizes the real part of $[S(f)e^{-j\theta}]$, where θ is known as the threshold line angle (TLA). The SNR of BPOFs can be improved by using ternary filters that take on three values, namely -1, 0, and $+1$. An efficient algorithm is available to determine the ROS for such ternary filters. It appears that four phases can provide average correlation peak intensities that are within 1 dB of the peak intensities from the full-phase POFs.

- Including the filter-independent detector noise in the SNR definition, and taking into account the minimal and maximal SLM magnitudes, results in saturated correlation filters. Saturated filters are the same as the matched filters except that the filter magnitude is saturated to either the maximal magnitude or the minimal magnitude.

- For a general SLM, amplitude and phase transmittances are cross-coupled. The filter that maximizes the SNR while taking on values only from Ω, the set of allowed SLM transmittance values, can be shown to be the minimum Euclidean distance optimal filter (MEDOF) of the matched filter; i.e., the optimal filter is obtained by mapping each pixel of the fully complex matched filter to the nearest value from set Ω. This still leaves two parameters (namely gain G and angle λ) to be determined, either by trying all meaningful values, or by a cross-correlation in a log-polar transform domain.

- In direct digital correlation, we can use the gray scale for both signals or use quantized versions of one or both signals for achieving hardware simplicity. The peak-to-sidelobe ratio (PSR) of gray-scale correlation can be shown to equal the space bandwidth product of the reference signal. For signals that can be modeled as sample realizations from a Gaussian random process, the PSR can be shown to be about 4 dB higher when both signals are binarized, compared to the gray-scale correlation. However, the noise behavior of the binarized correlations may be unsatisfactory in low-SNR situations.

- Additive noise is not a good model for non-overlapping background noise in observed scenes. Multiple hypotheses testing methods can be employed to locate the target in the presence of additive and non-overlapping background noise. For the low additive noise case, the optimal detection strategy is cross-correlation of the observed scene with the segmented target, except that we need to take into account a running scene energy. For the low background noise case, the optimal detection strategy is to cross-correlate the observed scene with the target signal in a constant background.

6

Advanced correlation filters

Ever since VanderLugt's pioneering work [5] on the implementation of matched filters (MFs) by coherent optical processing, there has been considerable interest in using correlators for recognizing patterns in images. The MF is of course optimal for finding a given pattern in the presence of additive white noise, and, as we have shown in Chapter 5, yields the highest output SNR. In radar signal processing and digital communications, matched filters have been very successful in many applications. For image processing, perhaps the greatest appeal of correlation filtering lies in its ability to produce shift-invariant peaks (because correlation filters are just a special class of LSI filter) and the resultant processing simplicity since we can avoid the need for image segmentation and registration.[1] Unfortunately, MFs are not adequate for practical pattern recognition since their response degrades rapidly when the patterns deviate from the reference [57]. Such pattern variations can be induced by scale changes, rotations or signature differences, all of which are common phenomena associated with the general pattern recognition problem.

One straightforward approach to this problem would be to apply a large number of MFs, each tuned to a particular variation. However, the enormous storage and processing requirements of this approach make it impractical. The alternative is to design robust correlation filters that can overcome the limitations of the MFs. Correlation filters are 2-D finite impulse response (FIR) filters whose output is expected to be stable and predictable in response to a known class of input patterns, and we can use many of the tools discussed in Chapter 3 for LSI systems. As in the case of the MF, the discrete time equation describing the output of the correlation filters is given by:

---

[1] Statistical pattern recognition, model-based techniques, and even neural networks often require the object to be segmented from the background and registered before matching the edges, dimensions, or other features.

$$g(m,n) = f(m,n) \otimes h(m,n)$$
$$= \sum_k \sum_l f(m+k, n+l) \cdot h(k,l)$$
$$= \sum_k \sum_l f(k,l) \cdot h(k-m, l-n) \tag{6.1}$$

where the symbol $\otimes$ represents two-dimensional correlation, and $g(m, n)$ is the correlation surface produced by the filter $h(m, n)$ in response to the input image $f(m, n)$. The analog version of the above equation is

$$g(\tau_x, \tau_y) = f(x,y) \otimes h(x,y)$$
$$= \iint f(\tau_x + x, \tau_y + y) h(x,y) \mathrm{d}x \mathrm{d}y \tag{6.2}$$

While the analog equation is useful for analysis, it is more convenient to deal with the discrete form since the filters are always synthesized using digital computers.

Strictly speaking, the entire correlation surface $g(m, n)$ is the output of the filter. However, the point $g(0, 0)$ is often referred to as the "filter output" because this is the value usually compared to a preset threshold to determine whether the input object is a desired pattern or not. In part, the reason for using this is to remain consistent with the terminology used for the MF. In any event, the reference imagery is expected to have its highest correlation at the origin. For this reason, a peak is expected in the correlation surface at $g(0, 0)$. With this interpretation in mind, the *peak* filter output is given by

$$g(0,0) = \sum \sum f(m,n)h(m,n) = \mathbf{f}^T \mathbf{h} \tag{6.3}$$

Of course, the position of the peak will shift with any movements of the pattern within the input image as well as with noise, and may not always be at the origin in practice.

Correlation filter theory deals with the design of $h(m, n)$ such that the filter output can be used to achieve the following three objectives:

1. recognize distorted versions of the reference pattern,
2. behave robustly in the presence of clutter and noise, and
3. yield a high probability of correct recognition while maintaining a low error rate.

Thus our goal in this chapter is to go beyond simple correlation filters, such as the matched filter and the phase-only filter that rely on a single reference image, and design advanced correlation filters that can build in tolerance to anticipated variability in the pattern of interest. Much of this chapter deals with the design of advanced digital correlation filters. Correlation filter designs for optical implementation are discussed in detail in Chapter 8.

While a complete treatment of the topic is beyond the scope of this book, several fundamental and key concepts for making correlation a viable approach for pattern recognition are covered in this chapter. The rest of the chapter is organized as follows. Section 6.1 deals with techniques developed to handle cases that are limited to in-plane distortions such as in-plane rotation and scale changes. The application of coordinate transform techniques and *circular harmonic functions* are discussed in this section. Section 6.2 is devoted to more general filter synthesis techniques that handle arbitrary distortions. Various early methods as well as more advanced techniques are reviewed in detail. A transform-based distance classifier technique using correlation is described in Section 6.3. This approach yields a quadratic classifier that has proved to be very useful for discriminating between similar classes of patterns in images. Section 6.4 introduces polynomial correlation filters (PCFs) that extend the concept of correlation filters to include nonlinearities. Section 6.5 shows how statistical methods can be used to predict the performance of simple correlation filters. In Section 6.6, we briefly introduce a recently developed approach [52] to optimizing advanced pattern recognition criteria such as the Fisher ratio and Bayes error rate. More details about advanced pattern recognition metrics can be found in Chapter 8. Finally, a chapter summary is provided in Section 6.7.

6.1 In-plane distortion invariance

Distortions in an image due to scale changes and in-plane rotations can be described mathematically in terms of coordinate transforms. To the extent that in-plane variations do not cause obscurations or changes in the general shape of an object, the pattern is completely characterized by a single image. It has been shown that these filters can also be used for estimating the distortion (scale and orientation) parameters. The image changes can be likened to what happens as a camera remains centered on, and orthogonal to, a planar object while being rotated about, or translated along, the line of sight.

6.1.1 A basic coordinate transform method

In this section, we describe a technique [58] to synthesize a correlation filter from a single image, assuming that the desired response of the filter has been specified for all possible rotation and scale variations of the reference pattern. Changes due to scale and rotations in a Cartesian coordinate system manifest themselves as shifts in a log-polar coordinate system. This property can be exploited to obtain a filter that behaves in a predictable fashion in response to in-plane rotation and scale variations. Consider $f(x, y)$ to be an image in Cartesian coordinates. Its log-polar transform (LPT) $f(\rho, \theta)$ is computed using mappings $\rho = \ln(r) = \ln\left\{ \sqrt{(x - x_o)^2 + (y - y_o)^2} \right\}$ and $\theta = \text{atan2}((y - y_o), (x - x_o))$, where $\text{atan2}(\cdot, \cdot)$ is the two-argument arctangent satisfying $\phi = \text{atan2}(k \sin\phi, k \cos\phi)$ with k being any positive number and $0 \leq \phi < 2\pi$. Thus the image is mapped to a new coordinate system measured by the natural logarithm of the radial distance and angle with respect to the origin $\{x_o, y_o\}$. Note that we use the two-argument arctangent to resolve the quadrant into which the point falls. The entire rotational span is captured within $0 \leq \theta < 2\pi$. Since $\log x \to -\infty$ as $x \to 0$, we cannot capture the entire image all the way to (x_o, y_o). If a is the minimum radius to be represented in the log-polar image and b is the largest, then an annular region in the original image is represented in $\log a \leq \rho \leq \log b$, $0 \leq \theta < 2\pi$.

It is often necessary to compute the LPT of a discrete image using a digital computer. If the transformed image is to be N pixels in ρ and M pixels in θ, then choosing

$$i = 1 + \frac{\log \rho - \log a}{\log\left(\dfrac{P}{\sqrt{2}}\right) - \log a}(N - 1)$$

$$\tag{6.4}$$

$$j = 1 + \frac{\text{atan2}(y, x)}{2\pi}(M - 1)$$

for an input image that is P pixels square, and $a = P \exp(-2\pi NM)/\sqrt{2}$, will result in a conformal version of the input image in the log-polar domain, and will also assure $1 \leq i \leq N$ and $1 \leq j \leq M$.

Although digital considerations must be made in practice, we shall treat images as continuous functions for the purposes of our discussion here. If the original image is scaled and rotated, the LPT essentially shifts and can be expressed as $f(\rho + \tau_\rho, \theta + \tau_\theta)$, where τ_ρ is the log of the scale factor, and τ_θ is the angle of rotation. The coordinates of the equivalent distorted image $f(x', y')$ in the Cartesian system can be obtained by the equation

$$\begin{bmatrix} x' \\ y' \end{bmatrix} = \mathbf{A} \begin{bmatrix} x \\ y \end{bmatrix} \tag{6.5}$$

where

$$\mathbf{A} = e^{\tau_\rho} \begin{bmatrix} \cos(\tau_\theta) & \sin(\tau_\theta) \\ -\sin(\tau_\theta) & \cos(\tau_\theta) \end{bmatrix} \tag{6.6}$$

It is true that scaled and rotated patterns may be recognized by first using the LPT to convert rotation and scale changes to shifts, and then applying an MF in the transformed coordinate system. This approach may prove cumbersome in practice since the images need to be centered and the on-line computation of the LPT is required. Real-time methods for getting the LPT image include (1) using an imager that samples the input irradiance pattern on the log-polar grid, and (2) programmed digital transformations exercised on special-purpose image-processing hardware. The correlation filter technique described here differs in that the LPT is only used for the design of the filter, but does not require the on-line use of any coordinate transform (i.e., the implementation involves only a straightforward correlation in Cartesian coordinates).

Let us now discuss how the LPT may be used to obtain a filter whose response can be fully controlled in the presence of rotation and scale changes. Recall that the filter output is the correlation value at the origin, which is the same as the inner product of the input image and the filter function. Mathematically, the filter output due to a distorted image is given by

$$\iint f(x', y') h(x, y) \mathrm{d}x \mathrm{d}y = c(\tau_\rho, \tau_\theta) \tag{6.7}$$

where (x', y') is related to (x, y) according to Eq. (6.5), and the notation $c(\tau_\rho, \tau_\theta)$ is used to emphasize the fact that, in general, the filter output changes with the distortion of the input image characterized by the parameters τ_ρ and τ_θ.

It follows from Eq. (6.7) that when the image is known, the filter can be uniquely determined if the output $c(\tau_\rho, \tau_\theta)$ is specified for all values of τ_ρ and τ_θ. We therefore treat $c(\tau_\rho, \tau_\theta)$ as a free parameter that can be selected by the user, and refer to it as the *signature control function* (SCF). In principle, the SCF, $c(\tau_\rho, \tau_\theta)$ can be any desired correlation peak signature (as a function of in-plane rotation) for the filter.

Integrals in Cartesian and log-polar coordinates can be related using the Jacobian (defined below) of the coordinate transform. If $f(x, y)$ and $g(x, y)$ are two functions in Cartesian coordinates with $f(\rho, \theta)$ and $g(\rho, \theta)$ as their respective LPTs, then

$$\iint f(x, y) g(x, y) \mathrm{d}x \mathrm{d}y = \iint f(\rho, \theta) g(\rho, \theta) |\mathbf{J}| \mathrm{d}\rho \mathrm{d}\theta \tag{6.8}$$

where the Jacobian of the transformation $|\mathbf{J}|$ is given by

$$|\mathbf{J}| = \begin{vmatrix} \dfrac{\mathrm{d}x}{\mathrm{d}\rho} & \dfrac{\mathrm{d}y}{\mathrm{d}\rho} \\ \dfrac{\mathrm{d}x}{\mathrm{d}\theta} & \dfrac{\mathrm{d}y}{\mathrm{d}\theta} \end{vmatrix} = \begin{vmatrix} \mathrm{e}^\rho \cos(\theta) & \mathrm{e}^\rho \sin(\theta) \\ -\mathrm{e}^\rho \sin(\theta) & \mathrm{e}^\rho \cos(\theta) \end{vmatrix} = \mathrm{e}^{2\rho} \tag{6.9}$$

Using this relation, the left side of Eq. (6.7) can be expressed as an integral in log-polar coordinates to yield

$$\iint f(\rho + \tau_\rho, \theta + \tau_\theta) h(\rho, \theta) e^{2\rho} d\rho d\theta = c(\tau_\rho, \tau_\theta) \qquad (6.10)$$

Defining $\hat{h}(\rho, \theta) = h(\rho, \theta) e^{2\rho}$, the equation for the filter output becomes

$$\iint f(\rho + \tau_\rho, \theta + \tau_\theta) \hat{h}(\rho, \theta) d\rho d\theta = c(\tau_\rho, \tau_\theta) \qquad (6.11)$$

or

$$f(\rho, \theta) \otimes \hat{h}(\rho, \theta) = c(\rho, \theta) \qquad (6.12)$$

In other words, given the LPT of the image $f(\rho, \theta)$ and the SCF $c(\rho, \theta)$ (i.e., the desired output for every combination of scale and rotation), it is possible to calculate the filter function using the relation in Eq. (6.11). Equivalently, the solution for the filter may be expressed as

$$h(\rho, \theta) = e^{-2\rho} \text{IFT}\left\{ \frac{\text{FT}\{c(\rho, \theta)\}}{\text{FT}\{f(\rho, \theta)\}} \right\} \qquad (6.13)$$

where FT{} and IFT{} denote the forward and inverse Fourier transform operations respectively. It should be noted that the expression in Eq. (6.13) cannot be evaluated at frequencies where the denominator is zero. In practice, it may be sufficient to set the filter response to zero at all such frequencies. The correlation filter h(x, y) can then be obtained by applying the inverse LPT to $h(\rho, \theta)$.

It is interesting to note that the system of equations in Eq. (6.11) is completely specified. As a consequence, only one solution for $h(\rho, \theta)$ exists that satisfies the constraints imposed on the output by the SCF. Therefore, the filter function cannot be further optimized with respect to any other performance criterion.[2] This is a consequence of the fact that the filter's output is fully specified over all possible distortions (scale and rotation). In the next section, we discuss techniques that deal only with rotation tolerance, and hence allow more degrees of freedom for optimization of the filter's performance.

6.1.2 Circular harmonic functions

Circular harmonic functions (CHFs) [59, 60] have proved to be an important concept for the design of in-plane rotation-invariant filters. To define CHFs,

---

[2] Several performance criteria for filters will be described in Section 8.2.

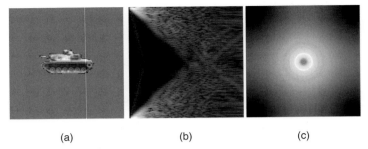

(a)   (b)   (c)

Figure 6.1 Images of (a) a tank, (b) its CHF expansion, and (c) the image reconstructed with only the zeroth order harmonic

we transform Cartesian coordinates $\{x, y\}$ to polar coordinates $\{r, \theta\}$ using the relations

$$r = \sqrt{(x - x_o)^2 + (y - y_o)^2} \quad \text{and} \quad \theta = \text{atan2}((y - y_o), (x - x_o)) \quad (6.14)$$

where $\{x_o, y_o\}$ is the center of rotation of the image in the Cartesian coordinate system. The polar transform $f(r, \theta)$ of an image is periodic in θ with period 2π, and consequently can be expressed by a Fourier series as follows:

$$f(r, \theta) = \sum_{m=-\infty}^{\infty} f_k(r) e^{jk\theta} \quad (6.15)$$

The term $f_k(r)$ is known as the kth *circular harmonic* component and is given by

$$f_k(r) = \frac{1}{2\pi} \int_0^{2\pi} f(r, \theta) e^{-jk\theta} d\theta \quad (6.16)$$

The numerical computation of circular harmonics requires Eq. (6.14) to be implemented using carefully written polar transform routines. Figure 6.1 shows the image of a tank from a broad-side view, its digitally computed CHFs and the reconstruction from just the zeroth-order CHF. Note that the reconstruction from the zeroth-order CHF is circularly symmetric, but loses much of the detail.

In some applications it may be convenient to apply the CHF expansion to an image in the frequency domain. In such cases, we let $F(r, \theta)$ be the polar transform of $F(u, v)$, the Fourier transform of $f(x, y)$. The corresponding circular harmonic functions, $F_k(r)$ are given by

$$F_k(r) = \frac{1}{2\pi} \int_0^{2\pi} F(r, \theta) e^{-jk\theta} d\theta \quad (6.17)$$

so that the image's Fourier transform (represented in polar coordinates as $F(r, \theta)$) can be expanded in terms of the CHFs as

$$F(r, \theta) = \sum_{k=-\infty}^{\infty} F_k(r) e^{jk\theta} \tag{6.18}$$

To see how CHFs can be used to obtain a rotation-invariant filter, let $C(\tau_\theta)$ represent the filter output when the input image is rotated by τ_θ degrees. When the image is at nominal orientation and $\tau_\theta = 0°$, the filter's output can be expressed in the frequency domain as follows:

$$
\begin{aligned}
C(0) &= \iint F(u, v) H^*(u, v) du dv \\
&= \iint F(r, \theta) H^*(r, \theta) r dr d\theta
\end{aligned} \tag{6.19}
$$

where $H(u, v)$ is the Fourier transform of the filter. Using the CHF expansion, this becomes

$$C(0) = \int_{r=0}^{\infty} \int_{\theta=0}^{2\pi} \left(\sum_{k=-\infty}^{\infty} F_k(r) e^{jk\theta} \right) \left(\sum_{l=-\infty}^{\infty} H_l^*(r) e^{-jl\theta} \right) r dr d\theta \tag{6.20}$$

It is now possible to derive the expression for the filter output when the input image is rotated by τ_θ degrees. The rotation produces a shift along θ so that the Fourier transform of the image is now given in polar coordinates by $F(r, \theta + \tau_\theta)$ with the CHF expansion

$$F(r, \theta + \tau_\theta) = \sum_{k=-\infty}^{\infty} F_k(r) e^{jk(\theta + \tau_\theta)} \tag{6.21}$$

The filter output for a rotation of τ_θ degrees is obtained by substituting Eq. (6.21) into Eq. (6.20) to yield the following:

$$
\begin{aligned}
C(\tau_\theta) &= \int_{r=0}^{\infty} \int_{\theta=0}^{2\pi} \left(\sum_{k=-\infty}^{\infty} F_k(r) e^{jk(\theta + \tau_\theta)} \right) \left(\sum_{l=-\infty}^{\infty} H_l^*(r) e^{-jl\theta} \right) r dr d\theta \\
&= \int_{r=0}^{\infty} \left[\sum_{k=-\infty}^{\infty} \sum_{l=-\infty}^{\infty} F_k(r) H_l^*(r) e^{jk\tau_\theta} \int_{\theta=0}^{2\pi} e^{j(k-l)\theta} d\theta \right] r dr
\end{aligned} \tag{6.22}
$$

Since

$$\int_0^{2\pi} e^{j(k-l)\theta} d\theta = \begin{cases} 0 & \text{if } k \neq l \\ 2\pi & \text{if } k = l \end{cases} \tag{6.23}$$

Eq. (6.22) simplifies as follows:

$$C(\tau_\theta) = 2\pi \int_{r=0}^{\infty} \sum_{k=-\infty}^{\infty} F_k(r) H_k^*(r) e^{jk\tau_\theta} r dr \tag{6.24}$$

Finally, this can be more compactly written as

$$C(\tau_\theta) = \sum_{k=-\infty}^{\infty} C_k e^{jk\tau_\theta} \tag{6.25}$$

where

$$C_k = 2\pi \int_{r=0}^{\infty} F_k(r) H_k^*(r) r dr \tag{6.26}$$

Thus, the filter output $C(\tau_\theta)$ as a function of the image in-plane rotation angle τ_θ can be expressed in terms of the coefficients C_k, defined in Eq. (6.26). If the filter is made up of only one CHF, i.e.,

$$H_k(r) = \begin{cases} H_{k_o}(r) & k = k_o \\ 0 & k \neq k_o \end{cases} \tag{6.27}$$

then there is only one non-zero coefficient C_{ko} and the filter output is given by

$$C(\tau_\theta) = C_{k_o} e^{jk_o\tau_\theta} \tag{6.28}$$

and

$$|C(\tau_\theta)| = |C_{k_o}| \tag{6.29}$$

which proves that the magnitude of the filter output is invariant to the angle of rotation. Without placing any other restrictions on the filter, it is thus possible to ensure a completely rotation-invariant response by allowing only one non-zero CHF in the composition of the filter. If, additionally, it is required that the filter be circularly symmetric, i.e., $H(r, \theta) = \hat{H}(r)$, then the circular harmonics of the filter are

$$H_k(r) = \frac{1}{2\pi} \int\limits_{\theta=0}^{2\pi} H(r,\theta)\mathrm{e}^{-jk\theta}\mathrm{d}\theta$$

$$= \frac{1}{2\pi} \int\limits_{\theta=0}^{2\pi} \hat{H}(r)\mathrm{e}^{-jk\theta}\mathrm{d}\theta = \begin{cases} \hat{H}(r) & k=0 \\ \\ 0 & k \neq 0 \end{cases} \qquad (6.30)$$

which shows that the only non-zero CHF occurs for $k=0$, and consequently the filter output is given by

$$C(\tau_\theta) = C_0 \qquad (6.31)$$

which is always independent of the angle or rotation. It is obvious from the circular symmetry of the zeroth-order CHF in Figure 6.1(c) that it will always yield a rotation-invariant response.

In this section, we have seen that a rotationally invariant filter can be obtained by using only one circular harmonic of the reference image. However, such filters usually exhibit poor discrimination properties, since a considerable amount of image information is lost when we ignore the remaining circular harmonics. Advanced techniques make use of all CHF harmonics to obtain a prescribed filter response. In fact, the theory has been extended to relate the CHF formulation to well-established techniques in digital signal processing for the design of finite impulse response (FIR) filters. The interested reader is referred to the references [58–61] for additional details.

6.2 Composite correlation filters

Composite correlation filters were developed to handle the more general types of distortions that cannot be mathematically modeled by coordinate transforms or CHFs. Composite filters are derived from several *training images*, which are representative views of the object or pattern to be recognized. In principle, such filters can be trained to recognize any object or type of distortion as long as the expected distortion can be captured by the training images. Thus, the proper selection of training images is an important step in the design of composite filters.

The objective of all composite filters is to be able to recognize the object on which they are trained, at any of a suite of viewpoints of the object, while being able to reject everything else. In other words, the filters should exhibit a high correct recognition rate while simultaneously keeping the false acceptance rate low, a concept later explained in more detail – the ROC curve. As will be shown, the optimization of key performance criteria offers a methodical approach to achieving these objectives. However, the importance of having a properly chosen training set cannot be overstated. Generally speaking, the training set should characterize the expected variations whether such variations are due to spectral changes in the object's signature, sensor phenomena, or imaging geometry. Ideally, the data should be acquired in the relevant spectral band with the sensor to be used. However this may not always be feasible in practice and one may resort to synthetic imagery and computer-generated models. It is also not practical to expect a large number of training images to capture all possible variations, since such an image set may be too large to store and use in field applications where storage and computation are at a premium. Thus, it is desirable that composite filters trained on a limited set of training images should be able to generalize and find objects under conditions not explicitly seen in the training process. This is termed the *generalization* ability of the composite filter. There are extensive issues relating to the fidelity and the source of training information, much of which is beyond the scope of this book. It will be assumed for the purposes of the following discussions that an adequate training set is available for filter synthesis. The training set, plus additional information such as the log-polar representation to cover scale and rotation distortions, must span the range of desired distortion, otherwise there will be insufficient information to provide to the filter.

6.2.1 *Early synthetic discriminant function (SDF) filters: The projection SDF filter*

One of the earliest composite correlation filters is known as the *synthetic discriminant function* (SDF) filter [7]. In this approach, the filter is designed to yield a specific value at the origin of the correlation plane in response to each training image. For example, in a two-class problem, the correlation values at the origin may be set to 1 for training images from one class, and to 0 for the training images from the other class. The hope is that the resulting filter will yield values close to 1 for all images from class 1 and close to 0 for all images from class 2, and thus we can tell which class the input belongs to by looking at the value at the origin.

The above idea works well if all of the images (including non-training images) are always centered and thus we look only at the correlation value at the origin. However, one of the main advantages of using a correlation filter is its shift-invariance so we need not require that the input image be centered. However, if the image is not centered, we will need to determine to which pixel location the controlled values (of 1 and 0) have moved. This can be done easily if the controlled values stand out (e.g., 1 should be the peak in the correlation output and 0 should be the minimum in the correlation output). Of course, the premise was that the training data are adequate for characterizing all expected variations (including the non-training images).

To develop the framework for SDF filters, we assume that a set of N training images is available. Let u_i be the value at the origin of the correlation plane $g_i(m, n)$ produced by the filter $h(m, n)$ in response to a training image $x_i(m, n)$, $1 \le i \le N$. Therefore,

$$u_i = g_i(0, 0) = \sum_{m=1}^{d_1} \sum_{n=1}^{d_2} x_i(m, n) h(m, n), \quad 1 \le i \le N \tag{6.32}$$

The original SDF filter approach was to design the filter to satisfy hard constraints on u_i. Let $\mathbf{x}_i$ and $\mathbf{h}$ be vector representations (obtained by scanning a 2-D array from left to right and from top to bottom, and placing the resulting sequence of numbers into a column vector) of $x_i(m, n)$ and $h(m, n)$, respectively. This permits Eq. (6.32) to be written as

$$\mathbf{x}_i^T \mathbf{h} = u_i, \quad 1 \le i \le N \tag{6.33}$$

The above N linear equations can be written as a single matrix-vector equation:

$$\mathbf{X}^T \mathbf{h} = \mathbf{u} \tag{6.34}$$

where $\mathbf{h}$ is the filter vector containing d values, $\mathbf{X} = [\mathbf{x}_1, \mathbf{x}_2, \ldots, \mathbf{x}_N]$ is a $d \times N$ matrix with the N training image vectors (each with d pixels) as its columns, and $\mathbf{u} = [u_1, u_2, \ldots, u_N]^T$ is an $N \times 1$ vector containing the desired peak values for the training images. However, since the number of training images N is generally much smaller than the number of frequencies in the filters, the system of equations is under-determined and many filters exist that satisfy the constraints in Eq. (6.34). To find a unique solution, $\mathbf{h}$ is assumed to be a linear combination of the training images,[3] i.e.,

$$\mathbf{h} = \mathbf{X}\mathbf{a} \tag{6.35}$$

where $\mathbf{a}$ is the vector of weights for linearly combining the columns of the data matrix $\mathbf{X}$. To determine $\mathbf{a}$, we substitute for $\mathbf{h}$ in Eq. (6.34), which yields

$$\mathbf{X}^T \mathbf{X} \mathbf{a} = \mathbf{u} \implies \mathbf{a} = \left(\mathbf{X}^T \mathbf{X} \right)^{-1} \mathbf{u} \tag{6.36}$$

---

[3] Another reason for the linear combination requirement is that the resulting filters could be obtained in an optics lab by exposing a holographic plate to different exposure levels/times of Fourier transforms of different training images.

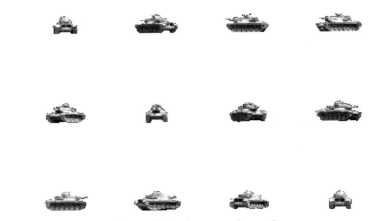

Figure 6.2 Representative training images of a tank

Figure 6.3 Projection SDF filter synthesized from images shown in Figure 6.2

Substituting this solution for **a** into Eq. (6.35) leads to the following *projection* SDF filter solution:

$$\mathbf{h} = \mathbf{X}(\mathbf{X}^T\mathbf{X})^{-1}\mathbf{u} \qquad (6.37)$$

where **h** is the $d \times 1$ filter vector expressed in the space (not transform) domain.

As an example, consider the training images of a tank shown in Figure 6.2. These images are used to synthesize the projection SDF filter shown in Figure 6.3. The composite nature of the filter is quite evident. Each of the training images is required to yield a value of 1.0 at the origin of the correlation plane. The resulting correlation planes are shown in Figure 6.4.

As can be seen in Figure 6.4, the projection SDF filter yields a correlation peak whose amplitude is guaranteed to be 1.0 by design when the input is a

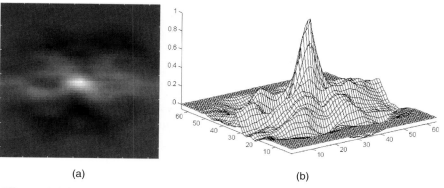

(a)   (b)

Figure 6.4 Output correlation plane produced by the projection SDF filter in response to one of the training images, shown as (a) an intensity image, and (b) as a mesh plot

training image. The process for determining the class of a test pattern requires the peaks in the correlation surface to be located. The position of the pattern at the input is indicated by the location of the peak. The filter is said to "recognize" the pattern when the peak value exceeds a certain threshold. However, the peak is surrounded by large sidelobes, which can lead to errors if they exceed the main peak. In fact, this occurs frequently in practice with a projection SDF filter since the filter design does not control any points in the correlation plane other than the origin.

6.2.2 Minimum average correlation energy filter

In practice, it is desirable to suppress the sidelobes to ensure a sharp and distinct correlation peak and reduce the chances of error. One way to achieve this is to minimize the energy in the correlation plane (which naturally includes the sidelobes). The average correlation energy (ACE) for the N training images is defined as follows:

$$\text{ACE} = \frac{1}{N} \sum_{i=1}^{N} \sum_{m}^{d_1} \sum_{n}^{d_2} |g_i(m,n)|^2 \tag{6.38}$$

Using Parseval's theorem, ACE can be expressed in the frequency domain as

$$\text{ACE} = \frac{1}{d \cdot N} \sum_{i=1}^{N} \sum_{k}^{d_1} \sum_{l}^{d_2} |G_i(k,l)|^2 \tag{6.39}$$

where $G_i(k, l)$ is the 2-D Fourier transform of $g_i(m, n)$. Since $G_i(k, l) = H(k, l) X_i^*(k, l)$, the frequency domain expression for ACE becomes

$$\text{ACE} = \frac{1}{d \cdot N} \sum_{i=1}^{N} \sum_{k}^{d_1} \sum_{l}^{d_2} |H(k,l)|^2 |X_i(k,l)|^2 \qquad (6.40)$$

The formulation of a frequency domain expression for ACE using Parseval's theorem was the cornerstone in the further development of correlation filters. To facilitate the analysis, we express $H(k, l)$ as a vector, $\mathbf{h}$, and define a diagonal matrix, $\mathbf{X}_i$, whose elements along the main diagonal are $X_i(k, l)$. Thus $\mathbf{h}$ and $\mathbf{X}_i$ represent the filter and the training images in the frequency domain respectively. The expression for ACE then becomes

$$\text{ACE} = \frac{1}{d \cdot N} \sum_{i=1}^{N} (\mathbf{h}^+ \mathbf{X}_i)(\mathbf{X}_i^* \mathbf{h}) = \mathbf{h}^+ \left[\frac{1}{d \cdot N} \sum_{i=1}^{N} \mathbf{X}_i \mathbf{X}_i^* \right] \mathbf{h}$$
$$= \mathbf{h}^+ \mathbf{D} \mathbf{h} \qquad (6.41)$$

where $\mathbf{D} = \frac{1}{d \cdot N} \sum_{i=1}^{N} \mathbf{X}_i^* \mathbf{X}_i$ is a $d \times d$ diagonal matrix.

As is the case with the projection SDF, the correlation peak is controlled using hard constraints. By minimizing ACE, we expect to reduce the sidelobes and sharpen the peak. To complete the analysis, the constraints on the correlation peak must be expressed in the frequency domain as well. Since inner products in the space domain are directly proportional to inner products in the frequency domain, the constraint equation becomes

$$\mathbf{X}^+ \mathbf{h} = d \cdot \mathbf{u} \qquad (6.42)$$

where $\mathbf{X}$ is now a matrix whose columns $\mathbf{x}_i$ are vector representations of the Fourier transforms of the training images. The *minimum average correlation energy* (MACE) filter [62] minimizes ACE in Eq. (6.41) subject to the hard constraints in Eq. (6.42). This is equivalent to a constrained quadratic optimization problem where the quadratic function $\mathbf{h}^+ \mathbf{D} \mathbf{h}$ is minimized subject to the linear conditions $\mathbf{X}^+ \mathbf{h} = d \cdot \mathbf{u}$. As discussed in Chapter 2, this can be achieved using the method of Lagrange multipliers, which yields the optimum solution

$$\mathbf{h} = \mathbf{D}^{-1} \mathbf{X} (\mathbf{X}^+ \mathbf{D}^{-1} \mathbf{X})^{-1} \mathbf{u} \qquad (6.43)$$

The solution in Eq. (6.43) is in the frequency domain and $\mathbf{h}$ is therefore the column vector containing the frequency-domain filter array $H(k, l)$.

An example of the output from a MACE filter is shown in Figure 6.5. As can be seen in the figure, the peak is very sharp with low sidelobes. MACE filters have been shown to be effective for finding training images in background and clutter, and they generally produce very sharp correlation peaks. They are the

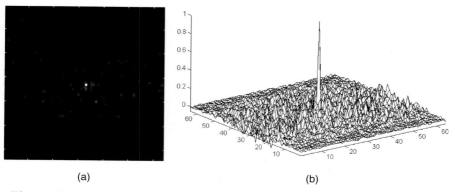

Figure 6.5 MACE filter correlation outputs: (a) an intensity image, and (b) mesh plot

first filters that attempted to control the entire correlation plane. However, there are two main drawbacks. Firstly, there is no in-built immunity to noise. Secondly, the MACE filters are often excessively sensitive to intra-class variations. Nevertheless, the MACE filters paved the way for the frequency domain analysis and development of correlation filters, and set the stage for subsequent developments which are discussed later in this chapter.

6.2.3 Minimum variance synthetic discriminant function

Noise and clutter can severely affect the performance of a filter. Therefore, it is important to characterize the behavior of the filter in the presence of noise and clutter to optimize its response. For now, we assume that all quantities are in the space domain and purely real. Also for simplicity of discussion, we adopt the discrete 1-D notation for random processes although the same concepts apply to 2-D signals and images. The filter's output in response to a training vector $\mathbf{x}_i$ corrupted by the additive noise vector $\mathbf{v}$ is given by:

$$(\mathbf{x}_i + \mathbf{v})^T\mathbf{h} = \mathbf{x}_i^T\mathbf{h} + \mathbf{v}^T\mathbf{h} = u_i + \delta \tag{6.44}$$

Clearly, fluctuations in the filter output occur because of the noise component δ. In fact, the input noise is often a time-varying process $\mathbf{v}(n)$ so that the output noise is also time-varying and may be denoted as $\delta(n)$. In the following discussions, the time-dependent notation is dropped for simplicity (and because of the stationarity assumption for the input noise) but invoked as necessary for clarification. The *minimum variance synthetic discriminant function* (MVSDF) [63] minimizes the variance of δ in order to minimize the fluctuations in the filter output.

Without loss of generality, let us assume that the noise is a zero-mean process. In order to derive the expression for the optimum filter, consider the *output noise variance* (ONV) given by:

$$\text{ONV} = E\{\delta^2\} = E\left\{(\mathbf{v}^T\mathbf{h})^2\right\}$$

$$= E\{\mathbf{h}^T\mathbf{v}\mathbf{v}^T\mathbf{h}\} = \mathbf{h}^T E\{\mathbf{v}\mathbf{v}^T\}\mathbf{h} = \mathbf{h}^T\mathbf{C}\mathbf{h} \tag{6.45}$$

where $\mathbf{C} = E\{\mathbf{v}\mathbf{v}^T\}$ is the input noise covariance matrix. Since ONV does not depend on the data, its expression is the same for all training vectors. The MVSDF is designed to minimize the ONV while satisfying the peak constraints on the training images in Eq. (6.34). This is equivalent to optimizing a quadratic function subject to linear constraints. The method of Lagrange multipliers readily yields the following MVSDF filter solution [63]:

$$\mathbf{h} = \mathbf{C}^{-1}\mathbf{X}(\mathbf{X}^T\mathbf{C}^{-1}\mathbf{X})^{-1}\mathbf{u} \tag{6.46}$$

We will now discuss some practical considerations for MVSDF filter synthesis. The projection SDF filter is a special case obtained when the noise is white (i.e., $\mathbf{C}$ is the identity matrix). Thus, the projection SDF filter is the optimum filter for recognizing the training images in the presence of additive white noise. In general, however, one difficulty in using the MVSDF is that $\mathbf{C}$ (a matrix of size $d \times d$, where d is the number of pixels in a training image and can be rather large) is difficult to estimate and computationally difficult to invert. In cases where the noise process is stationary, the covariance matrix is Toeplitz [17, 18] and its elements can be characterized as

$$\mathbf{C} = \{c_{i,j}\} = C_\upsilon(|i - j|) \tag{6.47}$$

where $C_\upsilon(\tau) = E\{\upsilon_i\upsilon_j\}$ is the *auto-covariance* function of the noise process, υ_i and υ_j are any two elements of the random vector $\mathbf{v}$, and $\tau = |i - j|$ is a discrete number representing the separation between the RVs, or the difference between the row and column location of the covariance matrix element. For instance, c_{ii} (elements on the main diagonal corresponding to the *variance*) are all equal to $C_\upsilon(0) = E\{\upsilon_i^2\}$. Essentially, all elements along a diagonal of a Toeplitz matrix are identical. This structure can be exploited to invert the covariance matrix and synthesize the MVSDF.

For the stationary noise case, a convenient form of the solution can be formulated in the frequency domain. The *power spectral density* $S_\upsilon(k)$ of a zero-mean, stationary random process is defined as the Fourier transform of the auto-covariance function, i.e.,

$$S_\upsilon(k) = \sum_{\tau=0}^{d-1} C_\upsilon(\tau) e^{-\frac{j2\pi k\tau}{d}} \tag{6.48}$$

and

$$C_\upsilon(\tau) = \frac{1}{d}\sum_{k=0}^{d-1} S_\upsilon(k) e^{\frac{j2\pi k\tau}{d}} \tag{6.49}$$

We have shown in Chapter 2 that when a random process with power spectral density $S_\upsilon(k)$ passes through an LTI system with frequency response $H(k)$, the power spectral density of the output is

$$S_\delta(k) = |H(k)|^2 S_\upsilon(k) \tag{6.50}$$

Hence the variance of the output process can be expressed in the frequency domain as

$$\text{ONV} = C_\delta(0) = \frac{1}{d}\sum_{k=0}^{d-1} S_\delta(k)$$

$$\tag{6.51}$$

$$= \frac{1}{d}\sum_{k=0}^{d-1} |H(k)|^2 S_\upsilon(k) = \mathbf{h}^+\mathbf{P}\mathbf{h}$$

where $\mathbf{P}$ is a diagonal matrix with samples of the noise power spectral density $S_\upsilon(k)$ along its main diagonal, and $\mathbf{h}$ is the filter represented as a vector in the frequency domain. There are many techniques to estimate the power spectral density of a stationary random process and the inversion of the diagonal matrix $\mathbf{P}$ is trivial.

The frequency domain version of the MVSDF can now be readily derived by defining the columns of the data matrix $\mathbf{X}$ with columns that are the vectorized versions of the 2-D Fourier transforms of the training images. We minimize the ONV in Eq. (6.51) subject to the frequency domain constraints in Eq. (6.42). Again, using the method of Lagrange multipliers we find that

$$\mathbf{h} = \mathbf{P}^{-1}\mathbf{X}(\mathbf{X}^+\mathbf{P}^{-1}\mathbf{X})^{-1}\mathbf{u} \tag{6.52}$$

The MACE filter yields sharp peaks that are easy to detect while the MVSDF filter is designed to provide robustness to noise. When there is only one training image, the MACE filter becomes the inverse filter, whereas the MVSDF filter becomes the matched filter. Since both attributes are important in practice, it is desirable to formulate a filter that possesses the ability to produce sharp peaks and behaves robustly in the presence of noise. It was realized early on [64] that both characteristics may be incorporated into a filter of the following type:

$$\mathbf{h} = (\mathbf{D} + \mathbf{C})^{-1}\mathbf{X}\left[\mathbf{X}(\mathbf{D} + \mathbf{C})^{-1}\mathbf{X}\right]^{-1}\mathbf{u} \tag{6.53}$$

and that the properties of such a filter would provide an optimal tradeoff between the MACE filter at one extreme and the MVSDF at the other. A more systematic development of this concept, and techniques for optimally trading off between various performance criteria will be discussed in Section 6.2.6.

6.2.4 *Designing distortion tolerant filters without hard constraints*

The SDF filters discussed up to now have assumed that the distortion tolerance of a filter could be controlled by explicitly specifying desired correlation peak values for training images. For example, we use correlation peak values of 1 for true class and 0 for false class. Another approach is to remove the hard constraints altogether. There are several observations that motivate this approach. Firstly, non-training images always yield different values from those specified and achieved for the training images. Secondly, no formal relationship exists between the constraints imposed on the filter output and its ability to tolerate distortions. In fact, it is unclear that even intuitively satisfying choices of constraints (such as the equal correlation peak [ECP] filter condition whereby the training images from the desired class give values of 1 and training images from the other class yield values of 0) have any significant impact on a filter's performance. In this section, we describe an approach that addresses the distortion tolerance problem without imposing such *hard* constraints on the filter.

The key idea is to treat the correlation plane as a new pattern generated by the filter in response to the input image. We start with the notion that the correlation planes are *linearly transformed* versions of the input image, obtained by applying the filter. Thus, attention should be paid not only to the correlation peak, but also to the entire correlation surface.

With the above discussion in mind, a metric for distortion is defined as the average variation in images after filtering. If $g_i(m, n)$ is the correlation surface produced in response to the i th training image, the average variation between the training image correlation surfaces in a mean square error (MSE) sense is

$$\text{distortion metric} = \frac{1}{N(N-1)}\sum_{i=1}^{N}\sum_{j=1}^{N}\sum_{m}\sum_{n}\left[g_i(m, n) - g_j(m, n)\right]^2 \tag{6.54}$$

We are using squared values rather than magnitude squared since our training images are assumed to be real and the resulting correlation outputs would be real (if we ensure proper symmetry for the filter). This metric is also referred to as the *average similarity measure* (ASM) [65], since it is a measure of the average similarity (or more correctly dissimilarity) between the filtered images. We now show that a simpler expression for ASM is

$$\text{ASM} = \frac{1}{N} \sum_{i=1}^{N} \sum_{m} \sum_{n} [g_i(m,n) - \bar{g}(m,n)]^2 \tag{6.55}$$

where $\bar{g}(m,n) = \frac{1}{N} \sum_{j=1}^{N} g_j(m,n)$ is the average of the training image correlation surfaces, and Eq. (6.54) and Eq. (6.55) are mathematically equivalent. The benefit of the form in Eq. (6.55) is that it requires fewer calculations and is easier to compute.

To show that the metrics in Eq. (6.54) and Eq. (6.55) are the same, we first switch to vector notation and rewrite Eq. (6.55) as follows:

$$\text{ASM} = \frac{1}{N} \sum_{i=1}^{N} |\mathbf{g}_i - \bar{\mathbf{g}}|^2 = \frac{1}{N} \sum_{i=1}^{N} \mathbf{g}_i^T \mathbf{g}_i - \bar{\mathbf{g}}^T \bar{\mathbf{g}} \tag{6.56}$$

where $\mathbf{g}_i$ and $\bar{\mathbf{g}}$ are the vector representations of $g_i(m, n)$ and $\bar{g}(m, n)$ respectively. Similarly in vector notation Eq. (6.54) becomes

$$\text{distortion metric} = \frac{1}{N(N-1)} \sum_{i=1}^{N} \sum_{j=1}^{N} |\mathbf{g}_i - \mathbf{g}_j|^2$$

$$= \frac{1}{N(N-1)} \sum_{i=1}^{N} \sum_{j=1}^{N} \left[\mathbf{g}_i^T \mathbf{g}_i + \mathbf{g}_j^T \mathbf{g}_j - 2\mathbf{g}_i^T \mathbf{g}_j \right] \tag{6.57}$$

$$= \frac{1}{N(N-1)} \sum_{i=1}^{N} \left[N\mathbf{g}_i^T \mathbf{g}_i + \sum_{j=1}^{N} \mathbf{g}_j^T \mathbf{g}_j - 2N\mathbf{g}_i^T \bar{\mathbf{g}} \right]$$

This can be further simplified to obtain

$$\text{distortion metric} = \frac{1}{(N-1)} \left[\sum_{i=1}^{N} \mathbf{g}_i^T \mathbf{g}_i + \sum_{j=1}^{N} \mathbf{g}_j^T \mathbf{g}_j - 2N\bar{\mathbf{g}}^T \bar{\mathbf{g}} \right]$$

$$= \frac{2}{(N-1)} \left[\sum_{i=1}^{N} \mathbf{g}_i^T \mathbf{g}_i - N\bar{\mathbf{g}}^T \bar{\mathbf{g}} \right] \tag{6.58}$$

$$= \frac{2N}{(N-1)} \left[\frac{1}{N} \sum_{i=1}^{N} \mathbf{g}_i^T \mathbf{g}_i - \bar{\mathbf{g}}^T \bar{\mathbf{g}} \right]$$

The right-hand sides of Eqs.(6.56) and (6.58) are identical except for a scale factor. Hence the pair-wise MSE distortion metric in Eq. (6.54) and the ASM in Eq. (6.55) are equivalent.

Thus, the ASM is an MSE measure of distortions (variations) in the correlation surfaces relative to an average shape. In an ideal situation, all correlation surfaces produced by a distortion-invariant filter (in response to a valid input pattern) would be the same, and the ASM would be zero. In practice, minimizing the ASM improves the stability of the filter's output in response to distorted input images.

We will now discuss how to formulate the ASM as a performance criterion for filter synthesis. This discussion is best developed in the frequency domain using matrix-vector notation for convenience. Using Parseval's theorem, the ASM can be expressed in the frequency domain as

$$\text{ASM} = \frac{1}{N \cdot d} \sum_{i=1}^{N} \sum_{k} \sum_{l} \left| G_i(k, l) - \bar{G}(k, l) \right|^2 \tag{6.59}$$

where $G_i(k, l)$ and $\bar{G}(k, l)$ are Fourier transforms of $g_i(m, n)$ and $\bar{g}(m, n)$ respectively, and d is the total number of pixels in each image. Then, in matrix-vector notation, Eq. (6.59) can be expressed as

$$\text{ASM} = \frac{1}{N \cdot d} \sum_{i=1}^{N} \left| \mathbf{g}_i - \bar{\mathbf{g}} \right|^2 \tag{6.60}$$

with $\mathbf{g}_i$ and $\bar{\mathbf{g}}$ now as vector representations of the $G_i(k, l)$ and $\bar{G}(k, l)$ respectively.

Let $\mathbf{x}_i$ be a vector that represents the Fourier transform of the ith training image, and $\mathbf{m} = \frac{1}{N} \sum_{i=1}^{N} \mathbf{x}_i$ be the average of the training images. We define the diagonal matrices $\mathbf{M}$ and $\mathbf{X}_i$ with the elements of $\mathbf{m}$ and $\mathbf{x}_i$ along the main diagonal. The Fourier transform of the correlation plane produced in response to the ith training image can be obtained as

$$\mathbf{g} = \mathbf{X}^* \mathbf{h} \tag{6.61}$$

where $\mathbf{h}$ is the filter vector in the frequency domain. Thus, multiplying the filter vector $\mathbf{h}$ by the diagonal matrix $\mathbf{X}^*$ to obtain $\mathbf{g}$ achieves the same result as the equation $G_i(k, l) = X_i^*(k, l) H(k, l)$. Similarly, $\bar{\mathbf{g}}$ is given by

$$\bar{\mathbf{g}} = \mathbf{M}^* \mathbf{h} \tag{6.62}$$

which is equivalent to $\bar{G}(k, l) = M^*(k, l) H(k, l)$, with $M(k, l)$ being the average training image Fourier transform. The expression for the ASM in Eq. (6.60) can now be written using the relations in Eqs. (6.61) and (6.62) as

$$\text{ASM} = \frac{1}{N \cdot d} \sum_{i=1}^{N} \left| \mathbf{X}_i^* \mathbf{h} - \mathbf{M}^* \mathbf{h} \right|^2 = \frac{1}{N \cdot d} \sum_{i=1}^{N} \mathbf{h}^+ (\mathbf{X}_i - \mathbf{M})(\mathbf{X}_i - \mathbf{M})^* \mathbf{h}$$

$$= \mathbf{h}^+ \left[\frac{1}{N \cdot d} \sum_{i=1}^{N} (\mathbf{X}_i - \mathbf{M})(\mathbf{X}_i - \mathbf{M})^* \right] \mathbf{h} = \mathbf{h}^+ \mathbf{S} \mathbf{h} \qquad (6.63)$$

The matrix $\mathbf{S} = 1/N \cdot d \sum_{i=1}^{N} (\mathbf{X}_i - \mathbf{M})(\mathbf{X}_i - \mathbf{M})^*$ is also diagonal, and therefore its "inversion" is trivial.

In addition to being distortion-tolerant, a correlation filter must yield large peak values to facilitate detection of the pattern and to locate its position. Towards this end, we maximize the filter's average response to the training images. However, unlike the SDF filters discussed in Section 6.2.1, no hard constraints are imposed on the filter's response to training images at the origin. Rather, we simply desire that the filter should yield a high peak on average over the entire training set. This condition is met by maximizing the *average correlation height* (ACH) criterion defined as follows:

$$\text{ACH} = \frac{1}{N} \sum_{i=1}^{N} g_i(0,0)$$

$$= \frac{1}{N \cdot d} \sum_{i=1}^{N} \sum_{k}^{d_1} \sum_{l}^{d_2} G_i(k,l) \qquad (6.64)$$

$$= \frac{1}{N \cdot d} \sum_{i=1}^{N} \sum_{k}^{d_1} \sum_{l}^{d_2} X_i^*(k,l) H(k,l)$$

Again, using matrix-vector notation to represent quantities in the frequency domain, Eq. (6.64) can be succinctly written as

$$\text{ACH} = \frac{1}{N} \sum_{i=1}^{N} \mathbf{x}^+ \mathbf{h} = \mathbf{m}^+ \mathbf{h} \qquad (6.65)$$

Finally, it is of course desirable to reduce the effect of noise and clutter on the filter's output by reducing the ONV. To make ACH large while reducing the ASM and ONV, the filter is designed to maximize

$$J(\mathbf{h}) = \frac{|\text{ACH}|^2}{\text{ASM} + \text{ONV}} = \frac{|\mathbf{m}^+ \mathbf{h}|^2}{\mathbf{h}^+ \mathbf{S} \mathbf{h} + \mathbf{h}^+ \mathbf{C} \mathbf{h}} = \frac{\mathbf{h}^+ \mathbf{m} \mathbf{m}^+ \mathbf{h}}{\mathbf{h}^+ (\mathbf{S} + \mathbf{C}) \mathbf{h}} \qquad (6.66)$$

As discussed in Chapter 2, the optimum filter that maximizes the ratio of quadratics (also called the Rayleigh quotient) is the dominant eigenvector of $(\mathbf{S} + \mathbf{C})^{-1} \mathbf{m} \mathbf{m}^+$, or

$$\mathbf{h} = \gamma(\mathbf{S} + \mathbf{C})^{-1}\mathbf{m} \qquad\qquad (6.67)$$

where γ is a normalizing scale factor. The filter in Eq. (6.67) is referred to as the *maximum average correlation height* (MACH) filter [65].

At this point, it is appropriate to make a few comments about the optimality of the MACH filter from a statistical point of view. Based on the central limit theorem (CLT), it can be argued that the correlation filter output can be approximated by a Gaussian random process. The "randomness" arises from the unknown background and clutter, as well as from variations in the target's signature caused by sensor and viewing geometry among other factors. While the "real world" may give rise to more complex cases, correlation over large areas often tends to satisfy the requirements of the CLT and, consequently, the distribution of a correlation plane is usually well approximated by a Gaussian model. The Gaussian approximation requires statistical independence among contributing RVs; and thus our Gaussian assumption becomes more valid for higher space–bandwidth product images.[4] Even so, it should be noted that the behavior of real correlation planes may be non-stationary and that simplifying assumptions are being made for the purpose of an insightful discussion.

As shown in Figure 6.6, the distribution of the correlation peak which is nominally at the origin for centered training images may be considered to be bimodal (with two Gaussian components): one component representing the clutter correlation (including noise and background) and the other the target's response. Assuming that the input noise/clutter is a zero-mean process, the distribution of the clutter correlation is also centered at zero. The variance of the output due to clutter with power spectral density matrix $\mathbf{C}$ is $\mathbf{h}^{+}\mathbf{Ch}$. Similarly, if $\mathbf{m}$ is the mean of the Fourier transform of the target's signature, then the target component of the correlation has a mean of $\mathbf{m}^{+}\mathbf{h}$. Its variance is a measure of the variation in the correlation patterns produced by the target after filtering, and is measured by ASM which is given by $\mathbf{h}^{+}\mathbf{Sh}$. When the mean of the noise and clutter is not zero, both distributions in Figure 6.6 will be shifted by the same amount resulting in no change in their relative separation.

The ideal correlation filter should separate output noise and signal distributions as much as possible to reduce the probability of error. Then, a threshold can be used to easily separate correlation values produced in response to the target from those produced by the background and clutter. Hence the objective for a minimum-probability-of-error scheme should be to find the filter $\mathbf{h}$ that maximizes the separation between the distributions of the correlation

---

[4] The space–bandwidth product is the product of the spatial support of an image and its 2-D bandwidth. A larger space–bandwidth product usually provides an image with more detail.

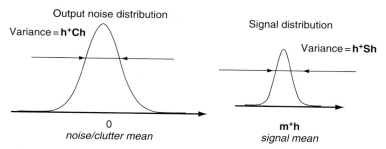

Figure 6.6 Gaussian distribution models for the output of a correlation filter

values produced in response to the background and the target. For Gaussian distributions, this is equivalent to maximizing the separation between the means while minimizing the net variance. Since the distance between the distribution means is simply $A(\mathbf{h}) = |\mathbf{m}^+\mathbf{h} - 0|^2$ and the net variance is $B(\mathbf{h}) = \mathbf{h}^+\mathbf{C}\mathbf{h} + \mathbf{h}^+\mathbf{S}\mathbf{h}$, the objective function to maximize is $J(\mathbf{h}) = (|\mathbf{m}^+\mathbf{h}|^2)/ (\mathbf{h}^+(\mathbf{c}+\mathbf{s})\mathbf{h})$, which is identical to the criterion in Eq. (6.66) maximized by the MACH filter. This interpretation of the MACH filter suggests that it should be statistically optimum and fairly robust for finding targets in clutter, at least when the Gaussian assumption holds [66].

Let us now discuss a strategy for analyzing the output of the MACH filter. We refer to this process as *post-processing*, i.e. the process for detecting correlation peaks and the criteria for making decisions. The amplitude of the correlation peak is often a poor metric because variations in the intensity of the input pattern will result in corresponding variations in the peak value. The correlation peak amplitude may be normalized with respect to the energy of the input pattern. Often however, the patterns of interest are embedded in the background making it difficult to estimate the energy. Another approach is to use the metric optimized by the MACH filter as the post-processing criterion. The statistical interpretation can be extended to show that, for valid targets, the MACH filter maximizes the *peak to sidelobe ratio* (PSR) defined as

$$\text{PSR} = N = \frac{\text{peak} - \mu}{\sigma} \tag{6.68}$$

where μ and σ are the mean and standard deviation of the correlation values in some neighborhood of the peak. The distribution of PSR values across the correlation plane may be treated as normal (Gaussian) with unit standard deviation and zero mean. From the analysis presented in Figure 6.6, the MACH filter separates the distributions of the PSR for the object of interest from that of the clutter. Therefore, the PSR is frequently used as a metric for post processing the output of the MACH filter.

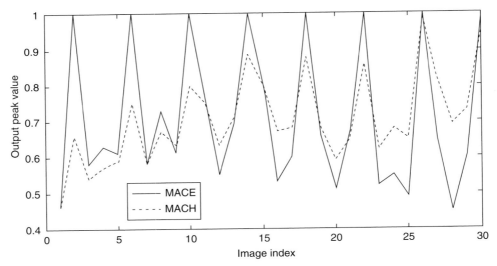

Figure 6.7 Comparison of the distortion tolerance capabilities of the MACH filter and the MACE filter

The improved distortion tolerance of the MACH filter is illustrated in Figure 6.7. A MACH filter and a MACE filter were synthesized using training images selected every 12° over a range of broad-side views between 36° and 120°, similar to those shown in Figure 6.2. The filters were then tested with images selected at 3° intervals over the same range of aspect angles, including both training and non-training views. The resulting correlation values are shown in Figure 6.7 for both cases. While the MACE filter yields an exact output of 1.0 for training images, its performance degrades considerably for the non-training images. On the other hand, the MACH filter maintains a more uniform performance over both training and test images, implying better distortion tolerance. Quantitatively speaking, the standard deviation of the MACE filter peak output over all of the images is $\sigma_{\mathrm{mace}} = 0.19$, whereas the standard deviation for the peak output of the MACH filter is $\sigma_{\mathrm{mach}} = 0.05$. The smaller standard deviation implies greater distortion tolerance. Another performance measurement (or "goodness") criterion is the ratio of the mean and standard deviation of the peak values. Higher values of this ratio indicate greater distortion tolerance. The mean peak value for the MACH filter is $\mu_{\mathrm{mach}} = 0.32$. The MACE filter yields a mean peak value of $\mu_{\mathrm{mace}} = 0.72$. The corresponding ratio for the MACH filter is $\mu_{\mathrm{mach}}/\sigma_{\mathrm{mach}} = 5.93$, whereas for the MACE filter the ratio is $\mu_{\mathrm{mace}}/\sigma_{\mathrm{mace}} = 3.81$, which again supports the contention that the MACH filter exhibits greater distortion tolerance.

6.2.5 Relationship between the MACH filter and SDF filters

As discussed in Section 6.2.1, the SDF filter's performance is controlled by imposing hard constraints on the filter output. In Section 6.2.4, we observed that the methodology for selecting constraint values for designing SDF filters has remained largely unaddressed. The MACH filter avoids this issue by merely requiring a large average peak without specifying exact constraints for every training image. However, an interesting relationship can be established between the MACH filter and SDF filters by showing that the two formulations can be equivalent if the SDF constraints are judiciously chosen to also minimize the performance criterion.

The SDF filter that minimizes the performance criterion

$$J(\mathbf{h}) = \mathbf{h}^+(\mathbf{S} + \mathbf{C})\mathbf{h} \qquad (6.69)$$

subject to the linear constraints $\mathbf{X}^+\mathbf{h} = \mathbf{u}$, is given by:

$$\mathbf{h} = (\mathbf{S} + \mathbf{C})^{-1}\mathbf{X}\left[\mathbf{X}(\mathbf{S} + \mathbf{C})^{-1}\mathbf{X}\right]^{-1}\mathbf{u} \qquad (6.70)$$

Substituting this solution for $\mathbf{h}$ in Eq. (6.69), the expression for the performance criterion becomes

$$J(\mathbf{h}) = \mathbf{u}^+\left[\mathbf{X}^+(\mathbf{S} + \mathbf{C})^{-1}\mathbf{X}\right]^{-1}\mathbf{u} \qquad (6.71)$$

We can now choose the constraint vector $\mathbf{u}$ to minimize $J(\mathbf{h})$ as well. However, in the absence of any other condition, this leads to a trivial solution. To avoid this trivial solution,[5] we simply require that the sum of the elements in $\mathbf{u}$ be non-zero. Let $\mathbf{z}$ be a vector of all 1s, the same length as $\mathbf{u}$. The sum of the elements of $\mathbf{u}$ is then given by $s = \mathbf{u}^+\mathbf{z}$. We may now find $\mathbf{u}$ by minimizing the quadratic term in Eq. (6.71) subject to this linear constraint. It is easy to show that the optimum choice $\mathbf{u}$ is given by

$$\mathbf{u} = \alpha\left[\mathbf{X}^+(\mathbf{S} + \mathbf{C})^{-1}\mathbf{X}\right]\mathbf{z} \qquad (6.72)$$

where α is a scale factor. The exact value of s is unimportant (as long as it is non-zero) since it merely scales the solution and is absorbed into α. Substituting the solution for $\mathbf{u}$ from Eq. (6.72) into Eq. (6.70) yields

---

[5] Other conditions such as maximizing the norm of $\mathbf{u}$ or minimizing the covariance of its elements are possible, and lead to somewhat different solutions.

$$\mathbf{h} = (\mathbf{S} + \mathbf{C})^{-1}\mathbf{X}\left[\mathbf{X}(\mathbf{S} + \mathbf{C})^{-1}\mathbf{X}\right]^{-1}\mathbf{u}$$

$$= \alpha(\mathbf{S} + \mathbf{C})^{-1}\mathbf{X}\left[\mathbf{X}(\mathbf{S} + \mathbf{C})^{-1}\mathbf{X}\right]^{-1}\left[\mathbf{X}(\mathbf{S} + \mathbf{C})^{-1}\mathbf{X}\right]\mathbf{z} \qquad (6.73)$$

$$= \alpha(\mathbf{S} + \mathbf{C})^{-1}\mathbf{X}\mathbf{z}$$

Recognizing that, except for a scale factor, the term $\mathbf{Xz}$ is identical to the average training image $\mathbf{m}$, the SDF expression may be re-written as

$$\mathbf{h} = \gamma \cdot (\mathbf{S} + \mathbf{C})^{-1}\mathbf{m} \qquad (6.74)$$

which is identical to the MACH filter in Eq. (6.67). Thus, the SDF filter is the same as the MACH filter in the special case when the correlation peak constraints (specified in the $\mathbf{u}$ vector) are optimally selected to minimize the combination of ASM and ONV performance criteria.

6.2.6 Optimal tradeoff filters

The prevalent correlation filter design techniques use several performance criteria such as ACE, ONV, and ASM as measures of goodness that relate to different properties of the filter. It is desirable to obtain a filter that achieves a good balance between these multi-criteria objectives. For example, a filter designed to minimize ACE (i.e., the MACE filter) would yield sharp peaks but is likely to have poor noise properties. On the other hand, the MVSDF filter, designed to minimize ONV, is robust to noise but does not yield sharp peaks. An intermediate filter that makes an acceptable compromise between the sharpness of the peak and the noise tolerance criteria might be preferable to either the MACE or MVSDF filter. Such a filter could be obtained by optimizing a weighted sum of the ACE and ONV metrics, and the resulting filter would have the following form:

$$\mathbf{h} = (\alpha\mathbf{D} + \beta\mathbf{C})^{-1}\mathbf{X}\left[\mathbf{X}^{+}(\alpha\mathbf{D} + \beta\mathbf{C})^{-1}\mathbf{X}\right]^{-1}\mathbf{u} \qquad (6.75)$$

The non-negative constants α and β can be chosen to tailor the filter's performance under noisy conditions. However, the question arises as to what is the optimum mix of the terms, and how does one prove the optimality.

The theory of optimal tradeoff (OT) filters [64] has shown that the best compromise between multiple quadratic performance criteria is obtained by optimizing their weighted sum. The weights are selected to make tradeoffs between different criteria. To understand the proof, consider the design of a filter that must satisfy a set of linear constraints and reduce two performance

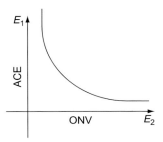

Figure 6.8 Optimal tradeoff between E_1 and E_2

criteria E_1 and E_2 (e.g., ACE and ONV). Since it is not possible to minimize both criteria, it is desirable to minimize E_2 for every possible choice of E_1. This can be done by forming the Lagrangian

$$\Phi(\eta) = E_2 - \eta E_1 - \mathbf{h}^+ \mathbf{X}\ell \tag{6.76}$$

where η is a single Lagrange multiplier that constrains E_1 to a fixed value, and ℓ is the vector of N Lagrange multipliers, corresponding to the linear constraints on the correlation peaks in response to N training images. It is easy to see that minimizing $\Phi(\eta)$ minimizes E_2 when E_1 is fixed to any value. Further, as η varies from 0 (corresponding to where E_2 dominates E_1) to $-\infty$ (where E_1 dominates E_2), E_2 varies from its best (smallest) value to its worst (largest) value. By defining $\eta = (\mu - 1)/(\mu), \mu \in [0, 1]$, the Lagrangian becomes

$$\Phi(\mu) = \mu\Phi(\eta) = \mu E_2 + (1 - \mu)E_1 - \mathbf{h}^+ \mathbf{X}\mathbf{p} \tag{6.77}$$

where $\mathbf{p} = \mu\ell$. Thus, the performance criterion to be minimized is a weighted linear combination of E_1 and E_2. In fact, as μ varies from 0 to 1.0, the emphasis shifts from minimizing E_1 to minimizing E_2.

To further understand the optimal tradeoff between E_1 and E_2, consider Figure 6.8. The solid line is generated by varying μ from 0 to 1.0 and plotting the resulting E_1 versus E_2. For every value of E_1, the most efficient filter is that which yields the smallest value of E_2. All points on the solid line are considered optimum in this sense. The filters represented by points above the line are sub-optimal, since we can find a filter with a smaller value for one of the two metrics and the same value for the other metric. The curve in Figure 6.8 can be used to select an operating point where the greatest reduction is obtained in one criterion for the least increase in the other. Typically, this occurs at the knee of the curve. The parameter μ thus provides a way to optimally trade the properties of the correlation filter to achieve an acceptable compromise between the two performance criteria. This approach is easily generalized to more than two performance criteria as well as other unconstrained correlation filters such as the MACH filter. In fact, the optimal tradeoff variant of the MACH filter has the form

$$\mathbf{h} = (\alpha\mathbf{D} + \beta\mathbf{S} + \gamma\mathbf{C})^{-1}\mathbf{m} \tag{6.78}$$

where α, β, and γ are the optimal tradeoff parameters associated with the ACE, ASM and ONV performance criteria. Once again, each parameter is varied until its best value is found while all others are held constant. This achieves the best balance between the filter's tolerance to noise, sharpness of the correlation peak, and distortion tolerance. Thus, the theory of OT filters provides a key technique for achieving the desired tradeoffs in designing robust correlation filters.

6.2.7 Lock-and-tumbler filters

Another interesting approach to correlation filters is the lock-and-tumbler approach pioneered by Schils and Sweeney [67]. Their initial idea was to design the filter so that any filter array obtained by rotating the original filter array in frequency domain around its center (namely, the (0,0) frequency point) would yield the same inner product with the target image. An input image is placed in the plane P_1, of the optical serial correlator (see Figure 5.5), the above filter is placed in plane P_2 and a correlation output is produced in plane P_3. Next, the filter in plane P_2 is rotated. If the filter is properly designed, the correlation output at the correct location (corresponding to the target's position in the input scene) will remain constant as the filter is rotated. All other points in the correlation plane will fluctuate. Then the target can be located by examining all locations in multiple correlation planes for points of constancy. Schils and Sweeney show that these points of constancy can be located using two buffers, each containing one output array. They have demonstrated the filter's success-ful optical implementation and have shown that it can successfully reject deterministic clutter.

This idea of looking for points of constancy is fundamentally different from seeking points of maximal magnitude. Schils and Sweeney generalized the notion of rotating filters by introducing *lock-and-tumbler* filters [68]. In the lock-and-tumbler approach, they design M filters so that they all give the same output (cross-correlation at the origin) magnitude value when the input image is a desired target. This constant value can be different for different training images. The presence of the training images is indicated by points of constancy in the M correlation planes. It is highly unlikely that an image from some other class will produce points of constancy in the M correlation outputs. The M filters are constructed by using an eigen-analysis of the training set. The training set is analyzed for its eigen-images, and the filters are assumed to be linear combinations of these eigen-images with complex weights. It is desirable to keep these weights of unit magnitude so that all filters have a similar number of eigen-images. These filters will provide complex cross-correlation outputs. Because we want all filters to give the same output magnitude, the phases are unconstrained. A spectral iteration algorithm is set up to iterate between complex weight domain (magnitudes unity, phases variable) and output-constraint domain (magnitudes constant, phases variable). Although there is no theoretical guarantee that this algorithm converges, good simulation results have been demonstrated by Schils and Sweeney [68].

A few other comments about the lock-and-tumbler approach are in order. It is an elegant procedure for filter design. To the best of our knowledge, it is the first attempt to break away from the traditional concept of pattern classification using correlation peaks. However, this approach uses multiple correlations and one needs a systematic way of figuring out how many filters are needed. The multiple filters in the approach are obtained by using different random starting points in the spectral iteration method, and thus are not necessarily orthogonal (i.e., they do not necessarily bring in completely new information). The computational complexity associated with the processing of multiple-correlation outputs in looking for points of constancy can be overwhelming.

6.3 Distance classifier correlation filters

The correlation filters described in the previous sections are linear systems whose response to patterns of interest is carefully controlled by the various optimization techniques. In this section, we present and use a somewhat different interpretation of correlation filters – as a means of applying transformations to the input data. It is well known that the correlation (or convolution) operation can be viewed as a linear transformation. Specifically, the filtering process can be mathematically expressed in the space domain as a transformation of the input data vector by a matrix. To see this, consider the correlation of two discrete sequences $x(n)$ and $h(n)$, each of length L. As we know, the result is a sequence of length $2L - 1$ given by:

$$g(n) = x(n) \otimes h(n) = \sum_{k=1}^{L} x(n+k)h(k), \quad -L+1 \leq n \leq L-1 \qquad (6.79)$$

The input sequence is $\mathbf{x} = [x(1), x(2), \ldots, x(L)]^{\mathrm{T}}$ and the correlation output sequence is $\mathbf{g} = [g(-L+1), g(-L+2), \ldots, g(0), \ldots, g(L-2), g(L-1)]^{\mathrm{T}}$. Then, the operation in Eq. (6.79) can be written in matrix–vector notation as

$$
\begin{bmatrix} g(-L+1) \\ g(-L+2) \\ \vdots \\ g(0) \\ \vdots \\ g(L-2) \\ g(L-1) \end{bmatrix} = \begin{bmatrix} h(L) & & & & & \\ h(L-1) & h(L) & & 0 & & \\ \vdots & \vdots & \ddots & & & \\ h(1) & h(2) & \cdots & h(L-1) & h(L) & \\ & & & \ddots & & \vdots \\ & 0 & & h(1) & h(2) & \\ & & & & h(1) & \end{bmatrix} \begin{bmatrix} x(1) \\ x(2) \\ \vdots \\ \vdots \\ x(L-1) \\ x(L) \end{bmatrix}
$$

$$
(6.80)
$$

or

$$
\mathbf{g} = \mathbf{H}\mathbf{x} \qquad (6.81)
$$

where $\mathbf{H}$ is a matrix whose rows are shifted versions of the sequence $h(n)$ as shown in Eq. (6.80). Thus the structure of $\mathbf{H}$ is such that its multiplication by the input data vector $\mathbf{x}$ implements the shift, multiply, and add operations of the correlation (or convolution) equation to yield the output vector $\mathbf{g}$. Matrix $\mathbf{H}$ is a *Toeplitz* matrix since all elements along any of its diagonals have the same value. The 2-D correlation equation can be similarly formulated as a matrix-vector equation. The input and output vectors are obtained by lexico-graphic re-ordering of the rows or columns, while the corresponding matrix $\mathbf{H}$ is *block-Toeplitz* (i.e., blocks of the matrix are arranged in the Toeplitz structure and each block is itself a Toeplitz matrix). Thus, the correlation operation (or filtering) can be interpreted as a linear tranformation of the data where the transform matrix is restricted to being Toeplitz.

The equivalent frequency domain relation is easily derived by defining $\tilde{\mathbf{g}} = [G(1), \ldots, G(L-1), G(L)]^{\mathrm{T}}$, and $\tilde{\mathbf{x}} = [X(1), X(2), \ldots, X(L)]^{\mathrm{T}}$, and

$$
\tilde{\mathbf{H}} = \begin{bmatrix} H(1) & & & & \\ & H(2) & & 0 & \\ & & \ddots & & \\ & 0 & & H(L-1) & \\ & & & & H(L) \end{bmatrix} \qquad (6.82)
$$

where the tilde is used to denote that the vectors and matrices refer to frequency domain quantities, and where $G(k)$, $X(k)$, $H(k)$ are the discrete Fourier transforms of $g(n)$, $x(n)$, and $h(n)$, respectively. Then the equation $\tilde{\mathbf{g}} = \tilde{\mathbf{H}}^{*}\tilde{\mathbf{x}}$ is equivalent to $G(k) = H^{*}(k)X(k)$. For the 2-D case, $\tilde{\mathbf{g}}$ and $\tilde{\mathbf{x}}$ are lexicographically re-ordered versions of the 2-D Fourier transforms $G(k, l)$ and $X(k, l)$, while the diagonal elements of $\tilde{H}$ are the elements of $H(k, l)$. Thus, the frequency domain interpretation of the correlation process is a linear transformation of the data vector by a diagonal matrix. We will now use the transform interpretation of correlation to develop a distance-based classifier [69].

Transform-based distance classifiers are well known in the pattern recognition literature. From now on, we will drop the tilde as we will be using only frequency domain quantities. Generally speaking, the distance of a test vector $\mathbf{x}$ to a reference $\mathbf{m}_k$ under a linear transform $\mathbf{H}$ is given by

$$d_k = |\mathbf{H}^*\mathbf{x} - \mathbf{H}^*\mathbf{m}_k|^2 = (\mathbf{x} - \mathbf{m}_k)^+\mathbf{H}\mathbf{H}^*(\mathbf{x} - \mathbf{m}_k) \qquad (6.83)$$

For instance, using the *Mahalanobis distance* applies a whitening transform (equal to the inverse square-root of the class covariance matrix) to normalize the feature space. Then, given an unknown input, its distance to all the class centers in the transformed space is computed. The input is assigned to the class to which the distance from the target is the smallest. For Gaussian distributions, this scheme yields the minimum probability of error.

However, the Mahalanobis distance and other transform domain techniques found in statistical pattern recognition are difficult to apply to images since they often require feature extraction or dimensionality reduction, image segmentation, and registration. The size of common images is usually so large that we cannot accurately estimate the required statistics. The needed inversion of large matrices is also impractical in most cases.

In this section, we concern ourselves with finding an expression for $\mathbf{H}$ suitable for processing images by assuming that the transform will be implemented as a linear system. This interpretation immediately affords us several benefits. Firstly, it imposes the Toeplitz structure on the transform matrix in the space domain and equivalently (but more importantly), it imposes the diagonal structure in the frequency domain. This alleviates matrix inversion problems. Secondly, the inherent shift invariance property eliminates the need for segmentation and image registration. Yet, the power of transform domain distance calculations can be obtained by suitably optimizing the properties of the linear transform matrix, in a similar manner to those used for designing correlation filters.

In other words, the filtering process transforms the input image to create new images. For the filter to be useful as a transform, we require that the images of the different classes become as different as possible after filtering, and that images from the same class be as similar as possible after the transform. Then, distances can be computed between the filtered input image and the references of the different classes that have also been transformed in the same manner. The input is assigned to the class to which the distance in the transformed domain is the smallest. One benefit of treating the filter as a transform for distance computations is that the resulting decision boundaries are quadratic (unlike conventional filters which produce linear decision boundaries).[6] Quadratic decision boundaries allow more "pickiness" in selecting portions of feature space to assign to the various classes. Secondly, the emphasis is shifted from just one point in the correlation output (i.e., the correlation peak) to comparison of the entire shape of the correlation plane. These facts along with the simplifying properties of linear systems lead to an interesting realization of a distance classifier with a correlation filter twist.

6.3.1 Designing the classifier transform

In this section, we concern ourselves with finding a solution for the transform matrix $\mathbf{H}$ in Eq. (6.82) using the properties of linear systems and correlation filter design techniques. It is assumed that the training images are segmented and appropriately centered (although test images are expected to contain background and not necessarily to be centered). An image $x(m, n)$ with d pixels can be expressed as a d-dimensional column vector $\mathbf{x}$, or as a $d \times d$ diagonal matrix $\mathbf{x}$ with the elements of $\mathbf{x}$ as its diagonal elements, i.e., diagonal $\{\mathbf{X}\} = \mathbf{x}$. Sometimes, the same quantity may be expressed both as a vector, say $\mathbf{m}_x$, and as a diagonal matrix $\mathbf{M}_x$. This implies that $\mathbf{H}^*\mathbf{m}_x$ and $\mathbf{M}_x^*\mathbf{h}$ are equivalent.

All analysis presented in this section is carried out in the frequency domain. As noted at the beginning of this section, the distance classifier uses a global transform $\mathbf{H}$ to separate the classes maximally while making them as compact as possible. In general, image vectors of length d can be considered as points in a d-dimensional hyperspace. For simplicity of illustration, consider signals of length $d = 2$ (i.e., data vectors with only two elements). In this notional signal space, Figure 6.9 depicts schematically the basic idea using a three-class example, where $\mathbf{m}_1$, $\mathbf{m}_2$, and $\mathbf{m}_3$ represent the class centers (obtained by averaging the Fourier transforms of the corresponding training images), and $\mathbf{z}$ represents an unknown input to be classified. The transformation matrix $\mathbf{H}$ is designed to make

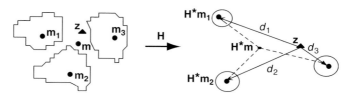

Figure 6.9 Transformation by $\mathbf{H}$ increases inter-class distance while making each class more compact, to improve both distortion tolerance and discrimination simultaneously

---

[6] Decision boundary refers to the boundary that separates different classes in a feature space. When these boundaries are hyperplanes, they are called linear decision boundaries.

the classes distinct by moving the class centers apart, while shrinking the boundaries around each class so that **z** can be more accurately identified with its correct class (class 3 in the figure, since d_3 is the smallest distance).

Although Figure 6.9 represents the concept in the $d=2$ hyperspace, the same arguments can be used for arbitrary hyperspaces. The general C-class distance classifier problem is formulated by stating that we require the transformed images to be as different as possible for each of the classes. At the same time, the classes should become as *compact* as possible under the transformation. Let $\mathbf{x}_{ik}$ be the d-dimensional column vector (which represents a point in a d-dimensional hyperspace) containing the Fourier transform of the ith image of the kth class, $1 \leq i \leq N$, and $1 \leq k \leq C$. We assume without loss of generality that each class has N training images. Let $\mathbf{m}_k$ be the mean FT of class k so that

$$\mathbf{m}_k = \frac{1}{N} \sum_{i=1}^{N} \mathbf{x}_{ik}, \quad 1 \leq k \leq C \tag{6.84}$$

Under the transform **H**, the difference between the means of any two classes is

$$\mathbf{v}_{kl} = \mathbf{H}^*(\mathbf{m}_k - \mathbf{m}_l) \tag{6.85}$$

Taking the expectation of the elements of $\mathbf{v}_k$ over all frequencies yields

$$\bar{\upsilon}_{kl} = \underset{\text{over } i}{E} \{\mathbf{v}_k(i)\} \cong \frac{1}{d}\mathbf{h}^+(\mathbf{m}_k - \mathbf{m}_l) \tag{6.86}$$

The quantity in Eq. (6.86) is a measure of the *spectral separation* between classes k and l, over all frequencies. The sign of $\bar{\upsilon}_{kl}$ is not important for the classes to be separated, but it should be as large as possible. Therefore, we want to make $|\bar{\upsilon}_{kl}|^2$ large. Taking all possible pairs of classes into consideration, we define the *average spectral separation* (ASS) criterion as

$$A(\mathbf{h}) = \frac{1}{C^2} \sum_{k=1}^{C} \sum_{l=1}^{C} |\bar{\upsilon}_{kl}|^2 = \frac{1}{C^2} \sum_{k=1}^{C} \sum_{l=1}^{C} |\mathbf{m}_k^+\mathbf{h} - \mathbf{m}_l^+\mathbf{h}|^2$$

$$= \frac{1}{C^2} \sum_{k=1}^{C} \sum_{l=1}^{C} \mathbf{h}^+(\mathbf{m}_l - \mathbf{m}_k)(\mathbf{m}_l - \mathbf{m}_k)^+\mathbf{h} \tag{6.87}$$

It should be noted that the terms corresponding to $l=k$ do not contribute to the sum. After algebraic manipulations, the expression for $A(\mathbf{h})$ can be simplified to

$$A(\mathbf{h}) = \mathbf{h}^+ \left[\frac{1}{C} \sum_{k=1}^{C} (\mathbf{m} - \mathbf{m}_k)(\mathbf{m} - \mathbf{m}_k)^+ \right] \mathbf{h} = \mathbf{h}^+\mathbf{T}\mathbf{h} \tag{6.88}$$

where $\mathbf{T} = \frac{1}{C}\sum_{k=1}^{C}(\mathbf{m} - \mathbf{m}_k)(\mathbf{m} - \mathbf{m}_k)^+$ is a $d \times d$ full (i.e., non-diagonal) matrix of rank $\leq (C-1)$, and $\mathbf{m} = \frac{1}{C}\sum_{k=1}^{C}\mathbf{m}_k$ is the mean of the entire data set. If $A(\mathbf{h})$ in Eq. (6.88) is maximized by appropriate choice of $\mathbf{h}$, the average spectral content of the classes will differ greatly and they will become well separated. At the same time, to improve distortion tolerance within a class, we want to minimize the criterion for compactness given by

$$B(\mathbf{h}) = \frac{1}{C}\sum_{k=1}^{C}\frac{1}{N}\sum_{i=1}^{N}\mathbf{h}^+[\mathbf{X}_{ik} - \mathbf{M}_k][\mathbf{X}_{ik} - \mathbf{M}_k]^*\mathbf{h} = \mathbf{h}^+\mathbf{Sh} \qquad (6.89)$$

We recognize that the term $\mathbf{h}^+\mathbf{Sh}$ is the same as ASM calculated over all classes, and therefore $B(\mathbf{h})$ is a measure of average class compactness after transformation by $\mathbf{H}$. Our objectives of maximizing $A(\mathbf{h})$ and minimizing $B(\mathbf{h})$ are met by maximizing the ratio of $A(\mathbf{h})$ and $B(\mathbf{h})$, i.e., we maximize

$$J(\mathbf{h}) = \frac{A(\mathbf{h})}{B(\mathbf{h})} = \frac{\mathbf{h}^+\mathbf{Th}}{\mathbf{h}^+\mathbf{Sh}} \qquad (6.90)$$

with respect to $\mathbf{h}$. As shown in Chapter 2, the optimum solution to Eq. (6.90) is the dominant eigenvector of $\mathbf{S}^{-1}\mathbf{T}$. We refer to the optimum $\mathbf{h}$ as the *distance classifier* correlation filter (DCCF).

6.3.2 Calculating distances with DCCFs

For testing purposes, the distance to be computed between the transformed input and the ideal reference for class k is

$$d_k = |\mathbf{H}^*\mathbf{z} - \mathbf{H}^*\mathbf{m}_k|^2 = p + b_k - (\mathbf{z}^+\mathbf{h}_k + \mathbf{h}_k^+\mathbf{z}), \qquad 1 \leq k \leq C \qquad (6.91)$$

where $\mathbf{z}$ is the input image, $p = |\mathbf{H}^*\mathbf{z}|^2$ is the transformed input image energy, $b_k = |\mathbf{H}^*\mathbf{m}_k|^2$ is the energy of the transformed kth class mean, and $\mathbf{h}_k = \mathbf{HH}^*\mathbf{m}_k$ is viewed as the *effective* filter for class k. For images that are real in the space domain, the expression for d_k simplifies to

$$d_k = |\mathbf{H}^*\mathbf{z} - \mathbf{H}^*\mathbf{m}_k|^2 = p + b_k - 2\mathbf{z}^+\mathbf{h}_k \qquad (6.92)$$

In general, the target may be anywhere in the input image. For shift-invariant distance calculation, we are interested in the smallest value of d_k over all possible shifts of the target with respect to the class references (i.e., the best possible match between the input and the reference for class k). In Eq. (6.92), since p and b_k are both positive and independent of the position of the target, the smallest value of d_k over all shifts is obtained when the third term (i.e., $\mathbf{z}^+\mathbf{h}_k$) is as large as possible. Therefore, this term is chosen as the *peak value* in the full space domain cross-correlation of $\mathbf{z}$ and $\mathbf{h}_k$. Since there are only C classes to which distances must be computed, we require only C such filters. It should be noted that for a given transform $\mathbf{H}$, all d_k, $1 \leq k \leq C$, have the same term p, which could be dropped if the only objective was to find the class to which the distance is the smallest, but this would yield only linear decision boundaries.

6.4 Polynomial correlation filters

In this section, we will discuss a generalization of the traditional correlation filtering approach known as the polynomial correlation filter (PCF) approach [70]. The fundamental difference between the PCF approach and traditional methods for correlation filters is that the correlation output from a PCF is a nonlinear function of the input. For a scalar input x, consider an output form given by

$$g(x) = a_o + a_1 x + a_2 x^2 + \cdots + a_N x^N \tag{6.93}$$

The polynomial $g(x)$ is a general form that can be used to represent any nonlinear function of x. For vector inputs (such as images), the most general nonlinear expression is more complicated. However, for reasons of analytical simplicity we assume point nonlinearities. The corresponding form of the output is then given by

$$\mathbf{g_x} = \mathbf{A}_1 \mathbf{x}^1 + \mathbf{A}_2 \mathbf{x}^2 + \cdots + \mathbf{A}_N \mathbf{x}^N \tag{6.94}$$

where $\mathbf{x}^i$ represents the vector $\mathbf{x}$ with each of its elements raised to the power i, and $\mathbf{A}_i$ is a matrix of coefficients associated with the ith term of the polynomial. It should be noted that the output $\mathbf{g_x}$ is also a vector.

We refer to the form in Eq. (6.94) as the PCF. Thus if $\mathbf{x}$ represents the input image in vector notation, then $\mathbf{g_x}$ is a vector which represents the output correlation plane as a polynomial function of $\mathbf{x}$. To ensure that the output is *shift-invariant*, all the coefficient matrices are required to be Toeplitz. Then, it can be shown that each term in the polynomial can be computed as a linear shift-invariant filtering operation, i.e.,

$$\mathbf{A}_i \mathbf{x}^i \equiv h_i(m, n) \otimes x^i(m, n) \tag{6.95}$$

or that filtering $x^i(m, n)$ by $h_i(m, n)$ is equivalent to multiplying $\mathbf{x}^i$ by $\mathbf{A}_i$. The output of the polynomial correlation filter can be mathematically expressed as

$$g_x(m, n) = \sum_{i=1}^{N} h_i(m, n) \otimes x^i(m, n) \tag{6.96}$$

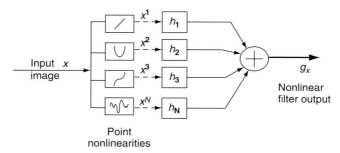

Figure 6.10 Nth-order polynomial correlation filter

The corresponding filter structure is shown in Figure 6.10.

6.4.1 *Derivation of the solution*

The objective is to find the filters $h_i(m, n)$ such that structure shown in Figure 6.10 optimizes a performance criterion of interest. We have shown earlier that, for correlation purposes, a useful approach is to maximize the OT performance criterion

$$J(\mathbf{h}) = \frac{|\mathbf{m}^+\mathbf{h}|^2}{\mathbf{h}^+\mathbf{B}\mathbf{h}} \tag{6.97}$$

where $\mathbf{h}$ is the filter vector in the frequency domain, $\mathbf{B}$ is a diagonal matrix related to a spectral quantity, and $\mathbf{m}$ is the mean image vector, also in the frequency domain. For example, MACH filter design involves maximizing the metric in Eq. (6.66). The polynomial correlation filter can also be designed in a similar way. Of course, the premise is that a higher-order (nonlinear) solution will yield higher values of $J(\mathbf{h})$ than the corresponding linear solutions.

For simplicity, we will firstly discuss the derivation of a second-order filter. In this case, the polynomial has only two terms and the output is given by

$$g(m, n) = x(m, n) \otimes h_1(m, n) + x^2(m, n) \otimes h_2(m, n) \tag{6.98}$$

The expression for $J(\mathbf{h})$ is obtained by deriving the numerator and the denominator of Eq. (6.97). In vector notation, the average intensity of the correlation peak for a second-order filter is given as follows:

$$|\text{average peak}|^2 = \left|\mathbf{h}_1^+\mathbf{m}^1\right|^2 + \left|\mathbf{h}_2^+\mathbf{m}^2\right|^2 + 2\mathbf{h}_1^+\mathbf{m}^1\mathbf{m}^{2+}\mathbf{h}_2 \tag{6.99}$$

where $\mathbf{h}_1$ and $\mathbf{h}_2$ are vector versions of the filters associated with the first and second terms of the polynomial, and $\mathbf{m}^k$ is the mean of the training images $\mathbf{x}_i$, $1 \le i \le L$, raised to the kth power. For illustration purposes, the denominator of the performance criterion in Eq. (6.97) is chosen to be the ASM metric, but it could easily be any other quadratic metric, such as the ONV, the ACE, or any combination thereof. The ASM for the second-order nonlinear filter is given by

$$ \text{ASM} = \frac{1}{L} \sum_{i=1}^{L} \left| \mathbf{h}_1^* \mathbf{X}_i^1 + \mathbf{h}_2^* \mathbf{X}_i^2 - \mathbf{h}_1^* \mathbf{M}^1 - \mathbf{h} \mathbf{M}^2 \right|^2 \tag{6.100} $$

where $\mathbf{X}_i^k$, $1 \le i \le L$, is the ith training image raised to the kth power expressed as a diagonal matrix, and $\mathbf{M}^k$ is their average. After algebraic manipulations, it can be shown that the expression for ASM is as follows:

$$ \text{ASM} = \mathbf{h}_1^+ \mathbf{S}_{11} \mathbf{h}_1 + \mathbf{h}_2^+ \mathbf{S}_{22} \mathbf{h}_2 + \mathbf{h}_1^+ \mathbf{S}_{12} \mathbf{h}_2 + \mathbf{h}_2^+ \mathbf{S}_{21} \mathbf{h}_1 \tag{6.101} $$

where

$$ \mathbf{S}_{kl} = \frac{1}{L} \sum_{i=1}^{L} \mathbf{X}_i^k \left(\mathbf{X}_i^l \right)^* - \mathbf{M}^k \left(\mathbf{M}^l \right)^*, \quad 1 \le k, l \le 2 \tag{6.102} $$

are diagonal matrices. Defining the block vectors and matrices,

$$ \mathbf{h} = \begin{bmatrix} \mathbf{h}_1 \\ \mathbf{h}_2 \end{bmatrix}, \quad \mathbf{m} = \begin{bmatrix} \mathbf{m}^1 \\ \mathbf{m}^2 \end{bmatrix}, \text{ and } \mathbf{S} = \begin{bmatrix} \mathbf{S}_{11} & \mathbf{S}_{12} \\ \mathbf{S}_{21} & \mathbf{S}_{22} \end{bmatrix} \tag{6.103} $$

the expression for $J(\mathbf{h})$ for the second-order filter can be succinctly expressed as

$$ J(\mathbf{h}) = \frac{|\text{average peak}|^2}{\text{ASM}} = \frac{\left| \mathbf{h}_1^+ \mathbf{m}^1 \right|^2 + \left| \mathbf{h}_2^+ \mathbf{m}^2 \right|^2 + 2\mathbf{h}_1^+ \mathbf{m}^1 \mathbf{m}^{2+} \mathbf{h}_2}{\mathbf{h}_1^+ \mathbf{S}_{11} \mathbf{h}_1 + \mathbf{h}_2^+ \mathbf{S}_{22} \mathbf{h}_2 + \mathbf{h}_1^+ \mathbf{S}_{12} \mathbf{h}_2 + \mathbf{h}_2^+ \mathbf{S}_{21} \mathbf{h}_1} = \frac{|\mathbf{m}^+ \mathbf{h}|^2}{\mathbf{h}^+ \mathbf{S} \mathbf{h}} \tag{6.104} $$

By now, we know that the solution that maximizes $J(\mathbf{h})$ in Eq. (6.104) is given by:

$$ \mathbf{h} = \mathbf{S}^{-1} \mathbf{m} \tag{6.105} $$

Using the definitions in Eq. (6.103), the solutions for the two filters of the second-order polynomial are

$$ \begin{bmatrix} \mathbf{h}_1 \\ \mathbf{h}_2 \end{bmatrix} = \begin{bmatrix} \mathbf{S}_{11} & \mathbf{S}_{12} \\ \mathbf{S}_{21} & \mathbf{S}_{11} \end{bmatrix}^{-1} \begin{bmatrix} \mathbf{m}^1 \\ \mathbf{m}^2 \end{bmatrix} \tag{6.106} $$

The inverse of the block matrix can be determined using matrix inversion lemmas (discussed in Chapter 2) to obtain the following explicit solutions for the two filter vectors:

$$
\begin{bmatrix} \mathbf{h}_1 \\ \mathbf{h}_2 \end{bmatrix} = \begin{bmatrix} \dfrac{\mathbf{S}_{12}\mathbf{m}^2 - \mathbf{S}_{22}\mathbf{m}^1}{|\mathbf{S}_{12}|^2 - \mathbf{S}_{11}\mathbf{S}_{22}} \\ \dfrac{\mathbf{S}_{21}\mathbf{m}^1 - \mathbf{S}_{11}\mathbf{m}^2}{|\mathbf{S}_{12}|^2 - \mathbf{S}_{11}\mathbf{S}_{22}} \end{bmatrix} \tag{6.107}
$$

The solution in Eq. (6.107) can be easily extended to the general Nth order case. The Nth order solution is given by:

$$
\begin{bmatrix} \mathbf{h}_1 \\ \mathbf{h}_2 \\ \vdots \\ \mathbf{h}_N \end{bmatrix} = \begin{bmatrix} \mathbf{S}_{11} & \mathbf{S}_{12} & \cdots & \mathbf{S}_{1N} \\ \mathbf{S}_{21} & \mathbf{S}_{22} & \cdots & \mathbf{S}_{2N} \\ \vdots & \vdots & \ddots & \vdots \\ \mathbf{S}_{N1} & \mathbf{S}_{N2} & \cdots & \mathbf{S}_{NN} \end{bmatrix}^{-1} \begin{bmatrix} \mathbf{m}^1 \\ \mathbf{m}^2 \\ \vdots \\ \mathbf{m}^N \end{bmatrix} \tag{6.108}
$$

The block matrix to be inverted in Eq. (6.108) can be large depending on the size of the images. However, because all $\mathbf{S}_{kl}$ are diagonal, and $\mathbf{S}_{kl} = (\mathbf{S}_{lk})^*$, the inverse can be efficiently computed using a recursive formula for inverting block matrices.

6.4.2 PCF extensions

While the power series representation of the polynomial correlation filter is initially used for deriving the solution in Eq. (6.108), there is no fundamental reason to limit ourselves to power nonlinearities. The analysis and the form of the solution remain the same irrespective of the nonlinearities used. Thus, we can use the more general form:

$$
\mathbf{g}_N = \sum_{i=1}^{N} \mathbf{A}_i f_i(\mathbf{x}) \tag{6.109}
$$

where $f(\cdot)$ is *any nonlinear function* of $\mathbf{x}$. For instance, possible choices for the nonlinearities include absolute magnitude and sigmoid functions. Clearly, guidelines must be established for choosing the most beneficial nonlinearities for specific applications. As a simple example, it may be detrimental to use logarithms when bipolar noise is present, since the log of a negative number is not defined.

The proposed algorithm can be used to correlate data from different sensors simultaneously. In this case, we may view the sensor imaging process and its transfer function itself as the nonlinear mapping function. The different terms of the polynomial do not have to be from the same sensor or versions of the same data. This is equivalent to data fusion in the algorithm. The concept is illustrated in Figure 6.11 where it is shown that data from different sensors

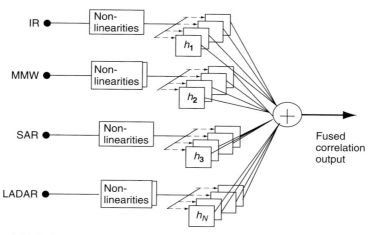

Figure 6.11 Polynomial correlation architecture for multi-sensor fusion

may be directly injected into the architecture, resulting in a fused correlation output. Again, the analysis and the form of the solution remain the same as those in Eq. (6.108).

The concept shown in Figure 6.11 can also be extended to multi-resolution inputs. In other words, the terms of the polynomial can be the data represented at different resolution levels. It is anticipated that this will result in a mechanism to achieve correlation in wavelet type transform domains.

6.5 Basic performance prediction techniques

There are many methods for designing correlation filters. To a certain extent, the choice depends on the application. Ultimately, what matters most is the performance of the filter in terms of the probability of correctly recognizing the desired pattern, and the probability of rejecting clutter.

The performance of correlation filters can be characterized, as in detection systems, in terms of the probabilities of correct detection (P_d), and false alarm (P_{fa}). Fundamentally, low detection thresholds improve the probability of correct recognition, while higher thresholds decrease false alarm probabilities by rejecting erroneous peaks. As discussed in Chapter 4, such a relationship between P_d and P_{fa} as the threshold T varies is represented in the form of a receiver operating characteristic (ROC) curve. These curves are useful for selecting a desired operating point, and for predicting whether a filter will achieve the desired performance level. In this section, we will provide a simple illustration for determining ROC curves for correlation filters.

To understand the process of obtaining a ROC curve, consider a set of K images, $x_1(m, n)$, $x_2(m, n)$, $\ldots$, $x_K(m, n)$ of a target. We refer to this as the set of *evaluation images* that in general may contain either test or training images. Let $h(m, n)$ be the impulse response of the filter that produces the correlation $g_i(m, n)$ in response to the ith target image $x_i(m, n)$. Assume that the input is corrupted by zero-mean additive white Gaussian noise with variance σ^2. If $x_i(m, n)$ is power-normalized (i.e., $\Sigma_i \Sigma_j |x(m,n)|^2 = 1$), the input SNR is given by

$$\text{SNR} \cong \frac{\text{signal energy/pixel}}{\text{noise variance}} = \frac{\displaystyle\sum_i \sum_j |x(m,n)|^2}{d\sigma^2} = \frac{1}{d\sigma^2} \tag{6.110}$$

where d is the total number of pixels on the target in the image. Equivalently, the input variance can be expressed in terms of the SNR as

$$\sigma = \sqrt{\frac{1}{d \cdot \text{SNR}}} \tag{6.111}$$

When the noisy input image is processed by the filter $h(m, n)$, a noisy output correlation is obtained. Using the central limit theorem arguments made earlier, the output correlation is assumed to be Gaussian-distributed. Since the input noise is zero-mean, the mean of the output is $g_i(m, n)$. It is easy to show that if the filter is also power-normalized (i.e., $\mathbf{h}^+\mathbf{h} = 1$), the output noise variance is equal to the input noise variance, σ^2. Therefore, the filter output is Gaussian distributed with mean $g_i(m, n)$ and variance σ^2. Without loss of generality, we shall assume that the evaluation images are nominally registered with respect to the training images used to design the filter, and the peak should occur at the origin. If the evaluation images are not centered at the origin, we simply move the detection window to the new location. The post-processing strategy is to declare a detection and accept the correlation peak as indicating the presence of the desired target if that peak exceeds a threshold T. The decision is said to be correct if the peak occurs within a *detection window* W around the origin. Conversely, the decision is an error if the peak is outside the detection window.

Strictly speaking, the power spectral density of the correlation output is proportional to $|H(k, l)|^2$ when the input noise is white. Thus, the output correlation pixels are not uncorrelated. However, for the sake of simplicity we assume that the points in the correlation plane are independent, so that the joint distribution is a product of the individual distributions. This approximation is more valid for high-frequency emphasizing filters such as MACE and MACH filters than for matched filters. As stated earlier, each point in the output is normally distributed with mean $g_i(m, n)$ and variance σ^2. The

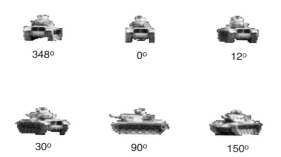

Figure 6.12 Tank images used for performance prediction example

probability of correct detection (that at least one point within W exceeds the threshold T) is then given as

$$P_{\mathrm{d}} = 1 - \prod_{\substack{m,n \\ \in \omega}} \left[\int_{-\infty}^{T} \frac{1}{\sqrt{2\pi\sigma^2}} \exp\left[\frac{-(\upsilon - g_i(m,n))^2}{2\sigma^2}\right] d\upsilon \right] \qquad (6.112)$$

Similarly, the probability of false alarm (that any point outside the detection window will exceed the threshold) is given by

$$P_{\mathrm{fa}} = 1 - \prod_{\substack{m,n \\ \notin \omega}} \left[\int_{-\infty}^{T} \frac{1}{\sqrt{2\pi\sigma^2}} \exp\left[\frac{-(\upsilon - g_i(m,n))^2}{2\sigma^2}\right] d\upsilon \right] \qquad (6.113)$$

The steps for computing the ROC curves for the ith evaluation image are:

1. For a specified SNR, obtain the corresponding σ^2 using Eq. (6.111).
2. Correlate the image with the filter and obtain $g_i(m, n)$.
3. Vary T over a range of values and calculate P_{d} and P_{fa} using Eqs. (6.112) and (6.113).

Consider the tank images shown in Figure 6.12. We will use these images to compare the predicted performance of a MACH filter and an MF. The MACH filter was synthesized using every other broad-side image between $30°$ and $150°$ (the data base has images separated by $3°$ in aspect). An MF was designed for the image at $90°$ only. While this may sound unfair, the precise definition of an MF is that it is matched to a single view. For each filter, ROC curves were generated for all images in the evaluation set and the results were averaged. The average ROC curves were computed for a square region, W of size ± 1, 2, and 3 pixels on each side of the expected correlation peak position.

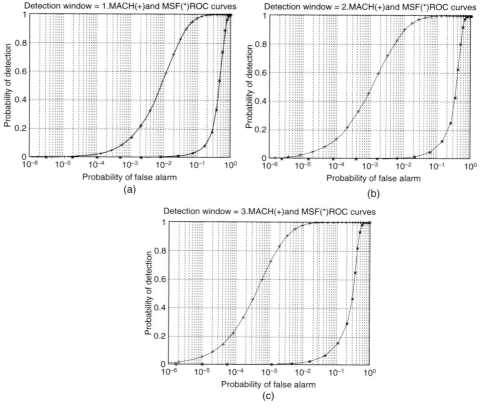

Figure 6.13 Average ROC curves for MACH $(+)$ and MSF $(*)$ filters for broad-side views (see Figure 6.12) using detection windows of: (a) $W = 1$, (b) $W = 2$, and (c) $W = 3$

The results are shown in Figure 6.13. The MACH filter offers a lower P_{fa} for any given P_d than that obtained with the MF. The ROC comparisons between the MACH and MF filters are fair in the sense that we are comparing the performance of a *single* correlation filter in each case, against the same set of input images.

The *power of the detector* (defined as the area under the ROC curve) offers a quantitative means of comparing the ROC curves. The closer the number is to 1.0, the more efficient the test and the better the filter. Table 6.1 shows the powers of the tests for the filters being compared. Like the ROC curves, these values attest to the significantly better performance of the MACH filter over the range of distorted broad-side tank images. The numbers are consistently higher for the MACH filter indicating that it outperforms the MF when distortions are taken into account.

Table 6.1. *Power of the detector values for the
MF and MACH filters*

Detection window	MF	MACH
1	0.5518	0.9815
2	0.6525	0.9960
3	0.7022	0.9986

Clearly ROC curves may be used to predict the performance of correlation filters. However, because of the simplifying assumptions it is possible that the observed values of P_d and P_{fa} may differ from the predicted values. Nevertheless, the curves provide useful estimates that can be used to gauge the chances of success. In any event, the ROC curves provide a method for comparing different correlators under the same set of conditions.

6.6 Advanced pattern recognition criteria

In this section we mention two ways in which correlation pattern recognition could be improved. One is by addressing the secondary peak problem, and the other discusses the ad hoc nature of most of the filters and the metrics they optimize. We describe these situations and their current status.

Among other problems with correlation pattern recognition, we do not have any guarantee that the presence of exactly the object the filter is set to detect will actually result in the largest correlation plane's value being at the designated center for that object. This fact is principally due to the effects of noise and clutter at the input plane as they propagate through the filter and onto the correlation plane, however there is a contribution resulting from the very nature of the reference object and its filter. Most objects that would be the subject of CPR have a degree of self-similarity; take an image of a man's shirt as an example. The button appears in several locations, and a filter that will respond to the shirt, including its buttons, will have some response to *each* of the buttons as translated to the position of any of the other buttons. This is an ineluctable consequence of the correlation algorithm, as contrasted with other pattern recognition algorithms that might search for all button-like objects and then evaluate the input scene as being shirt-like only if the detected button-like objects were in an expected orientation relative to each other. CPR produces a deterministic portion of the correlation surface's nature. A supposedly perfect correlation surface, looking like a point-up thumbtack, would only result from an object with no self-similarity and an inverse filter. The former condition is probably not very interesting, and the latter condition is very noise-sensitive. This is what is simultaneously a strength and weakness of CPR as compared with feature-based methods. We have begun to address this situation with filters that we term *apodizing* – that is, the filter is designed to reduce the effect of the deterministic sidelobes in the correlation surface. We do not yet have reportable results for apodizing filters, but for the moment we simply report this situation as a known problem that is being worked on. The stochastic nature of the correlation surface remains a problem, although the whitened matched filter generally deals well with this.

There are metrics from information theory that correspond to quantitative information (e.g., Bayes error, power of the test, and others; these are described in Section 8.5) that have long been used in the statistical pattern recognition community, and we have also begun to optimize these metrics within CPR. Section 8.5 will describe these metrics quantitatively within the correlation model. This material is developed in Chapter 8 (Limited-modulation filters) essentially for historical reasons – it was in the context of optical implementation that the information metrics were introduced to correlation pattern recognition [52]. For digital processing the optimization is noticeably easier than for optical processing for the usual reasons: there is less noise (no scattered light, no misaligned optics, no noisy detection), and the digital filter may take on a larger range of values than the optical filter.

Nevertheless, the model of how information flows through the correlation process and is extracted at the end is quite similar for optical and digital processing. The objective of pattern recognition is for the filter to cause the measurement densities to be separate for in-class and out-of-class targets, so that we can infer the presence or absence of a target from the occurrence of a given measurement value. Accordingly, we need a model for the density of measurement values. One critical development in Chapter 8 is that leading to the Bessel function expression (Eq. (8.8)) for measurement density, taking into account that for digital processing the detector noise σ_d^2 is zero. The optimizing digital filter can be used directly as computed in Eq. (8.51), without the projection step of Eq. (8.52) that produces the optimizing optical filter. Finally, the set of search parameters in Eq. (8.51), $\{\Gamma_i \exp(j\beta_i)\}_{\text{training set}}$ is simpler in form for digital processing with its unconstrained filter values. We can, in fact, arbitrarily normalize the search parameters by setting $\Gamma_1 \exp(j\beta_1) \equiv 1$, and conducting a search over the remainder of the set.

6.7 Chapter summary

In this chapter, we have introduced the following major concepts associated with advanced correlation filters:

- When the image distortion being considered obeys a mathematical relationship (e.g., in-plane rotation, radial scale change, etc.), algebraic mapping, such as Cartesian-to-polar transformations, can be used to design correlation filters with a specified response to such distortions.
- When images are subject to more general distortions (e.g., 3-D geometric distortions, occlusions, illumination changes, etc.), synthetic discriminant function (SDF) filters provide an attractive approach for designing correlation filters that are tolerant to such distortions. SDF filters are based on training images that contain examples of expected distortions.
- The first SDF filters (known as equal correlation peak SDF filters, or projection SDF filters) assumed that the correlation filter is a weighted sum of training images, and the weights are found so that correlation output at the origin takes on pre-specified values in response to training images.
- Projection SDF filters suffer from two problems. The first is that they do not have any in-built robustness to noise. The second is that they control only one value of the correlation output (namely the value at the origin), and thus sidelobes are often much larger than this value at the origin, making location of the peak difficult.
- Minimum variance SDF (MVSDF) filters minimize the output noise variance (ONV) while satisfying the hard correlation peak constraints, and thus provide maximum robustness to noise. If the input noise is white, MVSDF filters are the same as projection SDF filters.
- The problem of sidelobes can be attacked by the minimum average correlation energy (MACE) filters, which minimize the average of the correlation energies due to various training images, while satisfying the constraints of the correlation plane. MACE filters produce sharp correlation peaks, but exhibit noise sensitivity and may not provide high correlation peak values in response to non-training images from the desired class.
- As MVSDF filters typically emphasize low spatial frequencies and MACE filters emphasize high spatial frequencies, they provide conflicting attributes. Optimal tradeoff (OT) SDF filters provide a method to minimize one quantity (e.g., ONV) while holding the other (e.g., average correlation energy) constant.
- Using hard constraints for correlation peak values may be counter-productive in that these values may not be achievable for non-training images. Using softer constraints on peak values (e.g., maximizing average correlation height) leads to maximum average correlation energy (MACH) filters. MACH filters have the additional benefit that the only matrix inversions needed are of diagonal matrices and hence are trivial.

- Another step in the evolution of correlation filters is the use of the entire correlation output for pattern recognition. An example of this is the distance classifier correlation filter (DCCF), which transforms the training images so that different classes are better separated from each other, and are more compact in the transformed space than in the original space. For a test input, distance in the transformed space between prototypes from different classes is used for classification.
- Most correlation filter approaches are linear. However, the polynomial correlation filter (PCF) allows point nonlinearities such as power nonlinearity to achieve improved pattern recognition performance. In the PCF, each power nonlinearity introduces a new branch in correlation architecture, but all correlation filters are designed *jointly*. The PCF approach can be extended for multiple spectral bands (e.g., infrared, millimeter wave, etc.) and can thus provide automatic fusion in the multi-sensor case.
- Analytical performance prediction for advanced correlation filters is very difficult. However, simplifying assumptions such as independence of correlation output pixels can provide some high-level understanding of the filters' performance in noise.

The field of correlation filters is diverse and it is beyond the scope of this chapter to cover the various techniques in full mathematical detail. Over the years, many outstanding contributions have been made and some of the early ones have been covered in a tutorial survey [8]. In the field of circular harmonic function (CHF) filters, Arsenault and others [60] made fundamental contributions that spurred new research on rotation-invariant filter designs. These techniques have been extended by others [71, 72]. Optimal tradeoff CHF techniques [58] have recently been developed which can be used to produce any desired signature (e.g., ramp, parabola) at the correlation output in response to image-plane rotation.

In the field of SDF filters, much work has been done by many researchers and we have not covered it all in this chapter. Variants of the SDF filter such as the MACE and MVSDF filters have been extended by many teams. A notable extension is the introduction of nonlinear MACE filters by Fisher and Principe [73]. The generalized rotation-invariant MACE (GRI-MACE) filter by Ravichandran and Casasent [71] combines the MACE concept with CHF expansion to achieve sharp peaks while being invariant to rotations. Campos and others have developed rotation-invariant filters for color pattern recognition [74] and also explored combination with CHFs [72]. To handle in-plane rotations by using a training set of rotated images, Hassebrook [75] introduced the concept of the linear phase coefficient composite filters (LPCCF), which exploit the Toeplitz nature of the vector inner product matrix to obtain an orthogonal family of rotation-invariant filters. Another important development in trading off filter noise sensitivity to peak sharpness is the minimum noise and average correlation energy (MINACE) filter developed by Casasent [76].

A significant area of filter design investigated extensively by Javidi and others [54] deals with filter design for non-overlapping noise. While noise is commonly treated as an additive process, it is more realistic to model clutter and background noise as a process that is disjoint from the target.

The area of filter design for limited modulation devices is also very rich. There have been numerous techniques proposed for the design of filters for binary and ternary optical devices. Not all of these techniques take distortions or noise models into account, some are more focused on accommodating the limited dynamic range of the SLMs. However, with the advent of devices that are capable of complex modulation, the need for binary and ternary filters may diminish. We will discuss in Chapter 8 general filter designs for limited modulation devices. The *minimum Euclidean distance optimal filter* (MEDOF) is a very promising filter design technique that exploits the full range of complex modulation possible on modern devices. The MEDOF concept has been extended to implement the optimal correlation filters discussed in this chapter on limited modulation devices. This has paved the way for implementing the full-precision optimum filters in finite-precision optical systems.

7

Optical considerations

7.1 Introduction

In this chapter we will consider how light is used to carry and process information. To perform correlation pattern recognition optically, we begin with coherent light with the right beam characteristics. We impress a signal upon the beam with one spatial light modulator (SLM), alter the propagation of the beam with a second SLM so that information is preferentially passed through the system and gathered at a location in the output plane, and detect and identify any information that might have been on the input beam. Accordingly we shall examine coherent light sources, SLMs, noise, and detection methods. Polarization of light is a particular point of interest, since many of the extant SLMs operate by altering the polarization of the light. Light polarization is a proper subset of statistical optics, and for a more complete description we refer the reader to Goodman [77] and to O'Neill [78]. Diffraction phenomena are also examined in this chapter, as they are responsible for the Fourier transforming properties needed for an optical correlator.

A knowledge of the physics of light and light modulators can be used to place information onto a beam and to cause its propagation to convert the information to a usable (i.e., detectable) form. The light typically used in the correlators discussed in this book has idealized properties. It is almost monochromatic (i.e., of single wavelength), almost fully polarized (this concept will be explained in later sections), and often enters the processing system as almost planar waves. In earlier chapters we have seen the power of the Fourier transform in linear systems and correlation pattern recognition (CPR); the central concept in this chapter is that the diffraction of coherent light produces a Fourier transform. The diffraction of the coherent light is responsible for the Fourier transform that occurs, so that an optical correlator can be produced.

In this chapter we shall develop the necessary electromagnetic theory, including propagation, polarization, light modulation, calibrations, and noise. We shall see how modulators impose information on light beams and how that information propagates. We shall look at some methods of measuring the characteristics of the modulators so that they may be controlled to achieve the desired effect. We shall develop the meanings of these terms and some of the tools for analyzing departures from these ideal characteristics. Thus we shall become acquainted, if not fully familiar, with: light propagation and diffraction in Cartesian coordinates; the coherence matrix; the Jones and Mueller formalisms for fully and partially polarized light; the Poincaré sphere; phase, amplitude, and coupled modulation; interference, etc.

Except for some brief discussion on optical correlators in Chapter 5, we have previously developed correlation filters with essentially no restrictions on the values that objects, filters, etc. can have. In optical correlator implementations, though, we do not have the luxury of unrestricted filter values. The three major limitations on the optical implementation of CPR are: the restricted set of complex values a filter can take on when it is physically expressed on a modulator, the fact of only the correlation output magnitudes being observable, and the noise that is present in the optical system. In the next chapter we shall see the limitations on CPR imposed by optical systems, and discuss how best to live within these limitations. There are considerations (speed, volume, and power) that motivate us to use optical correlators, and we shall develop the best method to do this.

We shall develop sufficient familiarity with light as an electromagnetic wave phenomenon to grasp the concepts that are significant to performing correlation optically. We begin with light as an electromagnetic phenomenon – what are the disturbance's physically observable effects and how do they propagate? We show how a Fourier transform naturally occurs in the diffraction of light. Next we develop coherence and interference, since it is the constructive interference of coherent light originating at all locations in the filter plane that produces a correlation spot. Polarization is a property of coherent light that is often used for control by an SLM, so birefringence and polarization are reviewed in this chapter. We will principally use the Jones calculus to describe fully polarized light and its propagation, but the coherence matrix and the related Mueller and Poincaré formalisms will be mentioned as the most powerful description of polarization to include only partially polarized light. We shall work within the scalar diffraction model, which permits us to use the analytic signal representation of an electromagnetic wave. A large number of optical terms will be explained in sufficient, though not exhaustive, detail in this chapter.

7.2 Some basic electromagnetics

In this section we present a minimal amount of information on electromagnetic theory – just enough for the purposes of optical correlation. We use the analytic signal as developed in Section 7.5.

7.2.1 Description of plane electromagnetic waves

For shorter notation we will often use the angular frequency ω rather than the number of cycles per second ν. They are related by $\omega = 2\pi\nu$, of course, and sometimes (as when discussing Fourier transforms and related spectral quantities) it is notationally preferable to use $2\pi\nu$ since the 2π shows up in the normalization of the transformation.

Consider a function $f(x, y, z, t) = f(kz - 2\pi\nu t + \phi)$, where k is a constant, z is one of the Cartesian coordinates $\{x, y, z\}$, ν is frequency, t is time, and ϕ is a phase constant. If we restrict our attention to those places where $kz - 2\pi\nu t = \text{const.}$, then f is a constant. That is, of course, to say that f represents a plane wave propagating in the positive z direction, since the stated condition would have

$$z(t) = \frac{2\pi\nu}{k} t + \text{const.} \tag{7.1}$$

and furthermore the function has the same value at all (x, y) for such z. The wave propagates at speed $2\pi\nu/k$. This is for an arbitrary function f. Suppose that we restrict f to be a purely monochromatic (i.e., single sinusoid) function, then

$$f(x, y, z, t) = A\cos(kz - 2\pi\nu t + \phi) \tag{7.2}$$

is a fully general expression, with ϕ being the phase of the oscillation at $z = 0$, $t = 0$. If we pick a fixed value for z we observe f to be a sinusoid oscillating in time with frequency ν. As yet we do not have a physical requirement for k; let us remedy that.

The phase front of the wave advances at speed $c = 2\pi\nu/k$, giving rise to the definition of k as:

$$k = \frac{2\pi\nu}{c} \tag{7.3}$$

The reader should verify that k, called the wavenumber, is 2π times the number of wavelengths in a unit distance, whence its name. For light oscillating at 10^{14} Hz and propagating at approximately 3×10^{10} cm s^{-1}, k is about $20\,000$ cm^{-1}, so k is seen to be a large number. In fact it is so large (i.e., lightwaves are so short) that the small scale of optical diffraction was largely unobservable to the ancients, a literal shortsightedness that was a factor in allowing the corpuscular theory of light to hold pre-eminence over the wave theory until the nineteenth century. Newton himself favored the corpuscular theory.

Using the analytic signal notation discussed in Section 7.5,

$$f(x, y, z, t) = A[\exp jkz \exp(-j2\pi\nu t) \exp j\phi] \tag{7.4}$$

To obtain a little more generality than a wave propagating in the z direction, we replace kz by $\mathbf{k} \cdot \mathbf{r}$, where $\mathbf{k}$ is called the propagation vector and the light is a plane wave propagating in the direction of $\mathbf{r}$, and $\mathbf{r} = [x\ y\ z]^T$. In the complex notation we have a monochromatic plane wave f:

$$f = A \exp j(\mathbf{k} \cdot \mathbf{r} - \omega t + \phi) \tag{7.5}$$

that will be useful in the next section. In one more refinement, we can regard A to be complex and absorb the phase ϕ into it. Thus

$$f = A \exp j(\mathbf{k} \cdot \mathbf{r} - \omega t) \tag{7.6}$$

We emphasize that physically the time-varying phasor notation is actually used just for the convenience of complex notation, but in much of the discussion with linear operations it is not necessary to make the distinction between a physical quantity and its phasor representation. Real fields add, phasor fields add; real fields interfere, phasor fields interfere; etc. But the physical field is not complex, even though we speak of it as such. *Note* that there is not a uniform convention in the literature as to whether the wave is $\exp j(\omega t - kz)$, or $\exp j(kz - \omega t)$, whose real parts are equal. Uniformity is not required since those expressions have the same real part and each propagates in the positive z direction with increasing time. However, for the Jones representation of polarization (Section 7.2.4), the convention becomes important. In this book we use $\exp j(kz - \omega t)$ for a wave propagating in the positive z direction. When reading other material on polarization, check what convention is used. For example, of the five principal citations for this chapter, four use $\exp j(kz - \omega t)$ and one (Azzam and Bashara [1979], the most thoroughgoing with respect to polarization) uses $\exp j(\omega t - kz)$.

Typically in a monochromatic (and hence coherent) system we can understand the time-varying term, $\exp(-j\omega t)$, as a factor implicitly present in all terms, and thus delete it from the notation for simplicity. The justification is that ω for the light vibrations is far greater than the other temporal changes that we induce with SLMs and detect using conventional detectors (and especially imagers, as are needed to examine the whole spatial region of the output). Thus $\exp(-j\omega t)$ can be separated out as a common product in all of the processes, although when performing operations with the real part of the analytic signal it should be included again.

In using the analytic signal, we will find a couple of operator equivalences to be useful. Suppose we have a complex signal $u = B \exp j(kz - \omega t)$. Then we find $\partial u / \partial t = -j\omega u$, from which we make the association that taking the time derivative of the analytic signal is identical to multiplication by $-j\omega$.

$$\frac{\partial}{\partial t} \leftrightarrow -j\omega \tag{7.7}$$

Similarly we find

$$\frac{\partial^2}{\partial t^2} \leftrightarrow -\omega^2 \tag{7.8}$$

and

$$\frac{\partial}{\partial z} \leftrightarrow jk \tag{7.9}$$

Diffraction of light, as discussed in the next section, describes how light energy propagates.

7.2.2 Diffraction of light and the Fourier transform

In order for an optical correlator to function, we need to understand a little electromagnetic diffraction theory. Electromagnetic fields propagate quite differently from the way in which electrostatic and magnetostatic fields extend from their sources. As an example, compare the electric field strength at a meter and a kilometer from static and radiating point sources. The static field has a ratio of 10^6, since the electrostatic field strength is inversely proportional to the square of the source-to-destination distance. The electrostatic field is not depositing energy, since it is static. In contrast, the rate of energy deposition per unit area is proportional to the square of the time-varying field's magnitude, a condition that we can exploit in contemplating the rate of fall-off of a radiating field's amplitude. Consider that the electromagnetic wave is depositing energy onto the surface of a large centered sphere. If the wave propagates losslessly to the sphere's surface then, since the area of the sphere is proportional to r^2, the energy density falls off inversely with r^2. Electromagnetic energy density is proportional to the squared magnitude, so the magnitude of the electromagnetic field that is carrying energy falls off with r, not with r^2. When more than one source point is involved, their complex fields are added, not their independently determined intensities. The resulting intensity is the square of the magnitude of the sum of the fields, not the sum of the squares of the field magnitudes. This statement will be clarified later in the discussion of interference. The ratio of radiating electromagnetic field amplitudes at a meter and a kilometer is only 10^3, compared with the 10^6 for the electrostatic case. If we have a monochromatic source so that we can meaningfully talk about phase, then the electromagnetic wave propagates so that it lags the source with a phase that is directly proportional to the radius. This intuitively based discussion may serve to support the Rayleigh–Sommerfeld diffraction theory [32] that is the basis of Fourier optics.

The foundation of diffraction theory is that time-varying electromagnetic fields propagate themselves, with each point on an initial wave surface serving as a source for the subsequent appearance of the wave. Under the approximation of linearity – that fields add – we can compute the effect from an arbitrary element in the source as that effect is observed at a distant point. We then add together the contributions from each of the source elements. The interested reader may see Goodman [32], for example, for more detail.

In accordance with a physical arrangement (which we will examine later) of a spatial light modulator and a laser beam irradiating it, suppose that we have a planar object of complex transmissivity $\tau(\mathbf{r}_1)$ within a surface Σ, where $\mathbf{r}_1$ is the position of the object that is being irradiated by a uniform plane wavefront $\mathbf{E}_{\text{in}}$ from a fully coherent source, so that the propagated field is

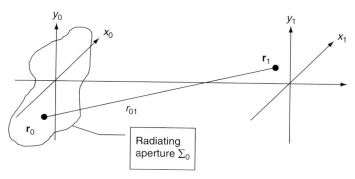

Figure 7.1 Superposition showing how $\mathbf{U}(\mathbf{r}_1)$ arises by propagation from $\mathbf{U}(\mathbf{r}_0)$. Each point in Σ_0 radiates into each point in Σ_1 with a propagation delay that is proportional to the distance r_{01} between them. The field $\mathbf{U}(\mathbf{r}_1)$ is the summation integral over all such contributing points in Σ_0. This diffraction can be made to be a Fourier transform of the pattern in Σ_0 (see text)

$$\mathbf{U}(\mathbf{r}_1) = \mathbf{E}_{in}\tau(\mathbf{r}_1) \tag{7.10}$$

The Rayleigh–Sommerfeld diffraction equation specifies that a radiating source of strength $\mathbf{U}(\mathbf{r}_0)$, at positions $\mathbf{r}_0$ on the surface Σ_0, is observed at a location $\mathbf{r}_1$ to have the effect:

$$\mathbf{U}(\mathbf{r}_1) = \frac{1}{j\lambda} \iint\limits_{\Sigma_0} \mathbf{U}(\mathbf{r}_0) \frac{\exp(jkr_{01})}{r_{01}} \cos(\mathbf{n} \cdot \mathbf{r}_{01}) \mathrm{d}s \tag{7.11}$$

where λ is the wavelength, k is the wavenumber, r_{01} is the distance between points $\mathbf{r}_0$ and $\mathbf{r}_1$, $\mathbf{n}$ is the normal to Σ, and the cosine is of the angle between the normal $\mathbf{n}$ and $\mathbf{r}_{01}$, the vector from $\mathbf{r}_0$ to $\mathbf{r}_1$. There are a couple of differences between the physically motivated discussion of the preceding paragraph and Eq. (7.11). One is the factor $1/j\lambda = -j/\lambda$ in front of the integral, which corresponds essentially to Eq. (7.7); it is the time-varying behavior of an electromagnetic field that allows it to propagate energy, and Eq. (7.7) and the definition of k give the connection to the time-varying aspect of $\mathbf{U}$ to create a propagating wave. The other item is the cosine term. Since it deals with contributions made at oblique angles to the surface normal it is called the obliquity factor, and it is justified in Green functions and boundary conditions that need not concern us here. The form of Eq. (7.11) is clearly indicative of the potential for diffraction's performing a Fourier transform, since that transform uses a kernel with nearly the same complex exponential.

Equation (7.11) is difficult to use as it stands, and approximations are commonly used to simplify it. The first such assumption is that the cosine term may be taken as unity and that, in the denominator, the term $\mathbf{r}_{01}$ may be taken as the distance z along the optical axis (the axis is normal to Σ). These approximations (as are the others that follow) are valid if z is much greater than the size of the aperture Σ, and if we stay close to the axis (the paraxial condition).

Two other approximations are commonly used for dealing with the complex exponential; they are the Fresnel and Fraunhofer approximations. Because the wavenumber is such a large factor in the complex exponential, and the effect of phase is so profound in the integral's accumulating a product, $\mathbf{r}_{01}$ has to be more closely approximated in the exponential than in the denominator. In the Fraunhofer approximation the phase is calculated as though the wavefront is planar. In the Fresnel approximation, which we treat first, the expanding wave is approximated as a quadratic surface rather than as a true sphere. The center of the sphere is at x_1, y_1 on the $z = 0$ surface.

$$\frac{1}{j\lambda}\frac{\exp(jkr_{01})}{r_{01}}\cos t(\mathbf{n},\mathbf{r}_{01}) \approx \frac{\exp(jkz)}{j\lambda z}\exp\left\{\frac{jk}{2z}\left[(x_0-x_1)^2+(y_0-y_1)^2\right]\right\}$$

(7.12)

which is valid for

$$z^3 \gg \frac{\pi}{4\lambda}\left[(x_0-x_1)^2+(y_0-y_1)^2\right]_{\max}^2$$

(7.13)

as a sufficient, but not wholly necessary, condition [78]. Under the Fresnel approximation, Eq. (7.11) becomes

$$\mathbf{U}(\mathbf{r}_1) = \frac{\exp(jkz)}{j\lambda z}\exp\left(\frac{k}{2z}(x_1^2+y_1^2)\right)\iint\limits_{\Sigma_0}\mathbf{U}(\mathbf{r}_0)\exp\left[j\frac{k}{2z}(x_0^2+y_0^2)\right]$$
$$\times \exp\left[-j\frac{2\pi}{\lambda z}(x_0x_1+y_0y_1)\right]ds$$

(7.14)

The Fresnel approximation does not yet allow the use of diffraction to perform a Fourier transform. If the first exponential within the integral were unity, we would have a Fourier transform. The next step – the Fraunhofer approximation – finally gets us there. In this approximation the expanding wave is considered in its dependence on x_0, y_0 to be planar as it emits from Σ, rather than in the Fresnel approximation where it is a quadratic approximation to a sphere. The Fraunhofer approximation is expressed as:

$$\frac{1}{j\lambda}\frac{\exp(jkr_{01})}{r_{01}}\cos(n,r_{01}) \approx \frac{\exp(jkz)\exp\left[\frac{jk}{2z}(x_0^2+y_0^2)\right]}{j\lambda z}\exp\left\{-j\frac{2\pi}{\lambda z}(x_0x_1+y_0y_1)\right\}$$

(7.15)

which holds for the far more stringent condition than Eq. 7.13,

$$z \gg \frac{k(x_1^2+y_1^2)_{\max}}{2}$$

(7.16)

in which the maximum is taken over the source aperture.

We digress for a moment for a physically intuitive approach to finding that the Fourier transform will appear at the focal plane of a lens. Note that the dependence on x_0 and y_0 is essentially angular, since the exponential term including them has z in the denominator. We could rewrite the last exponential term in Eq. (7.15) with angles ψ and ξ instead of spatial variables x_0 and y_0 by using

$$\psi = x_0/\lambda z, \xi = y_0/(\lambda z)$$

(7.17)

We observe that ψ and ξ are angles (divided by wavelength, a constant for monochromatic illumination as we have assumed) since the approximation in Eq. (7.16) applies and we remember that for small ρ, $\sin(\rho) \approx \rho \approx \tan(\rho)$. Then we have

$$\frac{1}{j\lambda} \frac{\exp(jkr_{01})}{r_{01}} \cos(\mathbf{n}, \mathbf{r}_{01}) \approx \frac{\exp(jkz) \exp\left[\frac{jk}{2z}\left(x_0^2 + y_0^2\right)\right]}{j\lambda z} \exp\{-j2\pi(x_0\psi + y_0\xi)\}$$

(7.18)

and the portion of the expression involving ψ and ξ has, for a large distance z, the kernel that takes the Fourier transform with respect to spatial frequencies in the x_1 and y_1 coordinates. (The conditions in Eq. (7.16) would cause the value of the second exponential term – the only other appearance of x_0 and y_0 – to be near unity.) A 2-D Fourier transform occurs for the spatial frequencies given by: $f_X = x_1/\lambda z$ and $f_Y = y_1/\lambda z$. (For more information see [32], for example, whose development we have abbreviated here.) In free space the stated conditions in Eq. (7.16) occur only at great distances from the diffracting aperture, but the imposition of a positive lens in essence draws infinity in to the focal distance, and we can use a lens to warrant using the Fraunhofer approximation in an optical correlator. Firstly, we consider just the geometric optics situation.

We see that the rays emitting from a point in the focal plane of the lens (the lens has focal length F) convert to rays having a single direction, and vice versa. Angle θ corresponds to a point in the focal plane lying at distance x from the optical axis, where $\tan\theta = q/F$. Following the preceding development we see that light originally arising in a spatial frequency f at the diffracting aperture comes to a location $q = F\lambda f$ in the focal plane. We can develop this in a more quantitative fashion and also examine a more general location of the lens that is responsible for the Fourier transform.

Returning to the main development, we say that the electromagnetic field $\mathbf{U}$ is expressed as the constant illumination field $\mathbf{E}_0$, being influenced by the complex transmittance $\tau(\mathbf{r}_1)$. Under the Fraunhofer approximation Eq. (7.11) now becomes:

$$\mathbf{U}(\mathbf{r}_0) = \mathbf{E}_0 \frac{\exp jkz \exp\left[j\frac{k}{2z}\left(x_0^2 + y_0^2\right)\right]}{j\lambda z} \int_\Sigma \int \tau(\mathbf{r}_1) \exp\left[-j\frac{2\pi}{\lambda z}(x_0x_1 + y_0y_1)\right] ds$$

(7.19)

and we see that, apart from the terms in front of the integral, we are obtaining the Fourier transform of the complex transmittance τ. We shall see how to use lenses to take care of these leading terms.

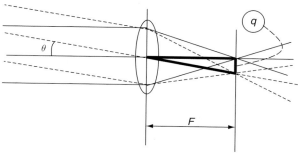

Figure 7.2 Drawing of a simple lens with three rays: chief (passing through center of lens, neither deviated nor displaced), top (from a point in focal plane to lens to parallel with chief), and bottom (the same). A lens converts angular behavior in the far field into positional behavior in the focal plane. The displacement q at the focal plane resulting from the off-axis angle θ of incoming light, is related to the focal length F by $\tan \theta = q/F$

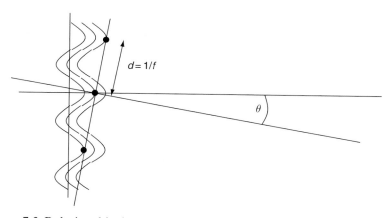

Figure 7.3 Relationship between Fourier transform and Bragg diffraction. The heavy dots are indicative of the Bragg condition, that reinforcement of light propagating into a given angle θ occurs when integer differences of wavelength correspond to the spatial frequency of periodic lateral structures. That is, when we move laterally through one cycle and longitudinally through integer wavelengths, the periodic structure interferes constructively in the direction θ, even though the principal direction of propagation is towards $\theta = 0$. Here, if f is the spatial frequency, then $\tan \theta = \lambda f$. But from Figure 7.2, $\tan \theta = q/F$. So we find that a laterally coherent periodic light wave with lateral spatial frequency f will come to a focus not only on the axis, but also at integral steps of $F \lambda f$, in the lens's focal plane. This is a heuristic adjunct to the formal Fresnel–Kirchoff diffraction integrals that show how a Fourier transform can occur in diffraction

A converging lens can be approximated as introducing a quadratic phase change to the arriving wavefront. To do this we assume that the wavefront is nearly perpendicular to the optical axis (the paraxial approximation), and that the lens is "thin" (i.e., a ray passing through it is deviated [in angle] without being displaced [in position]). Under these conditions the lens has an effect on an incident wavefront given by the phase change

$$\Delta\phi(x, y) = \exp\left(-j\frac{k}{2F}\left(x^2 + y^2\right)\right) \tag{7.20}$$

in which k is the wavenumber as usual, x and y are position coordinates within the lens, and F is the focal length of the lens. (The value $\Delta\phi$ is a phase change measured as an excess over the base phase thickness of the lens.) Let us examine the consequences of putting such a quadratic lens into the optical train. We assume that $\Delta\phi$ extends over the whole interesting extent of τ (in our practical sense, that the lens is larger than the spatial light modulator being used to affect $\mathbf{E}_0$, the incoming wavefront).

Goodman [32] develops three cases of interest in which the electromagnetic field is computed at a plane one focal distance behind the lens: (1) The complex object is placed directly against the lens; (2) the complex object is placed a focal length in front of the lens; and (3) the object is placed a distance d $(0 < d < F)$ behind the lens.

1. The lens cancels the quadratic term in the integrand of Eq. (7.14), with the result that within a constant phase factor, the electromagnetic field at x_f, y_f in the focal plane of the lens is

$$\mathbf{U}_1(\mathbf{r}_f) = \mathbf{E}_0 \frac{\exp\left(j\frac{k}{2F}\left(x_f^2 + y_f^2\right)\right)}{j\lambda F} \int\int_\Sigma \tau(\mathbf{r}_1)\exp\left[-j\frac{2\pi}{\lambda F}\left(x_f x_1 + y_f y_1\right)\right]ds \tag{7.21}$$

which differs from being the Fourier transform of $\tau(\mathbf{r})$ by only the quadratic phase factor before the integral.

2. Placing the object exactly a focal distance in front of the lens, and relying on no stronger an approximation than Fresnel to propagate the light from lens to object, results in cancellation of the quadratic term multiplying the integral in Eq. (7.21):

$$\mathbf{U}_2(\mathbf{r}_f) = \mathbf{E}_0 \int\int_\Sigma \tau(\mathbf{r}_1)\exp\left[-j\frac{2\pi}{\lambda F}\left(x_f x_1 + y_f y_1\right)\right]ds \tag{7.22}$$

in which a constant phase term has been omitted. We have now achieved a Fourier transform optically, using the diffractive properties of propagation.

3. With the object a distance d behind the lens, and again using the Fresnel approximation for propagating the light – this time from object plane to the focal plane – we have:

$$\mathbf{U}_3(\mathbf{r}_f) = \mathbf{E}_0 \frac{\exp\left(j\frac{k}{2d}\left(x_0^2 + y_0^2\right)\right)}{j\lambda d} \int_{\Sigma} \int \tau(\mathbf{r}_1) \exp\left[-j\frac{2\pi}{\lambda d}(x_0 x_1 + y_0 y_1)\right] ds$$

(7.23)

The Fourier transform of τ appears again, this time with the useful effect that the frequency ν in the plane of $\mathbf{r}_1$ is directed to a location $x_0 = \lambda \nu d$. That is, we can use d as a free parameter to scale the frequencies at the object plane to specified locations in the focal plane of the lens. (There is the quadratic phase term to be dealt with, but it is a small penalty for the convenience of being able to enlarge the spatial extent of the spectrum arbitrarily in its landing on the focal plane of the lens.)

7.2.3 Coherence, interference, and polarized light

With the tools from the previous sections we are able to get more particular about how light is used in an optical correlator. We shall see that coherence and polarization are quite intertwined subjects; coherence necessarily implies polarization, though the converse is not true. For further details we refer the reader to Chapters 8 and 9 of O'Neill [78] and Chapter 1 of Azzam and Bashara [79].

For optical correlation, the coherence of light is particularly important. In fact, the most frequently cited paper in the optical correlation literature [5] was published only slightly after the laser, with its notably coherent light, became available. The laser finally made it possible to have sufficient power in nearly monochromatic wavelengths for the diffraction integrals of propagation to be usable approximations of the exact Fourier transform. Interference occurs only with coherent light,[1] and constructive interference is exactly how detectable spots are created in the output plane of an optical correlator. We are also led to discuss polarization since it is used to control light with many of the current SLMs, and we shall see that a completely monochromatic light beam is unavoidably polarized.

---

[1] This would seem to call into question the concept of "white light interference," a term with which the reader may be familiar. What we mean by "white light" is that there is a breadth to the power spectrum when viewed at a given location as a function of time, and that the field de-correlates when we impose even small time differences. White light is not coherent but there is a finite bandwidth implicit in the experiment, or in the sensing and recording method, etc. A narrowing bandwidth implies changing character closer to coherence. However, if the white light beam is split into two beams, as by a partial reflecting mirror, and the two beams are brought back together after traversing identical path lengths, then interference can be observed. Each frequency component of the white light – being coherent with itself by virtue of arising in exactly the same oscillation – is able to interfere, and so the whole beam interferes. With very nearly identical path lengths the different frequencies interfere constructively or destructively simultaneously. The existence of white light interference fringes is a powerful indicator of very nearly equal path lengths in split beams.

In the two preceding sections we used a scalar electromagnetic wave; there was no mention of the vector nature of light's electric field. The equations related to a given component of the vector field. The vector nature now becomes significant, and the study of polarization is concerned with how the orthogonal components of the electromagnetic vector behave together. "Orthogonal" will be seen to have another connotation for polarized light rather than the familiar Cartesian meaning. It will denote polarization states that do not interfere (that is, their intensities strictly add).

A light wave for which the phase fronts are planes, perpendicular to the z axis and of uniform infinite extent in x and y, and propagating in an isotropic medium, is represented by the field vector:

$$\mathbf{U} = \mathbf{E} \exp(-j\omega t) \exp(jkz) \qquad (7.24)$$

in which we have absorbed the phase ϕ into the now complex $\mathbf{E}$. In isotropic media (and under some circumstances in anisotropic media) the electric vector $\mathbf{E}$ is perpendicular to the direction of propagation. We remind ourselves of the following features. The electromagnetic disturbance is sinusoidal both in space (along the z axis) and in time, with a phase $\phi = \arg \mathbf{E}$, at time $t = 0$ at the origin. The disturbance is a constant for $kz - \omega t = \text{const}$. That is, the wave propagates along z at a speed $\upsilon = \omega/k$. The electric field $\mathbf{E}$ has components E_x and E_y along x and y, and both E_x and E_y are complex in order to express the phase difference between them.

Coherence is quantified by how an x or y component behaves at different locations in space and/or time, and polarization is quantified by how the x and y components of $\mathbf{E}$ behave in combination at one location. For light that is not purely sinusoidal in its oscillations, Wolf's mutual coherence function [80] describes the correlation between parts of the wave at different times and locations; for one component, E, of the propagating field it is defined as:

$$\Gamma(\mathbf{x}_1, \mathbf{x}_2, \tau) = \langle E(\mathbf{x}_1, t + \tau) E^*(\mathbf{x}_2, t) \rangle \qquad (7.25)$$

in which $\langle \cdot \rangle$ indicates the time average taken over so long an interval that the exact length of the interval does not matter. Equivalently, we may assume a stationary process – sufficiently accurate for the environment within an optical correlator. Vectors $\mathbf{x}_1$ and $\mathbf{x}_2$ are position coordinates, and τ is a time difference that is usually implemented as a path length difference. The notation for Γ is shortened to $\Gamma_{12}(\tau)$. Let's look at some of the simplifications of the mutual coherence function. The mutual coherence function $\Gamma_{11}(\tau)$ tells us over what duration the source's light stays coherent with itself, and $\Gamma_{11}(\cdot)$ has its maximum value at zero. When τ is multiplied by the speed of light c, $\Gamma_{11}(\tau)$

tells us how the light is coherent with itself at different locations along its path. The maximum distance $c\tau_{coh}$ over which Γ remains near its peak is known as the coherence length of the beam. (Lasers have coherence lengths of centimeters to kilometers. They can conduct their internal electromagnetic vibrations in modes of slightly different frequencies, and the modes are not coherent with each other. As long as the laser is oscillating in one mode the light stays coherent, so the coherence length tells us how long the laser typically spends in one mode before hopping to another. Recent experiments in "counting light" have used molecular oscillations to stabilize light oscillations well enough for $c\tau_{coh}$ to be light-weeks.) Similarly, $\Gamma_{12}(0)$ tells us over what lateral spatial extent a light beam remains self-coherent. Clearly, an optical correlator must be built so that path length differences and beam widths are considered for their implications on coherence. These considerations lead us to install spatial filters to "clean up" a beam with respect to lateral spatial variations, and to select a source (such as a gas tube laser) whose coherence length suits the purpose. A spatial filter functions by focusing the beam down to a very small size and passing it through an even smaller aperture. (At this juncture we have begun to speak willy-nilly of Fourier domain processing of light, since the focal plane of a converging lens is where the angular spectrum of the incident light is displayed. That is, light from different angles focuses at different locations. By limiting the aperture in the focal plane we limit the set of angles into which the light will emit after it passes through the aperture.) The higher angular harmonic content of the beam is blocked by the aperture, and the emerging light, captured by a lens for its further use in the correlator, has a simpler behavior.

Generalizing from complex scalar to a vector field for $\mathbf{E}$, the mutual coherence function Γ becomes the coherence matrix $\mathbf{M}$, defined as

$$\mathbf{M} = \langle \mathbf{E}\mathbf{E}^{+} \rangle = \begin{bmatrix} \langle E_x E_x^* \rangle & \langle E_x E_y^* \rangle \\ \langle E_y E_x^* \rangle & \langle E_y E_y^* \rangle \end{bmatrix} = \begin{bmatrix} M_{xx} & M_{xy} \\ M_{yx} & M_{yy} \end{bmatrix} \qquad (7.26)$$

For purely monochromatic light (an assumption we shall make for the light in an optical correlator), the time averaging is simplified. The time-dependent term $\exp(-j\omega t)$ bears no relevance to the time averaging of the vector components, since under the strictly monochromatic assumption all the time variation occurs in the $\exp(-j\omega t)$ term. The coherence matrix will be used again in Section 7.2.5.

Let's look at some examples of specific polarizations. Suppose that the light is monochromatic and that as we look into the oncoming beam we observe the electric vector to oscillate in only the x direction. This light is linearly polarized in the x direction, and in the next section we shall develop some mathematics – the Jones matrix – for describing it and its propagation. At our observation point then we have

$$\mathbf{E}_a = E_a \exp(-j\omega t) \begin{bmatrix} 1 \\ 0 \end{bmatrix} \qquad (7.27)$$

in which we have momentarily retained the time variation just to remind ourselves that it is always present, usually only implicitly. Light that oscillates in only the y direction is

$$\mathbf{E}_b = E_b \begin{bmatrix} 0 \\ 1 \end{bmatrix} \qquad (7.28)$$

Similarly we would have

$$\mathbf{E}_c = \frac{E_c}{\sqrt{2}} \begin{bmatrix} 1 \\ \exp j\pi/2 \end{bmatrix} \qquad (7.29)$$

for light in which the y component lags $\pi/2$ radians behind the x component and the magnitudes are identical. Note that here we divided by a normalizing factor so that the norm of the vector $\mathbf{E}_b$ is the magnitude of the possibly complex E_b. We shall not always observe the normalizing nicety since the absolute magnitude (and the absolute phase, for that matter) are not usually of interest for our purposes. It is easily verified that the coherence matrices for these fields are

$$\mathbf{M}_a = E_a^2 \begin{bmatrix} 1 & 0 \\ 0 & 0 \end{bmatrix} \qquad (7.30)$$

$$\mathbf{M}_b = E_b^2 \begin{bmatrix} 0 & 0 \\ 0 & 1 \end{bmatrix} \qquad (7.31)$$

$$\mathbf{M}_c = \frac{E_c^2}{2} \begin{bmatrix} 1 & \exp(-j\pi/2) \\ \exp(j\pi/2) & 1 \end{bmatrix} \qquad (7.32)$$

Vectors as shown in Eqs. (7.27) to (7.29) are called the Jones vectors for the light of the described polarizations, and the polarization actions of optical elements on the Jones vectors are described using Jones matrices. When the light is not fully polarized, as is assumed in setting up these equations, then the appropriate vector to describe the state of polarization is the Stokes vector, and the operation of optical elements is described in the Mueller matrix. We shall spend the majority of our time on the Jones description.

We can examine how these components interact in producing the observed intensity. The intensity is $I = \mathbf{E}^+\mathbf{E}$, where the superscript $+$ indicates the conjugate transpose. Let's see what intensity results from adding light vibrating in the x and y directions.

$$\mathbf{E}_{\text{total}} = \begin{bmatrix} E_x \\ 0 \end{bmatrix} + \begin{bmatrix} 0 \\ E_y \end{bmatrix} = \begin{bmatrix} E_x \\ E_y \end{bmatrix} \tag{7.33}$$

$$I_{\text{total}} = \mathbf{E}_{\text{total}}^+\mathbf{E}_{\text{total}} = |E_x|^2 + |E_y|^2 = I_x + I_y \tag{7.34}$$

That is, the intensities add for x- and y-polarized light. In contrast, let's see what happens when similarly polarized light is added:

$$E_{\text{total}} = \begin{bmatrix} E_x \\ 0 \end{bmatrix} + \begin{bmatrix} E_x \\ 0 \end{bmatrix} = \begin{bmatrix} 2E_x \\ 0 \end{bmatrix} \tag{7.35}$$

$$I_{\text{total}} = \mathbf{E}_{\text{total}}^+\mathbf{E}_{\text{total}} = |2E_x|^2 = 4I_x \tag{7.36}$$

This looks like a problem! It appears that there is a net increase in energy from adding the similarly polarized light. It is true that for those points where the similarly polarized fields add exactly in phase, the intensity is quadrupled. However, suppose the fields add with a phase difference θ from being exactly in phase. It is easily verified that the resulting intensity is $(2 + 2\cos\theta)$ times the single-beam intensity. With uniform distribution of phase the cosine term averages to zero, so the intensity averages to twice the single-beam intensity. The point remains that for some phases the intensity of the combined beams is substantially different from the sum of the beams' individual intensities. This is the hallmark of interfering light. The x- and y-polarized light beams do not interfere under any phase relationship.

We say that non-interfering polarizations are *orthogonal*. It is also easily verified that more generally than for the x- and y-polarizations, two light beams having fields $\mathbf{E}_1$ and $\mathbf{E}_2$ are orthogonal in the present sense if

$$\mathbf{E}_1^+\mathbf{E}_2 = 0. \tag{7.37}$$

This topic is more fully developed in the next section, which deals with the Jones calculus for fully polarized light.

7.2.4 Jones calculus for fully polarized light

We begin with fully polarized light and develop the Jones formalism to describe it. Fully polarized light has the property that knowing one component of the field tells you what the other component is. In the simplest example, suppose that the light's electromagnetic vector vibrates only along the $x = y$ direction. Then the x and y components are equal. Similarly, if you know that the light's electromagnetic vector traces out a circle in a clockwise direction, you know that the y component leads the x component by $\pi/2$ radians. But if the light is not polarized at all, then knowing one component tells you nothing about the other. (Because of the definition of polarization this is an almost tautological statement.) In-between, we have partially polarized light, and the coherence matrix is the tool for handling this. We remark that if a beam is exactly coherent it is also fully polarized. This follows from each component's being monochromatic, so that if the phase relationship between components is once determined, it is fixed for all time, implying a state of full polarization.

There are various methods of producing polarized light; in some, like lasers, the light is initially emitted in fully polarized form, and there are optical devices ("polarizers") that produce polarized light by blocking one polarization state and passing the orthogonal state. The same physical device can be employed for either of two functional intents. The first is simply to provide a source of polarized light, in which case it is called a polarizer. The second is to determine how much light is in a particular arriving state, or to select only the light in that particular state for further processing; in either of these uses it is called an analyzer.

We let $\mathbf{E}_0$ be the incoming light wave's vector described in terms of its x and y Cartesian components. The complex relationship between the x and y components will describe the polarization of the wave. Suppose we have a wave propagating in the $+z$ direction:

$$\mathbf{E} = \mathbf{E}_0 \exp(jkz) \tag{7.38}$$

Writing $\mathbf{E}$ in terms of its Cartesian components,

$$\mathbf{E} = \begin{bmatrix} E_{0x} \\ E_{0y} \end{bmatrix} \exp(jkz). \tag{7.39}$$

The terms E_{0x} and E_{0y} are complex, and if the light is fully polarized they have a fixed ratio. The phase and relative magnitude between E_{0x} and E_{0y} describe the polarization state. The terms E_{0x} and E_{0y} are the components of the Jones vector (with respect to a Cartesian basis). Let's see how to find the Jones vector of a polarization state.

Even as a two-component vector can be represented with respect to any two independent vectors within the span of the vector's components, the polarization vector can be represented with respect to any two independent polarization states (we ordinarily take them to be orthogonal). For now we will be representing polarized light with respect to the x and y components, but later we shall represent such light with respect, for example, to left- and right-hand circular polarizations. That is enough for now, but remember this point for later.

Pick a position along z (0 is handy) and observe the vector $\mathbf{E}_0$ as an explicit function of time. In an isotropic medium the z-propagating electromagnetic vibration is perpendicular to the z-axis – that is, it has components in only x and y. (Young's coming up with the observation that propagating electromagnetic vibration is transverse saved the day for the wave theory of light, since the corpuscular ideologues had seized upon the failure of longitudinal vibration wave theory to explain polarization effects. Prior to Young's epiphany, the wave theorists had thought of light as a longitudinal wave similar to that of sound.) The perspective in Figures 7.4 to 7.8 shows the behavior as we look into the arriving wave. One cycle of the electric field is shown at the left. The temporal behavior of the x and y components is in the center. At the right, the complex phasor for the initial value of the electric field components is given in Jones vector and graphical form, with the phasor rotation at angular rate ω indicated. Consistent with the rest of the discussion on polarization, we have not bothered to normalize the vectors to unit length. In Figures 7.4 to 7.8, we see the significance of selecting the sign convention in Eq. (7.6); under the other convention, the phasor labeled ω would be rotating the other way around, and a different phase relationship between the x and y components of $\mathbf{E}$ (i.e., the real parts) would result.

The Jones vectors in Figures 7.4 to 7.8 are given with respect to the Cartesian basis. That is, writing $\mathbf{E} = \begin{bmatrix} E_x \\ E_y \end{bmatrix}$ is to say "I take E_x much of an E-field having zero phase at $t = 0$ and polarized to vibrate along the x-axis, and add it to E_y

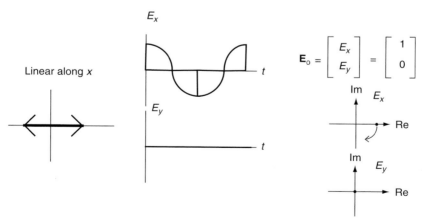

Figure 7.4 Linear polarization along x

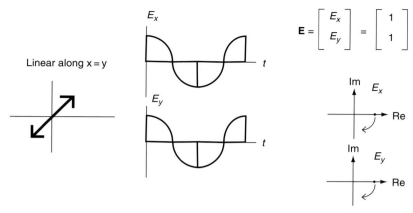

Figure 7.5 Linear polarization along $x = y$

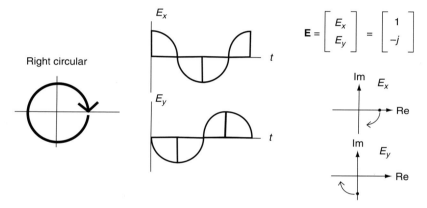

Figure 7.6 Right circular polarization

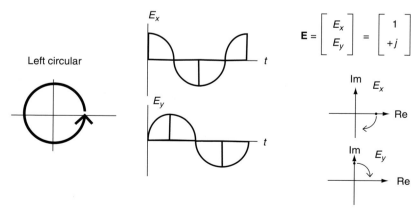

Figure 7.7 Left circular polarization

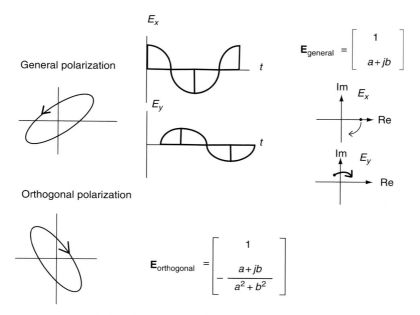

Figure 7.8 General elliptical polarization

much of an E-field having zero phase at $t = 0$ and polarized to vibrate along the y-axis. I express phase lag between the x and y components in the argument of the complex values of E_x and E_y." We can equally well use any pair of independent polarizations (e.g., non-collinear Jones vectors) as a basis. For example, if we add right-hand circularly polarized (RHCP) light and left-hand circularly polarized (LHCP) light together, we have the resulting vector

$$\mathbf{E}_{\text{total}} = \begin{bmatrix} 1 \\ j \end{bmatrix} + \begin{bmatrix} 1 \\ -j \end{bmatrix} = 2 \begin{bmatrix} 1 \\ 0 \end{bmatrix} \tag{7.40}$$

in which we have, for this instance, kept the normalization in addition to keeping the complex ratio between components. We see that adding RHCP light and LHCP light (of equal intensities and zero phase relative to each other) gives x-polarized light. Similarly to Eq. (7.40) we can formally subtract RHCP from LHCP and obtain y-polarized light:

$$\mathbf{E}_{\text{total}} = \begin{bmatrix} 1 \\ j \end{bmatrix} - \begin{bmatrix} 1 \\ -j \end{bmatrix} = 2 \begin{bmatrix} 0 \\ j \end{bmatrix} \tag{7.41}$$

(Subtraction is physically done by delaying the wave by π radians and then adding.) From these last two equations we can see the following relationship. Suppose we have a polarization state $\mathbf{S}$ that is expressed in terms of the

Cartesian x and y states, $\mathbf{S} = [a\ b]^\mathrm{T}$. We can build $\mathbf{S}$ by taking an amount $a/2$ each of LHCP light and RHCP light (i.e., the x polarization) and adding an amount $b/2$ of LHCP light, then subtracting an amount $b/2$ of RHCP light (i.e., the y polarization). As a matrix expression, $\mathbf{S}$ is given by:

$$\mathbf{S} = \begin{bmatrix} a \\ b \end{bmatrix} \begin{bmatrix} 1/2 & 1/2 \\ -1/2 & 1/2 \end{bmatrix} \begin{bmatrix} \text{light in RHCP} \\ \text{light in LHCP} \end{bmatrix} \qquad (7.42)$$

The matrix in the center expresses a change of basis; we can use a and b to tell how much RHCP light and LHCP light is needed to express state $\mathbf{S}$, although we initially expected to have $\mathbf{S}$ specified to us in Cartesian-orthogonal polarizations. The major point here is that we confirm the statement made earlier, that we may use independent polarizations as a basis for representing any polarization. In the next section, the Poincaré sphere description of a polarization state will be a function of the polarization basis set.

We remark again that orthogonal polarizations do not interfere. To understand this, notice that for the interacting Cartesian-polarized lightwaves, the Pythagorean theorem, restated, says that the total intensity of the electromagnetic field is the sum of the intensities in x and y. Interference classically is observed when the light intensity is other than the sum of the intensities of the two beams in question. Orthogonality extends to all other states as well as Cartesian polarization. The statement of orthogonality between general states $\mathbf{E}_1$ and $\mathbf{E}_2$ is that, $\mathbf{E}_1^+\mathbf{E}_2 = 0$. Thus linear polarizations at angles of $\pi/2$ with respect to each other are orthogonal; RHCP and LHCP are orthogonal; and elliptical polarization has an orthogonal elliptical polarization (see Figure 7.8).

Energy originally in orthogonal polarization states *can* be made to interfere by the imposition of optical elements that affect polarization. Suppose we have mutually coherent co-propagating light beams linearly polarized at 0 and $\pi/2$; they are orthogonal to each other and they might differ in their phase. If they pass through a linear analyzer aligned at $\pi/4$, then light from each beam emerges at $\pi/4$, although each with its magnitude reduced by a factor of $\sqrt{2}$. Their individual phases are maintained in relation to each other. Each beam has been analyzed for its linear component at $\pi/4$, which is to say that its orthogonal component has been discarded (absorbed or reflected). An exact vector analogy is that Cartesian x and y vectors are orthogonal (have zero projections onto each other), but if we take unit vectors in the x and y directions and take their projections separately onto a vector at 45°, those projections add directly (are themselves parallel and non-zero).

The analysis occurs so as to retain the individual phases of the originally orthogonal beams, and the phases of the original beams determine whether the light passing through the analyzer interferes constructively or destructively. Upstream of the analyzer, there is no such interference. A key point here is that an irreversible action occurs when the light passes through the analyzer, and some of the light has been lost. In terms of the Jones matrix to be developed now, we would say that the analyzer's rank is not full, so light cannot in general be reversibly propagated through it. Jones matrix devices of full rank and unity determinant are reversible (actually a unity determinant implies full rank), and states that are originally orthogonal will propagate through the full rank Jones matrix devices so that they continue to be orthogonal.

Now that we have the Jones vector description of the state of polarized light, let's describe how optically active elements change the polarization state of light. This description is the Jones matrix, which specifies how differing polarizations are altered and summed in passing through the element.

Inducing, altering, and measuring polarization states requires devices that block one polarization of light and pass the orthogonal polarization. We will try to be consistent in the use of "polarizer" and "analyzer" in describing these optical elements. As mentioned earlier, the differentiation is in the intended effect: the same physical device can be either a polarizer or an analyzer. A polarizer by intent delivers light in a particular polarization state. An analyzer by intent extracts light of a particular polarization state from the arriving light. Where both meanings are intended, "polarizer" suffices. The context will indicate which is the correct interpretation.

A linear polarizer blocks one Cartesian direction of electromagnetic vibration and passes the orthogonal one. Say it blocks the x component and passes the y component undiminished and undelayed. Then the input–output relationship is:

$$\mathbf{E}_{\text{in}} = \begin{bmatrix} E_x \\ E_y \end{bmatrix}, \quad \mathbf{E}_{\text{out}} = \begin{bmatrix} 0 \\ E_y \end{bmatrix}. \tag{7.43}$$

This action is expressed in a matrix operation as:

$$\mathbf{E}_{\text{out}} = \begin{bmatrix} 0 & 0 \\ 0 & 1 \end{bmatrix} \mathbf{E}_{\text{in}} \tag{7.44}$$

and the matrix $\begin{bmatrix} 0 & 0 \\ 0 & 1 \end{bmatrix}$ is the Jones matrix for a y polarizer. Similarly, $\begin{bmatrix} 1 & 0 \\ 0 & 0 \end{bmatrix}$ polarizes linearly along x. A linear polarizer at an angle ψ counterclockwise with respect to the x-axis is expressed as $\begin{bmatrix} \cos^2 \psi & \sin \psi \cos \psi \\ \sin \psi \cos \psi & \sin^2 \psi \end{bmatrix}$ as we see next.

Let **E** be the Jones vector in the original (x, y) coordinate system and let **E′** be the Jones vector of that same electromagnetic wave in the Cartesian system with $x′$ aligned at angle ψ counterclockwise with respect to x. Then the rotation matrix **R** converts $\mathbf{E}_{\text{in}}$ to $\mathbf{E}'_{\text{in}}$:

$$\mathbf{E}'_{\text{in}} = \mathbf{R}\mathbf{E}_{\text{in}} = \begin{bmatrix} \cos\psi & \sin\psi \\ -\sin\psi & \cos\psi \end{bmatrix} \mathbf{E}_{\text{in}} \tag{7.45}$$

The x polarizer's action is $\begin{bmatrix} 1 & 0 \\ 0 & 0 \end{bmatrix}$ in the primed system, so

$$\mathbf{E}'_{\text{out}} = \begin{bmatrix} 1 & 0 \\ 0 & 0 \end{bmatrix} \begin{bmatrix} \cos\psi & \sin\psi \\ -\sin\psi & \cos\psi \end{bmatrix} \mathbf{E}_{\text{in}} = \begin{bmatrix} \cos\psi & \sin\psi \\ 0 & 0 \end{bmatrix} \mathbf{E}_{\text{in}} \tag{7.46}$$

which is rotated back to the unprimed (original) coordinate system by $\mathbf{R}^{\text{T}}$:

$$\mathbf{E}_{\text{out}} = \begin{bmatrix} \cos\psi & -\sin\psi \\ \sin\psi & \cos\psi \end{bmatrix} \begin{bmatrix} \cos\psi & \sin\psi \\ 0 & 0 \end{bmatrix}$$

$$\mathbf{E}_{\text{in}} = \begin{bmatrix} \cos^2\psi & \cos\psi\sin\psi \\ \cos\psi\sin\psi & \sin^2\psi \end{bmatrix} \mathbf{E}_{\text{in}} \tag{7.47}$$

giving us the claimed form for the Jones matrix. An analyzer that transmits only RHCP light can be designed by similar concatenation; firstly we use a waveplate that converts RHCP light into x-polarized light, then we use an x polarizer (rejecting the y-polarized light that was previously orthogonal to RHCP light – i.e., LHCP light), and then inverting the action of the first waveplate.

(Although we shall not develop the fact here, you may note that the Jones matrices for linear polarization are all of rank one, whereas a full-rank Jones matrix is of rank two; the linear polarizers' determinant is zero. This is because when considered as a linear transform, it has a nullspace; the whole of the orthogonal polarization is mapped to zero by virtue of being blocked.)

Now let's see how physical optical devices can be assembled to produce the various Jones matrices. They depend on differential effects on polarizations of light.

The linear polarizer is perhaps simplest to envisage. A fine array of straight wires, close together in a grid, forms a polarizer. Somewhat simplistically, the conductive wires short out the electromagnetic field in the direction of the wires, so only the electric component perpendicular to the wires makes it through.

In addition to the polarizers, the other principal optical devices for controlling polarization are retarders (also called waveplates), both variable and fixed. The defining character of a waveplate is that it differentially retards light of one polarization compared to another. Retarders affect the state of polarized light passing through them by slowing light vibrating in one direction more than that vibrating in another, thus altering the phase relationship between the components as they propagate through the material. Crystallography provides appropriate materials for retarders. There are crystals (natural and artificial; solid and liquid) that have differing indices of refraction depending on the polarization of the light and the direction of propagation. In any direction of propagation there are generally two values of the phase velocity, depending on the polarization, so this property is called birefringence. In fact, in crystals, only those with cubic symmetry are isotropic; asymmetric stiffness in the binding of the charged atoms in other crystals causes the speed of light propagation (equivalently, the index of refraction) to be a function of polarization and propagation direction. The susceptibility tensor is the mathematical description that relates the material polarization in the crystal to the impressed electric field. As is manifest in Maxwell's equations, the material's electric polarization affects the propagation of light. Crystals for which there are three distinct values in the diagonalized susceptibility tensor are called biaxial; those with two distinct values, uniaxial; and those with one, isotropic. The names come from the biaxial crystal's having two directions in which polarizations propagate with equal phase velocity; and the uniaxial crystal, just one. This direction of equal phase velocity is called an optic axis of the crystal. The purpose of this paragraph has not been to fully elucidate all of these terms, but instead to show them in a logical sequence and context for further study if it becomes important to the reader (for example, in reading product literature for variable retarders).

To obtain polarization effects we require optical elements that affect the polarization states of incident light in different ways. Without going into greater crystallographic detail than already given in the preceding paragraph, we mention that it is possible to cut plates from uniaxial crystalline materials so that they have "fast" and "slow" axes at right angles to each other, and in the plane of the plate. As light passes through the plate, linear electromagnetic vibrations that are aligned with the fast axis are retarded less (i.e., travel faster) than those aligned with the slow axis. The indices of refraction for the two directions are n_e and n_o, the subscripts o and e standing for ordinary and extraordinary propagation. The *ordinary* propagation is that for which the speed of light is *independent of direction* in the bulk crystal. In a *positive* uniaxial crystal, $n_e > n_o$, and conversely $n_e > n_o$ in a *negative* uniaxial crystal. As in the isotropic retarder in monochromatic light, where we were mostly interested in fractional excess delay above integral wavelengths, we are now interested in fractional *differences* in the delay in fast and slow axes.

In the process invented by Edwin Land (US Patent 2,041,138), a mass of minute polarizing crystals is embedded in a plastic sheet that is then stretched.

The stretching orientates the small crystals, and the orientated mass together allows only a single linear polarization state to pass. The extinction ratio is the factor by which an unpolarized light beam is reduced in intensity by passage through a polarizer/analyzer pair that is so disposed as to minimize the transmittance. Quite high extinction ratios (intensity ratios in the order of a million to one) can be obtained with the stretched plastic polarizers. There are other methods that use the fact that reflection can cause polarization, a common example being the Brewster window found in lasers. A glass plate tilted with respect to an incoming light beam at an angle equal to the arctangent of the index of refraction of the glass will transmit only a single polarization.

Linear polarization and isotropic retarding are not sufficient to get to arbitrary polarization states. Thus we now discuss birefringence and optical activity. "Optical activity" is the property possessed by substances that rotate the plane of vibration for linearly polarized light passing through them; it is the result of differing speeds of propagation for right- and left-hand circularly polarized light, and many common substances are optically active. Crystalline quartz and certain sugars are examples, dextrose and levulose being sugars named for their optical rotary powers (the first rotates the plane of linear polarization to the right, as viewed looking into the beam, the latter to the left). Birefringence has more common application in optical processing than does optical activity. Birefringence is the quality of having two distinct indices of refraction. (A prime example is the calcite crystal, whose birefringence causes a double image that the adherents to the theory that light is a longitudinal wave were unable to explain.)

An isotropic retarder delays all polarizations of light by the same factor. Compare two parts of a light wave, one passing through an isotropic retarder and the other not. We can describe the lag of the first with respect to the second in three ways: time, distance, or phase. Of these, phase is the most commonly used. If d is the physical thickness and n is the index of refraction, the optical thickness is nd. The free-space optical thickness is $1d$, so the distance lag is $(n-1)d$. The phase lag ϕ is wavelength dependent and is found from the proportionality:

$$\frac{\phi}{2\pi} = \frac{(n-1)d}{\lambda} \tag{7.48}$$

The Jones matrix for the isotropic retarder is $\begin{bmatrix} \exp(-j\phi) & 0 \\ 0 & \exp(-j\phi) \end{bmatrix}$. Typically $(n-1)d/\lambda \gg 1$, but in a monochromatic system with slowly varying modulation of the carrier, we are interested in only the excess delay over $2m\pi$, for the largest m, such that $m \leq (n-1)d/\lambda$. Because of this, we can build waveplates that are physically large enough to be robust to handling; we can build the bulk of the device to be a multiple of 2π of the phase, and then add enough extra thickness to get the fractional effect we want. For example, the differential index between ordinary and extraordinary modes for quartz, a positive uniaxial crystal, is about 0.01. That is, for 100 wavelengths of propagation through quartz, one of the polarizations will lag behind the other by one wavelength. For $100N$ wavelengths, the lag is N wavelengths. To obtain a quarter-wave of relative lag we need 25 wavelengths of quartz, or about two and a half micrometers. Such a piece of quartz is unmanageably thin. With coherent light, a lag of N wavelengths is undetectable. So instead a quarter-wave quartz plate is made $100N + 25$ wavelengths thick, with N large enough such that the optical element can be handled. The coherence length of the light sources used in optical correlators is far greater than the thickness of the optical elements, so inducing $2N\pi$ of additional phase lag is inconsequential.

Suppose we take a retarder with a slow axis aligned with x and a fast axis aligned with y. Let δ be the phase by which the x component of the Jones vector is retarded compared with the y component. Then the effect is:

$$\mathbf{E}_{\text{out}} = \begin{bmatrix} \exp(-j\delta) & 0 \\ 0 & 1 \end{bmatrix} \mathbf{E}_{\text{in}} \tag{7.49}$$

which can be written in a balanced form as:

$$\mathbf{E}_{\text{out}} = \begin{bmatrix} \exp(-j\delta/2) & 0 \\ 0 & \exp(+j\delta/2) \end{bmatrix} \mathbf{E}_{\text{in}} \tag{7.50}$$

Here, as in other locations, we have implicitly acknowledged that we are not concerned with absolute phase. We have suppressed a factor $\exp(-j\delta/2)$ on all components. If we wish to leave the x component unchanged to provide the reference phase,

$$\mathbf{E}_{\text{out}} = \begin{bmatrix} 1 & 0 \\ 0 & \exp(+j\delta) \end{bmatrix} \mathbf{E}_{\text{in}} \tag{7.51}$$

Two common retarders are half- and quarter-wave plates in which δ is respectively $\pi/2$ and $\pi/4$. By putting $\delta = \pi/4$ into Eq. (7.51) we see that a quarter-wave plate has the Jones matrix

$$\mathbf{M}_{\lambda/4} = \begin{bmatrix} 1 & 0 \\ 0 & +j \end{bmatrix} \tag{7.52}$$

and converts $\theta = \pi/4$ linearly polarized light into LHCP. Similarly, $\psi = 3\pi/4$ leads to RHCP light. For example, suppose that we begin with light polarized at $\theta = -45°$, expressed as $[1 \ -1]^{\text{T}}$. Then

$$\mathbf{E}_{\text{out}} = \begin{bmatrix} 1 & 0 \\ 0 & +j \end{bmatrix} \begin{bmatrix} 1 \\ -1 \end{bmatrix} = \begin{bmatrix} 1 \\ -j \end{bmatrix} \qquad (7.53)$$

which is RHCP light.

The concept that the x and y vibrations are orthogonal to each other extends to other polarization states as well. We can verify that the given condition on polarization states expressed in the Jones calculus means that they do not interfere. An intuitive way to think of this is that an x component of a wave cannot cancel a y component. Two polarization states do not interfere if and only if the intensity of their sum is the sum of their intensities. For x- and y-linearly polarized light, this is easy enough:

$$I_{x+y} = \left\| \begin{bmatrix} 1 \\ 0 \end{bmatrix} + \begin{bmatrix} 0 \\ 1 \end{bmatrix} \right\|^2 = \left\| \begin{bmatrix} 1 \\ 1 \end{bmatrix} \right\|^2 = \left\| \begin{bmatrix} 1 \\ 0 \end{bmatrix} \right\|^2 + \left\| \begin{bmatrix} 0 \\ 1 \end{bmatrix} \right\|^2 = I_x + I_y \quad (7.54)$$

More generally, suppose state S_1 is $[a\ b]^{\mathrm{T}}$, and state S_2 is $[c\ d]^{\mathrm{T}}$, with complex elements in each vector. The statement of orthogonality expressed earlier is that, $S_1^+ S_2 = 0$, or $a^*c + b^*d = 0$. Using this condition we obtain:

$$I_{1+2} = \left\| \begin{bmatrix} a \\ b \end{bmatrix} + \begin{bmatrix} c \\ d \end{bmatrix} \right\|^2 = \left\| \begin{bmatrix} a+c \\ b+d \end{bmatrix} \right\|^2 = \left\| \begin{bmatrix} a \\ b \end{bmatrix} \right\|^2 + \left\| \begin{bmatrix} c \\ d \end{bmatrix} \right\|^2 = I_1 + I_2 \quad (7.55)$$

after expanding the terms and using the statement proposed for orthogonality. So we see that polarization states meeting the definition of orthogonality do not interfere. To emphasize: this statement is true for all sorts of completely polarized light, be it linearly, circularly, or elliptically polarized.

The Jones description of fully polarized light makes it simple to see that any polarization state (described by its Jones vector) can be converted to a linear state by an appropriate waveplate.

$$S_1 = \begin{bmatrix} A \exp j\phi_1 \\ B \exp j\phi_2 \end{bmatrix} \text{ is an arbitrary polarization}$$

$$S_2 = \begin{bmatrix} \exp\left(j\frac{\phi_2-\phi_1}{2}\right) & 0 \\ 0 & \exp\left(-j\frac{\phi_2-\phi_1}{2}\right) \end{bmatrix} \begin{bmatrix} A \exp j\phi_1 \\ B \exp j\phi_2 \end{bmatrix} = \begin{bmatrix} A \exp\left(j\frac{\phi_1+\phi_2}{2}\right) \\ B \exp\left(j\frac{\phi_1+\phi_2}{2}\right) \end{bmatrix}$$

$$S_2 = \exp\left(j\frac{\phi_1+\phi_2}{2}\right) \begin{bmatrix} A \\ B \end{bmatrix} \text{ which is seen to be linear polarization.}$$

It is also easily verified that any waveplate operating on orthogonal states produces orthogonal states. We leave it to the reader to verify that the arbitrary waveplate $\begin{bmatrix} \exp(j\delta) & 0 \\ 0 & \exp(-j\delta) \end{bmatrix}$ does not change the orthogonality of incident polarizations.

Table 7.1 *Jones matrices for various optical elements*

Optical element		Jones matrix
Linear polarizer	Horizontal	$\begin{bmatrix} 1 & 0 \\ 0 & 0 \end{bmatrix}$
	Vertical	$\begin{bmatrix} 0 & 0 \\ 0 & 1 \end{bmatrix}$
	$\pm 45°$	$\begin{bmatrix} 1 & \pm 1 \\ \pm 1 & 1 \end{bmatrix}$
Quarter-wave plate	Fast axis vertical	$\begin{bmatrix} 1 & 0 \\ 0 & -j \end{bmatrix}$
	Fast axis horizontal	$\begin{bmatrix} 1 & 0 \\ 0 & j \end{bmatrix}$
	Fast axis at $\pm 45°$	$\begin{bmatrix} 1 & \pm j \\ \pm j & 1 \end{bmatrix}$
Half-wave plate	Fast axis either vertical or horizontal	$\begin{bmatrix} 1 & 0 \\ 0 & -1 \end{bmatrix}$
Isotropic phase retarder		$\begin{bmatrix} \exp j\phi & 0 \\ 0 & \exp j\phi \end{bmatrix}$
Relative phase retarder		$\begin{bmatrix} \exp j\phi_x & 0 \\ 0 & \exp j\phi_y \end{bmatrix}$
Circular polarizer	Right	$\begin{bmatrix} 1 & j \\ -j & 1 \end{bmatrix}$
	Left	$\begin{bmatrix} 1 & -j \\ j & 1 \end{bmatrix}$

Following Wolf [80], we summarize in Table 7.1 the foregoing development of the Jones calculus for fully polarized light and linear optical elements. We do not, however, normalize the entries.

7.2.5 *Another formalism for polarized and partially polarized light*

There is more than one way to quantify the state of polarization of light, each method being suited to the purpose for which it was developed. In this section we briefly discuss the Stokes–Mueller formalism for partially polarized and unpolarized light. We shall contrast the Stokes–Mueller formalism with the Jones calculus, and state why we use the Jones method to describe light in optical correlation. We will discuss the Poincaré sphere, the Stokes parameters and the Stokes vector comprising them, the Mueller matrix, and the coherence matrix.

The Jones calculus is the most often used formalism in optical correlators. We have seen that it is well suited to fully polarized light both with respect to describing the state of polarization (the Jones vector) and the expression of a linear optical element's action on the light (the Jones matrix). Furthermore, and quite significantly, it describes how phase propagates. Optical correlators work by restructuring phase and amplitude among the harmonic components of a signal, with their resulting interference providing the correlation output. Another primary reason for concentrating so much attention on coherent light is that its coherence enables the optical Fourier transform. Fully monochromatic light is perforce polarized (why?) and is hence amenable to the Jones calculus and, in fact, quasi-monochromatic light that is fully polarized is indistinguishable from monochromatic light, by most instruments. In optical correlators, coherent light carries the information, and furthermore we often use a modulator whose action is described by a Jones matrix to filter the image borne on the light. As long as it is valid, we prefer using the Jones calculus for its descriptive economy. We discuss the Poincaré sphere as a graphical adjunct to the Jones calculus; positions on it describe the polarization state, and linear optical elements that alter the polarization state simply move the point around on the sphere.

As light becomes other than fully coherent (and may then become other than fully polarized, too), statistical methods are necessary to describe it. The Stokes parameters (and the Stokes vector built from them) characterize coherent, partially coherent, and incoherent polarized light. The Mueller matrix describes the conversion of the Stokes parameters by optical elements. The coherence matrix mentioned in Section 7.2.3 relates to the Stokes description. Again we refer the reader to Chapters 8 and 9 of O'Neill [78], and Chapter 1 of Azzam and Bashara [79] for details. Chapter 3 of Mandel and Wolf [81] gives a very thorough and readable treatment of the coherence matrix.

Often we do not need to include all of the information in the Jones description of fully polarized light. An example is when the absolute phase or absolute amplitude of the polarized light is unimportant, which is most often true in optical correlators. In this case the Poincaré sphere, adapted from complex function theory, is adequate to depict the state, although quantitative computations are still most conveniently performed in the Jones calculus. Movements of a point on the Poincaré sphere describe the action of linear polarizing elements. Let's see how to get to the Poincaré representation.

Firstly, we note that if the absolute phase and amplitude of fully polarized light are unimportant, we can normalize the Jones vector to the Cartesian x component, and then all of the information that describes the light's polarization is contained in the normalized y component. The result is that a point in the complex plane describes the polarization state; this is called the Cartesian complex plane representation:

$$\chi = \frac{E_y}{E_x} \tag{7.56}$$

Note that the absolute phase and amplitude are lost in this representation. In the limit as E_x goes to zero, χ becomes infinite in a phase direction that depends on how E_x and E_y behave in that limit. Thus, finite values of the electric field in polarized light require the entirety of the complex plane to be sure that all polarization states can be described. This is an unconscionably thin use of complex real estate, however cheap the land prices might be! However, in the theory of functions of a complex variable, Riemann has shown us how to compact the entire complex plane onto a sphere. A sphere of unit diameter is placed tangential to the complex plane at the origin. The point of contact we shall call the south pole, and the sphere's diametrically opposite point, the north pole or vertex. Zero longitude will be in the direction of the positive x-axis at the point of tangency at the south pole. A point in the plane is mapped onto the sphere along the line connecting the point to the vertex, a process known as a stereographic projection (it is sometimes used in mapping the Earth's nearly spherical surface onto a flat map, most commonly in the polar regions). We note with satisfaction that however E_x approaches infinity, the representation on the Riemann sphere approaches a single point, the north pole. When the values of χ from Eq. (7.56) are plotted on a Riemann sphere it is then called a Poincaré sphere.

We remind ourselves that the values of the Jones vector depend on the polarization basis vectors. In the Cartesian description of χ, linear

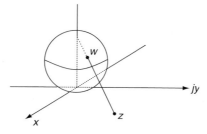

Figure 7.9 Diagram of Riemann sphere. Complex value $z = x + jy$ is mapped to point w on the unit-radius sphere tangent to the complex plane at the origin

polarization along the x-axis has the Jones vector $\mathbf{E}=[E_x\ E_y]^T=[1\ 0]^T$, linear along y is $[0\ 1]^T$, linear at an angle θ to the x-axis is $[\cos\theta\ \sin\theta]^T$, RHCP is $[1\ -j]^T$, LHCP is $[1\ j]^T$, and the vector

$$[\ (\cos\theta\cos\varepsilon+j\sin\theta\sin\varepsilon)\quad(\sin\theta\cos\varepsilon-j\cos\theta\sin\varepsilon)\ ]^T$$

describes the general elliptical state depicted in Figure 7.8. (As an exercise for the reader, the latter relationship is obtained from noting that for $\theta=0$, the Jones vector is $[(\cos\varepsilon)\ (-j\sin\varepsilon)]^T$, and then rotating the coordinate system by θ.) Let us depict these conditions graphically. The Cartesian point for light linearly polarized at θ is the ratio of the E_y to the E_x component, or $\tan\theta$. Thus the linear polarizations occur along the real axis of the complex plane, being linear along x at the origin and linear along y at the infinities. The real axis in the complex plane maps onto the great circle passing through the poles at $0°$ and $180°$ of longitude. Similarly we see that RHCP light is represented at $-j$ and LHCP light at $+j$. Since the diameter of the Poincaré sphere is unity, points in the plane at unit radius are mapped to the equator, and so RHCP light falls on the equator at $270°$ east longitude, and LHCP light falls at $90°$ east longitude. A quarter-wave plate aligned with its fast axis parallel to the x-axis has a Jones matrix $\begin{bmatrix} 1 & 0 \\ 0 & j \end{bmatrix}$ that differentially retards the y component by $\pi/2$, and so it would convert RHCP light $[1\ -j]^T$ into linearly polarized light $[1\ +1]^T$ at $45°$, halfway between the x and y axes. On the Poincaré sphere, this rotates the representing point along the equator through $90°$ of longitude. Other actions that alter the polarization state amount to similar motions of points on the Poincaré sphere. We trust that this meager introduction to the Poincaré sphere representation of fully polarized light and its propagation through optical instruments will whet the reader's appetite and encourage them to read further in Section 2.5 of Azzam and Bashara [79]. We remind the reader that the Poincaré representation is a function of the basis polarization (Azzam and Bashara often use RHCP and LHCP as the basis), and also that those authors use an alternative sign convention to ours in the analytic signal (see Section 7.5).

The Jones and Poincaré methods are adapted to fully polarized light. (In fact if we permit a sphere of smaller radius than unity, the Poincaré sphere is also sufficiently descriptive of partial polarization, with zero radius corresponding to entirely unpolarized light.) However, light in the real world is never exactly fully polarized. The Stokes parameters replace the Jones vector description of the polarization state, and the Mueller formalism, related to the coherence matrix, describes the propagation of partially polarized light. We now summarize the Stokes vector and Mueller matrix for operating with partially polarized light.

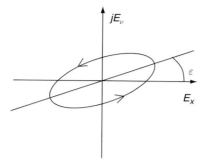

Figure 7.10 The general polarization ellipse

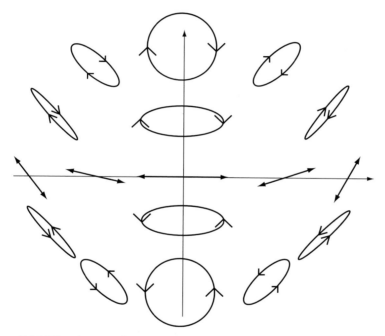

Figure 7.11 The shapes of polarization ellipses as a function of the complex value of χ, the ratio of y-component to x-component of the light's electric vector

The Stokes parameters are statistical descriptions of the polarization state of light for light that is not fully polarized – that is, for light that does not have complete predictability of one component of the light, given perfect knowledge of the other component. The four Stokes parameters are, S_0, S_1, S_2, and S_3. (They correspond to measurable intensities, if suitable elements are inserted

Table 7.2 *Stokes parameters*

Stokes parameter	Expression	Range of value	Inference
S_0	$\left\langle\left\|E_x(t)\right\|^2\right\rangle + \left\langle\left\|E_y(t)\right\|^2\right\rangle$	Positive	Light is present
		Zero	No light is present
		Negative	Can't happen!
S_1	$\left\langle\left\|E_x(t)\right\|^2\right\rangle - \left\langle\left\|E_y(t)\right\|^2\right\rangle$	Positive	More like x polarization
		Zero	Not more like either x or y linear polarization
		Negative	More like y polarization
S_2	$2\left\langle\left\|E_x(t)E_y(t)\right\|\cos\left[\delta_y(t)-\delta_x(t)\right]\right\rangle$	Positive	More like linear at $+45°$
		Zero	Not more like either $\pm45°$ linear polarization
		Negative	More like linear at $-45°$
S_3	$2\left\langle\left\|E_x(t)E_y(t)\right\|\sin\left[\delta_y(t)-\delta_x(t)\right]\right\rangle$	Positive	More like RHCP
		Zero	Not more like either circular polarization
		Negative	More like LHCP

into the beam. However, we do not make measurements within the optical train of a correlator during its operation, but instead mathematically describe the light in terms of its complex amplitude as it propagates. This leads us to favor the Jones formalism for describing the propagation of light through the correlator.) These Stokes components, arranged as a vector, $\mathbf{S} = [\, S_0\, S_1\, S_2\, S_3]^{\mathrm{T}}$, are called the Stokes vector; S_0 is the total intensity of the light, S_1 is the degree to which the light resembles linear polarization along the x or y axes, S_2 is the degree to which light resembles polarization at $\theta = \pi/4$, S_3 is the resemblance to circular polarization. As for the coherence matrix briefly introduced earlier, expectation quantities are time-averaged over sufficiently long periods that the exact value of the period is unimportant. Table 7.2 gives more information.

For completely polarized light, $S_0^2 = S_1^2 + S_2^2 + S_3^2$. For partially polarized light, the inequality $S_0^2 > S_1^2 + S_2^2 + S_3^2$ occurs, and for completely unpolarized light, $S_1^2 + S_2^2 + S_3^2 = 0$. This leads naturally to the definition of the degree of polarization as:

$$P = \frac{\sqrt{S_1^2 + S_2^2 + S_3^2}}{S_0} \tag{7.57}$$

which is zero when there is no correlation between the Cartesian components of the electromagnetic field, and unity when there is complete correlation. The charm of the Stokes parameters is that: they are real quantities unlike the Jones complex quantities; that their values have a somewhat more direct and intuitive connection with mental images of polarization behavior, as outlined in Table 7.2; that they deal with partially polarized light (this is really the big deal here); and that they are directly calculable from physical measurements. In the latter connection, suppose we have measurements of intensity without any analyzer (I_{bare}), following analyzers that select light in the linear polarizations at $0°$, $45°$, $90°$, and $135°$, and also in RHCP and LHCP. Letting these intensities be subscripted in the obvious manner, the Stokes parameters can be computed as:

$$S_0 = I_{bare} = (I_0 + I_{90}) = (I_{45} + I_{135}) = (I_R + I_L)$$
$$S_1 = I_0 - I_{90}$$
$$S_2 = I_{45} - I_{135} \tag{7.58}$$
$$S_3 = I_R - I_L$$

The above-described method uses six measurements to obtain the Stokes parameters in a natural and straightforward manner, easily amenable to laboratory practice. Mandel and Wolf [81], in Section 6.2 of their book, show how to obtain the elements of the coherence matrix – and hence the Stokes parameters – using only four intensity measurements.

Here is the conversion between Stokes parameters and the coherence matrix:

$$\left. \begin{aligned} S_0 &= M_{xx} + M_{yy} \\ S_1 &= M_{xx} - M_{yy} \\ S_2 &= M_{xy} + M_{yx} \\ S_3 &= j(M_{yx} - M_{xy}) \end{aligned} \right\} \leftrightarrow \left\{ \begin{aligned} M_{xx} &= \tfrac{1}{2}(S_0 + S_1) \\ M_{yy} &= \tfrac{1}{2}(S_0 - S_1) \\ M_{xy} &= \tfrac{1}{2}(S_2 + jS_3) \\ M_{yx} &= \tfrac{1}{2}(S_2 - jS_3) \end{aligned} \right. \tag{7.59}$$

The conversion from coherence matrix to Stokes vector is patent here, and the formulas are also shown inverted to give the coherence matrix in terms of the Stokes vector. We have shown how to make physical measurements that yield the Stokes vector, so we can achieve the coherence matrix from the selfsame physical measurements.

7.2.6 *Which formalism to use?*

We now have all the elements: Jones vector and matrix, Stokes vector, Mueller matrix, and coherence matrix. We use these tools in ascertaining the action of SLMs to control light and in characterizing the SLMs themselves. For optical correlators, though, we much prefer to use the Jones calculus. The principal reason is that the Jones calculus retains phase information on the propagated light and does so in easily retrievable form. (Note that the measurements behind the Mueller–Stokes description are all intensities in which phase is not immediately evident.) Phase is a very important descriptor in computing how the spectral components of a lightwave propagating through the optical correlator will add together.

7.3 Light modulation

The role of an SLM is to encode a desired spatially varying effect (in particular an input image or a filter) onto a lightwave. In Chapter 8 we shall examine more of the physics of light propagation, but now we discuss some of the ways in which light is modulated as a function of position on the face of the SLM.

7.3.1 *Architecture and diffraction from a Cartesian SLM*

The SLMs that we shall consider are modeled as pixels in a Cartesian (often square) array. There are two major diffraction effects necessary to understanding the effect of an SLM. The first is the diffraction pattern of a single pixel, and the second is the diffraction from the array structure itself. The array of pixels can be regarded as a convolution of a single pixel with an array of impulse functions. As is apparent from an earlier section we can compute the diffraction of light from a single pixel by using the pixel as the aperture and performing the diffraction integral. (Accurate modeling of this effect from a single pixel is tedious and fraught with problems that we shall not go into in any depth here.) Earlier in this chapter we saw that under the Fraunhofer approximation the far-field diffraction pattern of a radiator is given by the Fourier transform of the field. The Fourier transform of the convolution between two functions is the product of the transforms of the two functions. What this means here is that the diffraction pattern of light coming from an SLM is the product of the Fourier transforms of light from a single pixel and the transform of an array of impulse functions corresponding to the pixel locations. The transform of the array of impulses is another array of impulse functions; the spatial array is transformed into an angular array by diffraction, and the angular array is the set of diffraction orders. In this section we shall get quantitative about these concepts.

7.3.2 *Birefringence (Jones matrix) SLMs*

We have put so much attention into polarization because a large number of SLMs operate by polarization effects; the liquid crystal SLMs and the magneto-optic SLMs are examples. They operate by having a voltage-dependent Jones matrix, with each pixel having its own drive voltage. These SLMs are to be operated in a particular pair of polarization states (input and analyzed), and the SLM's effect on the light is to be understood in terms of the polarization states. We shall refer to these SLMs as *indirect* in their action; the SLM does not care what the input or analyzed polarization state is, it simply alters its own Jones matrix according to the applied voltage. The ultimate operating characteristics of the SLM depend generally on both the input and analyzed polarization states between which the SLM operates. Manufacturers of these devices include Boulder Nonlinear Systems, Displaytech, Kopin, Litton, Sharp, and others. Many of the LCD devices for the display industry fall into this category.

7.3.3 Direct action SLMs

In addition to the polarization-operated SLMs, there are the *direct*-action SLMs, such as the deformable mirror surface devices produced by Texas Instruments (TI), by Optron, and by Hamamatsu. In the case of the TI devices, small flat mirrors either tilt or undergo piston (up and down) motion in response to the control voltage. In the case of the Optron and Hamamatsu devices, a flat membrane surface is locally dimpled according to a spatially varying voltage applied to the reverse side. The light diffracts differently from the dimpled surface than from the flat, and light from the dimples is ordinarily scattered out of the light path and trapped. The direct action SLMs do not typically have as strong a polarization-sensitive variation in their action as the indirect action SLMs. To first order, we can think of the tilting of a small mirror element as redirecting the beam according to the laws of reflection from planar surfaces, so if the light is redirected so as to miss the entrance aperture of subsequent optics, it is observed as a reduction in amplitude. (A more accurate picture is that the tilted mirror element is "blazed," in spectroscopic terms, which affects the efficiency with which it directs light into its various diffraction orders. Pixelated holographic optical elements in general do not so much steer beams around in a continuous fashion as simply re-balance the light among the orders. Optical correlators are usually set up to capture a single diffraction order.) Also to first order, and with more precision than for the effect of tilting, translation of a pixel along the optic axis introduces a phase change in the diffracted light. The optical path, computed in units of distance, is a constant function of wavelength; hence the phase change, measured in radians, is a varying function of wavelength.

7.3.4 Optically addressed and electrically addressed SLMs

In another binary partition of the SLM population, there are optically addressed
SLMs (OASLMs) and electrically addressed SLMs (EASLMs). The OASLMs
are often referred to as light valves, much as vacuum tubes are referred to as
electronic valves. Vacuum tubes amplify the effect of small electronic signals to
control large results. The light values use a low-intensity "write" beam to control
the reflective or transmissive properties of the SLM, and a high-intensity "read"
beam is then spatially controlled by the input light image. The development of the
OASLMs is most strongly dictated by the display industry, which uses them for
projecting bright images onto large screens. The EASLM market is also highly
responsive to commercial interests in projection display, but it is more strongly
influenced by optical processing than the OASLM business. An example of
incompatibility between optical processing and projection requirements is that
for display to a human observer you can easily afford to duty-cycle a binary-
intensity display, in order to attain the effect of continuously variable intensity
modulation. Digital circuitry is fast enough to duty-cycle the image faster than
the flicker fusion response of the human eye. Optical processing is not so tolerant,
however; the correlation image is formed in literally nanosecond-scale time, far
faster than the SLMs change state, so if the effect of analog modulation is to be
had, then analog modulation it has to be. Analog modulation is particularly
significant in optical correlation.

7.4 Calibration of SLMs and their drive circuitry

In using an SLM, an image (usually from a television camera, if it is the input
SLM, or digital storage, if it is the filter SLM) is converted to a drive signal and
applied to the SLM's pixels. Then the coherent light is directed to the SLM and
the SLM alters the light's local characteristics. The original electronic image is
thereby encoded on the wavefront. The encoding is determined by the behavior of the SLM, and there is no single form of that behavior. The set of
complex values that a pixel takes on as the drive goes through its domain is
called the operating curve, and to ascertain the operating curve is called
characterizing the SLM. The general rule is that the SLM must be characterized in the same sort of optical environment in which it will operate. We shall
see several methods of characterization. We start with a case where strong
simplifying assumptions hold, specifically that the SLM is completely uniform
over its face and noise is negligible. We progress to cases in which the fringes
are non-uniform, the SLM varies in its behavior from pixel to pixel, and noise
must be accounted for.

We emphasize that it is the *complex amplitude*, not the *intensity*, of the SLM's action that is primary to optical correlation. That is because the complex amplitude (i.e., the phased magnitude) is what adds linearly inside a correlator, not the intensity. The things that happen inside a correlator are described by linear system theory, but not if one were to make the observed intensity (rather than magnitude) have first-order dependence on the drive signal.

7.4.1 Interference fringe analysis for uniform and non-uniform SLMs

In order to make best use of an SLM we must know what it does when a given drive value is applied to it at any point. In the preferred case the SLM is uniform in its response across its face. We can build spatial interference patterns and analyze them for the SLM's action. If the action is different from location to location, however, the analysis is more difficult and we do what we call a "depth" analysis. That analysis is done, pixel by pixel, on a set of interference values collected for a set of reference phases at each pixel.

We earlier discussed the interference of light in terms of effects at a point. In general when using or characterizing SLMs we have interference over the spatial extent of a beam, not just at a point. The electromagnetic field of the kth beam incident at a location may be written as $E_k(x, y)$, the combined electromagnetic field is $E(x, y) = \sum_k E_k(x, y)$, and the observable intensity is $I(x, y) = |E(x, y)|^2$. It will be implicit that we are analyzing for the interference of light that is in a single polarization state (see the discussion of how different polarization states do or do not interfere in Section 7.2.3).

The magnitude portion of an SLM's action is measurable in a very straightforward fashion: simply change the drive applied to the SLM and measure the intensity of the light that passes through. The magnitude is proportional to the square root of the measured intensity.

Phase is not ordinarily directly visible, and to infer phase we rely on the analysis of observed intensity. Phase interferometry is quite a broad topic, but we summarize here some of the essential elements and their application to optical correlation. We limit our observation and analysis to the interference of two beams.

Suppose we have a voltage-dependent Jones matrix SLM operating in the presence of an input polarization state S_1, and that the output polarization S_2 is analyzed for state S_3. The input polarizer (if present – we might instead rely on the inherent polarization of the coherent source to produce S_1), the SLM, and the output analyzer form a system (as shown schematically in Figure. 7.12) that delivers an image encoded on the lightwave. The action of the SLM is regarded as the magnitude and phase changes induced in the light in state S_3.

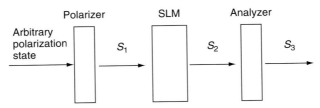

Figure 7.12 Arbitrary coherent light is polarized into state S_1; modulated by the SLM into state S_2; and analyzed for the component of S_2 that is in state S_3. The SLM is characterized by the phase and magnitude of that light in state S_3 according to various drive conditions of the SLM

As an aside, we might choose to operate the SLM without a following analyzer, and indeed there are laboratories in which that is done. If the input polarization state is such that the Jones matrix of the SLM produces a single output polarization, all may be well and good since the light from all pixels will interfere. If, however, the output polarization state varies with voltage, then the light from pixels driven at different voltages will not fully interfere, and the effectiveness of the SLM is reduced. This is because the scalar diffraction integrals, so manipulated as to produce the Fourier transform of a lightwave, assume that the light all across the source aperture is able to interfere. The operating characteristics of a Jones matrix SLM will generally be strongly dependent on the selected states S_1 and S_3, and it will be beneficial to examine the effects of various combinations of those states.

The action of the system is to change all or part of the light in state S_1 into S_2 with varying amplitude and/or phase in that state, and we model the system as though there are no other effects. (Other such modulation effects that actually do occur, and which are sometimes the basis of an SLM, are a change in the frequency of the light or time-modulation of light.) SLMs are typically linear in their propagation of light – i.e., they are not like frequency-doubling crystals – so we will find that the light is little changed in frequency. SLMs certainly time-modulate the light, but the SLMs used in optical correlators do not change state quickly in comparison with the optical processing time. We can regard the SLM as essentially static while the light is bearing the image through the processor. The retardation and the absorption remain as the significant modulation effects. As long as we recognize that we are measuring only the light in state S_3, we can proceed as though the sole effect of the SLM is to retard and absorb the light as a function of local position on the SLM. We can express the absorption and retardation as a field of complex numbers, one number per pixel. At a pixel in question, denote the amplitude transmittance as ρ and the time delay as τ. The phase delay corresponding to τ is, $\phi = 2\pi\, c\tau/\lambda$, and in fact the phase delay (or simply "the phase") of the SLM is what we analyze directly for, rather than τ. We construct the complex phasor, $s = \rho \exp(j\phi)$ as the encoded signal. It remains only to decide what to use as reference for zero phase and perhaps also to decide on an amplitude normalization. Otherwise, once the SLM is characterized, we can proceed just as though all we have is a field of complex numbers to diffract, or take the DFT of, or whatever.

We will mention four architectures for characterizing an SLM. All depend on interference with a reference beam to visualize the complex effect, since phase is an important part of the action of an SLM and the phase is not directly measurable. The conventional architectures are the Michelson interferometer, the Mach–Zehnder interferometer, and the dual-beam interferometer. A new architecture uses the assembled correlator itself as an interferometer, and it is called an *in situ* interferometer.

The interferometers are analyzed from their imaged interference patterns. For the moment, suppose that the action of the SLM is uniform across its face, which is a common enough assumption in the discipline. We activate the SLM with a value V for the drive over its whole face. The illuminated SLM is imaged onto a pixelated imager having x as its coordinate in the direction perpendicular to the interference fringes. The reference beam and the beam from the SLM would optimally be uniform, although as we shall see there are compensations for non-uniformities. Ordinarily a slight slant is put between the reference and object beams so that the reference beam has a spatially affine phase variation across it when its phase is compared with the object beam from the SLM. Consider that we arrange the object beam and reference beam so that they arrive at angles equally disposed from the normal onto an imager's surface where we measure the intensity of their combination.

We now examine the *in situ* characterization setup. The other interferometers are well described in the literature but the *in situ* method is newer in the way that it relates to an optical correlator. The fringe analysis is common to all of the setups.

When a periodic disturbance is written onto the input SLM its transform appears at symmetric locations in the transform (or filter) plane. This is similar to what is indicated when the incoming beam strikes the grating in the dual-beam setup in Figure 7.13. The correlator automatically selects the -1, 0, and $+1$ orders of the diffracted beam. Optimally the zero order (the DC component of the input image) is blocked.

To do the *in situ* characterization of the filter SLM, we begin with the magnitude. Easily enough, we keep a stable pattern on the input SLM, drive the whole filter SLM with a uniform value, and observe the light collected at the correlation plane as the filter drive changes. The magnitude of the filter SLM's response is proportional to the square root of the measured correlation

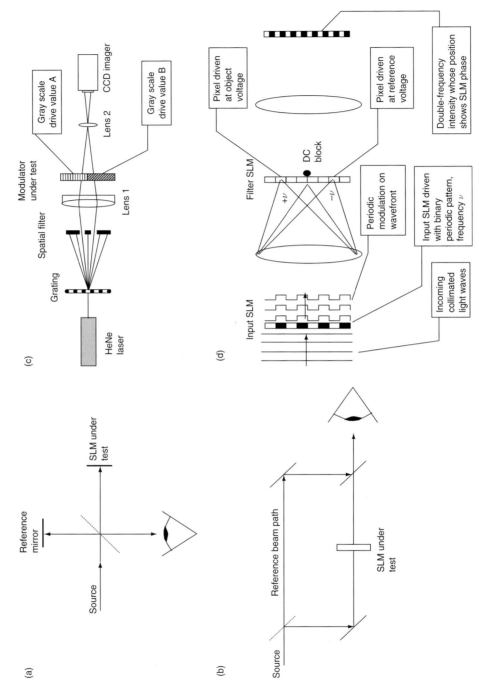

Figure 7.13 Four interferometric setups: (a) Michelson, (b) Mach–Zehnder, (c) dual-beam, and (d) *in situ*

intensity. Similarly, by holding the filter drive steady and varying the input SLM's drive we can characterize the input SLM's magnitude.

Phase is a little more troublesome. We begin by putting a periodic signal onto the input SLM. In order to minimize the effects of coupled-input SLM behavior, we use only two values of the input drive, and we repeat a value a sufficient number of times so that its doubled-frequency appearance in the correlation plane will be resolvable. For example, we might use a sequence such as: [0 0 0 0 255 255 255 255 0 0 0 0 ...] as input bars. The periodicity of the input creates matching transform plane locales (at positive and negative frequencies) where there is appreciable energy density. The halves of the filter SLM (and thus the positive and negative frequencies of the bar pattern) are driven with different values, and the differential effect of the drives is analyzed for the filter SLM's phase.

We approximate the input pattern as a sinusoid:

$$u(x_{in}) = \frac{1}{2}[1 + \cos(kx_{in})] = \frac{1}{2}\left\{1 + \frac{1}{2}[\exp(jkx_{in}) + \exp(-jkx_{in})]\right\} \quad (7.60)$$

The input signal u is decomposed under diffractive Fourier transform into a DC term, a positive frequency term, and a negative frequency term, at the filter modulator. We now apply signals to each half of the filter modulator, one of which (the positive f_x side, say) we hold constant (at some convenient value such as either maximum transmittance or zero voltage) and refer to as unity. It is the SLM's action on the other half that we change by altering the applied drive, V. It multiplies the negative frequencies by the complex factor, $\tau(V) = M(V)\exp[j\theta(V)]$, while we also block the DC component with a physical stop (or minimal-magnitude SLM value) at the filter plane. Variables M and θ are, of course, respectively the magnitude and phase of the SLM action. In the retransform (correlation) plane, we have the complex amplitude s derived from u by having its DC component removed and the positive and negative frequencies differently affected:

$$s_{no\ DC}(x) = \exp(+jkx) + M\exp(j\theta)\exp(-jkx) \quad (7.61)$$

(We suppress constant factors. Also, k and x refer, as appropriate, to input or correlation plane quantities without explicit notice of the scale difference when the correlation plane imager has a different resolution than that of the input SLM.) We wish to extract the values for M and θ by observing the intensity of $s_{no\ DC}$. When $s_{no\ DC}$ has its intensity $I_{no\ DC}$ detected in the CCD imager:

$$I_{no\ DC}(x) = [\exp(jkx) + M\exp(j\theta - jkx)][\exp(-jkx) + M\exp(-j\theta + jkx)]$$
$$= 1 + M^2 + 2M\cos(\theta - 2kx) \quad (7.62)$$

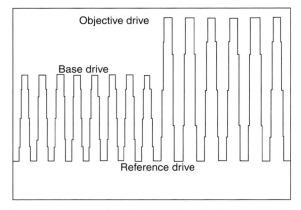

Figure 7.14 The drive pattern for the input SLM when ascertaining its phase. Left and right halves are driven at differing frequencies and between different limits. Bandlimiting followed by detection at the correlation plane yields an analytical signal phase. The objective drive value is changed, and the difference in analytical signal phase gives the change in SLM phase caused by the change in objective drive value

Note that the phase θ of the SLM appears directly in the argument of the cosine term, and that it appears with a term $2kx$ that is double the input frequency. Removing DC light balances the cosine about zero, and detecting it doubles the frequency. (Remember the trigonometry identity, $2\cos^2\theta = (1 + \cos 2\theta)$.) The analytical signal is ideally adapted for measuring θ. We compute the analytical signals, $s_a(V, x)$, and $s_a(V_{ref}, x)$ as a function of position x in the correlation plane, and of the drive voltages V (the objective) and V_{ref} (the reference). The complex ratio τ is the transmittance of the SLM, and the phase is directly read from τ:

$$\tau(V, x) = \frac{s_a(V, x)}{s_a(V_{ref}, x)} \tag{7.63}$$

At the time of writing, we have had quite good results from this means of characterizing the filter SLM in place within the correlator.

It has proved more difficult to characterize the input SLM's phase than the filter SLM's and as of this time we have only initial laboratory results. Theory and simulations, plus preliminary laboratory results, encourage us to describe the method here.

Input a signal of vertical bars across the whole input SLM. The left half is driven at frequency f_1, and the right half at f_2, as in Figure 7.14. The left half alternates between the base value and the reference value, the right half alternates between the reference value and the object value. When the filter SLM blocks all but narrow one-sided passbands around the two fundamental frequencies on one half of the frequency plane, the effect at the correlation plane is that each of the two resulting complex sinusoids extends across the whole correlation plane, and interference at the difference frequency results. The interference is not exactly sinusoidal, because the passbands are inevitably not each at a pure single frequency and the source patterns are not each sinusoidal and uniform across the input SLM. The drives produce complex values, z_{base}, z_{ref}, and z_{obj}. The signal arising from the left side is modeled, with acknowledged imprecision, as the sinusoid, $(z_{base} - z_{ref}) \cos(2\pi f_1 x)$, and the right-side signal as $(z_{base} - z_{obj}) \cos(2\pi f_2 x)$. If we have set up the filter SLM so as to have a low-transmittance point on its operating curve, we use that value and a high-transmittance value to create passbands only in the vicinities of $+f_1$ and $+f_2$. The complex sinusoids, $(z_{base} - z_{ref}) \exp(j2\pi f_1 x)$, and $(z_{base} - z_{obj}) \exp(j2\pi f_2 x)$ will reach the correlation plane. The detected intensity is then:

$$\text{intensity} = \left| z_{base} - z_{obj} \right|^2 + \left| z_{base} - z_{ref} \right|^2$$
$$+ 2 \, \text{Re}\left\{ (z_{base} - z_{obj})(z_{base} - z_{ref})^* \exp[\, j2\pi(f_1 - f_2)x] \right\} \quad (7.64)$$

Just as for the filter SLM, we use digital processing of the intensity pattern to create the analytical signal from a digital bandpass around the difference frequency, $(f_1 - f_2)$.

$$\text{Sig}_{an} = (z_{base} - z_{obj})(z_{base} - z_{ref}) \exp[j2\pi(f_1 - f_2)x] \quad (7.65)$$

We then change the drive for the objective part of the input pattern and recompute the analytical signal. The ratio of the two analytical signals is nominally a constant function of position x:

$$\rho = \frac{\text{Sig}_{an,1}}{\text{Sig}_{an,2}} = \frac{(z_{base} - z_{obj,1})}{(z_{base} - z_{obj,2})} \quad (7.66)$$

The "cookbook" description is this:

1. Write two high-frequency bar patterns at the input. One of these patterns oscillates between two drives we call "reference" and "base," the other oscillates between "reference" and "objective." The reference, base, and objective complex values are ideally well separated from each other. The high frequencies are so chosen that their difference frequency is observable at the correlation plane.

2. Use the filter SLM for spectral isolation of the low-frequency difference between the the two high-frequency input components. That is, except for a window surrounding the beat frequency, the filter SLM is set to low transmittance. (We assume that the filter SLM is capable of a low-magnitude transmittance – and if the correlator is not set up to permit such a low-magnitude filter transmittance, in our opinion it should have been!)

3. Use the analytical signal to extract a value for the phase of the interference pattern.
4. Change the "objective" drive value and repeat Steps 1 to 3.
5. Infer the relative phase between the two objective drive values.
6. Repeat Steps 1 to 5 for a sufficient number of objective pairs, to construct the whole operating curve.

The foregoing analysis assumes that the action of the SLM is uniform over its face. The display industry is the major financial force in SLM development, and it places a high value on visual uniformity in the manufacture of display SLMs, since even a single misbehaving pixel out of millions can attract the attention of the human visual system (the eye). As previously mentioned, the preferences for optical signal processing and for display are different. In optical signal processing we can stand considerably more non-uniformity and errant pixels than the display application, and we prefer a different kind of analog behavior as well (see the later discussion on an SLM's "coverage" of the unit disk). Practical experience has shown that SLMs can come with non-uniformities that are significant problems to optical correlation but of no consequence to display, so we need a method to deal with them. We call this method a "depth" analysis of the fringing, as opposed to the "sideways" analysis we just developed. By depth we imply that measurements at one location are taken at differing times using differing reference beam values, whereas in sideways analysis we examine spatial patterns in an image from a single snapshot.

7.4.2 *"Depth" fringe analysis for spatially variant SLMs*

In the foregoing analysis, fringes that occur simultaneously across the face of the imaged SLM are captured in a frame grabber and fitted with various parameters. If the SLM has significant spatial variability in its action, the fringes will not be cleanly delineated and the method fails; we then revert to fundamentals. What is important later on, when we are computing the drive value for each pixel in order to assemble a filter, is the set of complex values that are accessible to the pixel. That is, we fundamentally want to know the action of each pixel, and we can use this information regardless of whether or not the SLM is uniform. To obtain this information, we re-order the data set that we obtain according to the prescription for the uniform SLM given above. There is one difference, though: we will preferably land the reference beam onto the imager so as to minimize the number of fringes across its face. The interference will be present, but we will not wish to have many fringes as they will unnecessarily complicate the analysis. In practical use, the filter SLM is ordinarily positioned perpendicular to the beam (or as nearly perpendicular as the possibly warped face allows) – i.e., it is mounted in a minimum-fringe orientation.

Suppose that we image the SLM exactly, pixel for pixel, onto the imager. We begin this process in the same way as for the uniform SLM; we block in turn the reference and modulated beams to obtain $M(x, y)$ and $A(x, y)$. For each value of V we image the interference between the modulated and reference beams, obtaining the intensity data set, $I(x, y, V)$. With as near zero fringe as possible across the modulator, and with the reference and modulated amplitudes in appropriate ratio (near unity), then at each SLM pixel (x, y) we have the data set I that we can analyze for $\theta(x, y, V)$, just as before. It is just that the fringe appears as a function of V at each (x, y), not as a function of x at each V. There is a much greater amount of data to be handled, since we potentially have a different operating curve for each pixel. However, those are the breaks of the game when we deal with non-uniform SLMs. Clustering of operating curves [82] reduces the mass of data, and it can be tuned to the problem at hand. A processing problem that requires a high degree of filter precision (typically the discriminant filters that are trying to tell apart similar objects) can have its clusters set by criteria that have little variation amongst the members, although this is at the expense of handling more clusters.

7.4.3 Establishing synchronism of D/A/D mappings

We compute a filter drive based on the set of complex values that the filter SLM's pixels can produce, in the hope that the drive value we assign will be applied to each pixel. This is easiest if a digitally stored filter drive is directly displayed on the filter SLM. However, when one uses an SLM that is taken from a projection display system, the electronics that drive the SLM are often configured to take its signal from an analog video waveform such as RS-170. In this sort of correlator system, a digital filter drive is stored on a board that converts its internal discrete array into a string of analog video lines that are resampled for display on the filter SLM's pixels. Here we risk being led astray, with a filter SLM's drive value being some sort of neighborhood average of what was intended. The problem is that the display pixel sampling is not likely to be synchronized with the digitally stored values; this is because in the usual display of video imagery for visual consumption there is little or no need for sub-pixel registration precision. We have to provide that for ourselves. An example of the problem is illustrated in Figure 7.15, where we want the ith pixel to have a value V_i, but it ends up having a value associated with the average of V_{i-1} and V_{i+1}. This is not a problem where the filter is slowly changing in value as a function of frequency, but the advanced filters are ordinarily found to be very busy functions of frequency. This puts strong demands on both optical and electronic alignment.

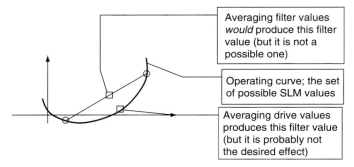

Averaging filter values *would* produce this filter value (but it is not a possible one)

Operating curve; the set of possible SLM values

Averaging drive values produces this filter value (but it is probably not the desired effect)

Figure 7.15 Illustrating the ill effect of incorrect assumptions in mapping from digitally stored filter drive values to the affected filter. If the assumed relationship between storage and filter pixels is wrong, then the incorrect filter value is likely to result

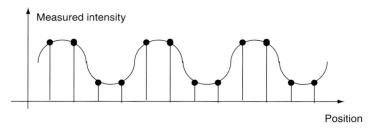

Measured intensity

Position

Figure 7.16. A very sensitive way of determining the mapping between storage cells and SLM pixels is to draw storage cell values from a sine wave with varying phase and spatial frequency, until two-on, two-off pixels are observed when imaging the filter SLM

In aiming for a discrete filter drive value to be displayed on a given filter pixel, we must pre-compensate for the discrete-to-analog-to-discrete electronic processing that will follow.

Figure 7.16 illustrates how to learn the spatial mapping between storage cells from which the filter SLM is driven, and the pixels. We have found a digital moiré method to be satisfactory. It assumes linearity in the rate of change of board location with rate of change of position on the SLM. There is ordinarily not a problem in mapping from lines on the drive board to lines on the SLM. At the center of the problem, then, is finding the affine mapping constants k and b (scale and offset) so that

$$n = km + b \qquad (7.67)$$

where n is the pixel number on the SLM, and m is the pixel number on the drive board. The value of b produces an offset in the analog line so that we can hit

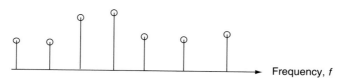

The set of drive values that are determined as optimal for a given problem. The set of frequency indices are those at the filter plane pixels.

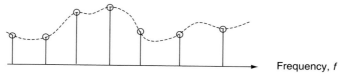

Here we have drawn a bandlimited video signal that will be sampled by the SLM circuitry to produce the desired drive values

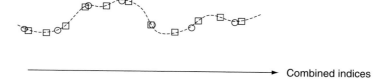

Once the relationship is determined between the pixel grid (circles are data values) and drive storage location (squares are data values), an interpolation on the circles produces the values to be stored as filter drives at the storage locations (squares)

Figure 7.17 Interpolating to compute stored drive values that construct desired drive voltages at the SLM pixels

the first pixel on the SLM, and the value of k stretches the analog line so that we end on the last pixel. For the moment it will be convenient to let x be the continuously variable version of discrete n, and to let y be the similar version of m. We wish to apply the drive value $v(n)$ to pixel n, and we have to determine the set of stored values $d(m)$ in the vicinity of $m = (n - b)/k$ that will produce it, realizing that the value of m so computed will not generally be an integer. When we know k and b, we can interpolate storage cell values from, say, sinc $(\cdot)$ interpolation on the optimal drive values.

7.5 Analytic signal

It is extremely convenient to use complex numbers to describe quasi-periodic phenomena (light, in particular). In this section we justify using the complex description of real time-varying quantities. If the reader is comfortable with complex descriptions of electromagnetic phenomena the section can be skipped.

The analytic signal depends on the famous Euler's identity:

$$e^{j\theta} = \cos\theta + j\sin\theta \tag{7.68}$$

From Euler's identity we also have

$$\cos\theta = \frac{1}{2}\left(e^{j\theta} + e^{-j\theta}\right) \tag{7.69}$$

and

$$\sin\theta = \frac{1}{2j}\left(e^{j\theta} - e^{-j\theta}\right) \tag{7.70}$$

Many manipulations involving trigonometric functions are made easier by using this identity. We can use $\exp j\theta$ where we would otherwise use $\cos\theta$, and then appropriately take the real part when we are through. As long as we are careful in performing only linear operations, we can interchange these operations with taking the real part of the expressions. We frequently find complex notation to be very convenient in describing optical processes.

Our present objective is to begin with a real nearly periodic signal and determine the appropriate complex form. We want to regard a nearly monochromatic oscillation as the real part of a complex time-varying phasor, and we will further require that the rotation of the phasor be unidirectional. That is, if we look in the three-dimensional space with time as one axis and the real and imaginary axes as the other two, the complex phasor describes a spiral about the time axis. A consequence is that the negative frequency component of the spiral (which we call the analytic signal) is zero; this condition specifies how to compute the imaginary part of the analytic signal from the given real part. The clue is to compare among the previous three equations. The trick is to do the operations in the frequency domain. Let us replace θ with $2\pi\nu t$ to emphasize that we are dealing with time-varying quantities.

An intuitive view of the analytic signal is that the analytic signal, $f_a(2\pi\nu t)$ is to $f(2\pi\nu t)$, as $\exp(j2\pi\nu t)$ is to $\cos(2\pi\nu t)$, and we build the analytic signal from $f(2\pi\nu t)$ in the same fashion as we would build $\exp(j2\pi\nu t)$ from $\cos(2\pi\nu t)$ using the forms above. Suppose we have a strictly monochromatic signal, e.g., the signal is a pure cosine. We convert $\cos(2\pi\nu t)$ to $\exp(j2\pi\nu t)$ by going through the frequency domain. The cosine's transform is two delta functions lying at $\pm\nu$. We double the coefficient of the positive frequency coefficient and replace the negative frequency's value by zero, and then inverse transform – thus converting $\cos 2\pi\nu t = \frac{1}{2}\left(e^{j2\pi\nu t} + e^{-j2\pi\nu t}\right)$ to $e^{j2\pi\nu t}$. That is, we transform,

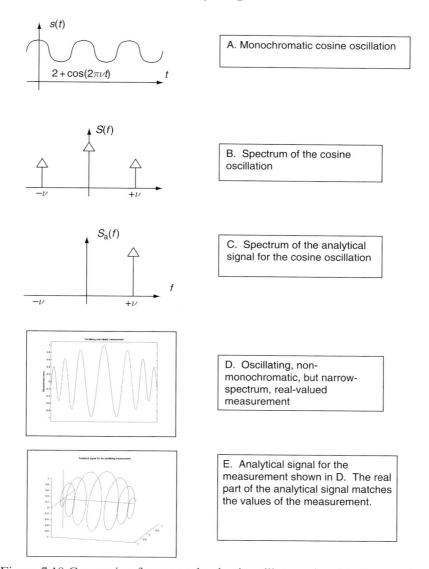

Figure 7.18 Conversion from a real-valued oscillatory signal to its complex-valued rotary analytical signal

multiply in the frequency domain by $2\,\mathrm{sgn}(\nu)$ (where $\mathrm{sgn}(\nu)$ is -1 for $\nu < 0$, 0 for $\nu = 0$, and $+1$ for $\nu > 0$), and then inverse transform. (The convention is that the zero-frequency term is unaltered in ascertaining the analytic signal, but we most often want to suppress DC in our analyses.) Multiplication in the transform domain by the sgn (ν) term is closely related to taking the Hilbert transform, $\hat{f}(\nu) = 1/\pi \int_{-\infty}^{\infty} f(t)/(t-\nu)\,dt$ of $f(\cdot)$, since taking the Hilbert transform is seen to be a convolution with $1/\pi t$, whose transform is $j\,\mathrm{sgn}\,\nu$. Thus the analytic signal, $f_a(2\pi\nu t)$ is obtained from the real-valued $f(2\pi\nu t)$ by:

$$f_a(2\pi\nu t) = f(2\pi\nu t) + \mathrm{FT}^{-1}\{\mathrm{sgn}\,\nu\,\mathrm{FT}\{f(2\pi\nu t)\}\} \tag{7.71}$$

This equation displays the computational method of obtaining the analytic signal.

Now instead of being monochromatic, suppose $f(t)$ is a carrier of constant frequency $2\pi\nu$, modulated by a slowly varying envelope $A(t)$, and a slowly varying phase $\phi(t)$ according to

$$f(t) = A(t)\cos[2\pi\nu t + \phi(t)] \tag{7.72}$$

We invoke the term "slowly varying" to mean, somewhat imprecisely, that the amplitude spectra of the positive and negative frequencies do not overlap – that they are confined well away from zero. The transform of the signal (in optical signals, a comparatively slowly varying modulation of a 10^{14} Hz carrier) consists of narrow bands centered on $\pm\nu$, and we would like to focus our attention on just the positive frequency component. The analytic signal in Eq. (7.72) is seen to be, $f_a(t) = A(t)e^{j\phi(t)}$, in which expression we have suppressed the carrier frequency, ν, as is customary since it is common to all terms and carries no significance for the optical processing system (any practical modulator or sensor will average over many cycles of the 10^{14} Hz carrier). In the mathematical descriptions of electromagnetic waves and the effects of optical components, we shall use the analytic signals and functions without further ado.

For a deeper discussion including the analytic function for constants, pairs of Hilbert transforms, even/odd Hilbert transform characteristics, etc., see the excellent references by Bracewell [45] and Goodman [32]. We shall use the complex form for the electromagnetic wave and the optical elements without further comment.

8

Limited-modulation filters

In this chapter we shall optimize statistical correlation pattern recognition under the constraints of being implemented optically. We shall also trace their genealogy and look at some of their predecessors.

8.1 Introduction

The objective of this chapter is to treat optical correlation pattern recognition (OCPR) by considering signals, correlation metrics, noise, and limited filter domains.

Digital correlation is computationally more flexible and less noisy than optical correlation. On the other hand, optical correlation can be much faster, and have less weight and volume and power consumption, etc., which motivates us to give it a try. The constraints are very different in the two processes; in some digital correlation filter designs we have seen the necessity of *introducing* constraints (such as that the filter should have unit energy). In optics we are thoroughly constrained already, without introducing any artificial constraints. Unfortunately, in contrast to the digital version, the form of the optical constraint does not usually provide the solution for an optimizing filter. In this chapter we will nevertheless see how to operate optimally within the limitations imposed by optical practicalities.

The objective of pattern recognition is to recognize the presence of the reference object in the input signal or scene. Optical correlation aims to make a comparatively bright spot of light that is detectable against a notably dimmer background when the desired object is present in the input image. We shall work with various criterion functions that measure the optical distinctness.

In Chapter 5 we developed correlation basics, and in Chapter 7 we introduced the basic optics necessary to do the correlation. In this chapter we bring these two chapters together and show how best to go about the pattern

295

recognition in physically realizable systems. We will show that a simple and attractive mathematical form – a version of the matched filter – optimizes a large number of the metrics that we would choose to optimize.

In this chapter, we will base our metrics on what we can observe in the output – the intensity. This will have us using slightly different statistics. For example, we will need to use the variance of intensity, whereas in previous chapters we used the variance of magnitude.

There are three principal stages in the historical development of OCPR discussed in this chapter.

1. The history of OCPR has seen many ad hoc approaches (some of which we introduced in Chapter 5) to filter optimization, in which generally good ideas are used to guide the selection of a filter, but not always on the basis of analytical rigor.
2. In the second stage, we compute a fully competent digital filter that is correct according to the precepts developed earlier, and then find a way to adapt it to the devices in the optical correlator.
3. In the third stage, we begin by expressing a metric that is motivated by the concepts already developed, but altered to acknowledge the facts of physical implementation. We then optimize the metric expressly in view of the limited set of filter values we can make.

Another design philosophy we espouse is the use of every degree of freedom we can lay our hands on. A prime example is in the optical phase at the correlation plane; that phase is unobserved, so it becomes a free parameter that can be used to good advantage in computing an optimal filter.

A good deal of signal and image processing concerns itself with processing a transmitted and noisy signal to reconstruct the original signal. One wishes to reduce error in the reconstruction. We have a similarity and a difference here. The similarity is that we are trying to minimize an error but, differently, we are using the signal's harmonic content to build statistically distinguishable optical results for in-class and out-of-class objects. We use the statistical pattern recognition (SPR) metrics to quantify the discriminability, just as the expected error's energy quantifies the quality of reconstruction. The reconstruction error – or, better in our application, the *con*struction error – is unavoidably larger with limited-modulation filters. The light that passes through the filter unabsorbed and not contributing favorably to the pattern recognition problem goes *some* where; it is distributed hither and thither in the correlation plane. In doing so it is detected and contributes to the background against which the correlation is to be detected.

Historically there have been a number of approaches to operating a correlator using choice of filter SLM type and method of driving that SLM. The historical filters have not typically been tied rigorously to the SPR metrics, and with respect to the stated aim of OCPR we shall refer to them as physically based rather than statistically based. They include the phase-only filter (POF), maximizing the correlation intensity with a coupled SLM (its phase and magnitude vary together), binarized filters, etc. Optimizing the SPR metrics within optical correlation is a more recent development. There are information metrics by which we might judge (and then optimize) the correlator's performance, and they include the Bayes error and cost [1, 2], the area under the ROC curve [29], the Fisher ratio [1], expected information (averaged log likelihood ratio) [83], the Kullback–Liebler "distance" [83], the Bhattacharyya and Chernoff bounds [22], and others.

Our preferred approach to filter optimization is analytical optimization, in which it is shown that the metric is at an extremum obtained by choice of filter. Another approach is found in the various ad hoc optimizations in which justification is by analogy rather than by strict analysis. An example of the ad hoc approach is choosing to match a signal's phase on an SLM that has coupled phase and magnitude behavior. Although phase is a stronger conveyor of image information than magnitude, just matching the phase may not necessarily optimize a pattern recognition metric of interest. We shall develop the conditions for which phase matching is optimal.

It is easy to compute competent digital or optical filters that are not expressible on extant SLMs, and the challenge becomes to use the optical filter values at hand to approach the performance of the ideal fully competent SLM. In this chapter we will see how a correlation metric will help us select an SLM for a stated purpose, to run a modulator as it stands, or to select from the set of operating characteristics of the SLM we are using.

There are at least two principal environments for constructing optimal filters. Firstly, suppose you already *have* a correlator and you want to operate it most effectively. You must know the whole system accurately and quantitatively. Along with the range of the filter SLM, you must know the signal that is presented to the correlator and how it is encoded to appear at the filter SLM. The noise environment is important information. Your knowledge of all of these things will guide you to the optimal filter.

Secondly, if you are *designing* a correlator, you would do well to know the problem you are trying to work with, and you should quantitatively describe the metric by which you will judge the result. The system, including the filter range and the filter itself, limits the achievable value of the metric. The metric becomes the tool by which a system design is optimized.

A powerful and general method applicable to both of these objectives is to design a metric that suits your correlator's pattern recognition problem, and then to maximize that metric by filter selection (in the first case), or by system design plus filter selection (in the second case). For every system design there is a filter range, and for every filter range and metric there is an optimal filter; we can see that we do not independently consider correlators, metrics, and optimal filters.

In comparison with digital processing, OCPR has some serious limitations to recognize. Most significant among them are:

1. The limited set of filter values that the filter SLM can take on. The physics of light modulation and convenient architectures admit only certain subsets of the complex plane – and, more particularly, of the unit disk in the case of passive modulators.
2. Larger amounts of noise in the detection process, including some that is not filterable. Optical detection is subject to noise, which does not happen in digital computations. Scattered light in the correlator's optical path, shot noise in the detector, etc., are examples of physical noise that are not experienced in computational pattern recognition methods.
3. Imprecision, including spatial variability of an SLM, non-uniformity of illumination, incorrect alignment, temporal changes in the behavior of an SLM, etc.
4. The loss of phase in detecting the correlation plane. Optical correlation detects the intensity of the light field, and distinctions such as $+1$ versus -1 that are plainly possible in the digital sense are impossible in optics.

To expand on the first point, there are strong differences between the limitations imposed optically and the limitations apropos digital processing (e.g., that a filter should have unit energy). The digital Rayleigh quotient is a clear example. In order to optimize it digitally we use a degree of freedom: that the Rayleigh quotient is invariant to a complex factor applied to the filter. Because of this, filter energy can be restricted to unity to reduce the solution space. However, the constraints imposed on optical filters are substantially different. We cannot plan to multiply all filter values by the same complex factor and still have a realizable filter. The most fundamental optical limitation is that all filter values must be realizable, and this is a very different situation than for the digital condition.

We are not able to produce optically most of the ideal filters developed in Chapter 5, because all modulators are limited in the combinations of phase and magnitude that they can work on the light, whereas the digital values assumed by the filters of Chapter 5 are limited only by the digital machinery on which they are computed. Optical filter values are drawn from a limited set of complex values. In digital processing of an image a vastly larger set of complex numbers, both in dynamic range (magnitude), phase, and computation precision is possible. The most competent passive optical filter SLM would be one that can take on exactly any desired value in the unit disk of the complex plane, with no representation error, and with infinite precision. Unless we use a lasing medium for the filter SLM (and to our knowledge, none do), the complex filter values are limited to the unit disk, since the processed light necessarily has the same or smaller magnitude after it leaves the SLM. If we are using a filter medium driven by a single analog parameter (such as a liquid crystal responding to a voltage), we are restricted to a curvilinear subset of the unit disk. Some SLMs have binary or ternary response to the drive and so are even further restricted. Enlarging the set of complex values is possible by combining SLMs, but to do so complicates the optical architecture to a degree that may be unpalatable after considering cost versus benefit.

The second major limitation is that the optical system invariably inserts noise to the correlation. Some of that noise is affected by the filter (e.g., noise in the input signal) and some is not (e.g., light that leaks around the active part of the filter pixels, or arises directly in the detection process). Some is input signal energy that the filter is not able to direct into the correlation spot. Each can be accommodated in the optimal filter theory, and demonstrably better results attain when the noise is kept in mind as a filter is computed.

Regarding the third limitation, both digital and optical correlation methods have limits to their resolution. The limit may lie in the finite precision with which a signal will reproduce upon repeated presentations to the system; in the noise accompanying the input signal as it enters the correlator: in the finite precision of digital computation; in the optical noise of scattered light; in the electronic noise of detecting the light at the correlation plane; or elsewhere. The point is that a proper metric will take into account the physical nature of the observables upon which it is based. The metrics in this chapter show an evolution from an ideal system (noiseless, infinite resolution) to more practical forms.

We make the historical observation that most of the quantitative work to be found in the OCPR literature as of this writing is performed for statistical evaluation of the magnitude of the correlation electromagnetic field. We think it preferable to use statistics of the measurable quantity, the detected optical intensity. (This intensity is, of course, proportional to the magnitude squared.) The distinction is not insignificant; as a quick elucidation, note that the mean of the square of a quantity is not usually the square of the mean of the quantity. Because the perspective of using statistics of the measured intensity is a relatively new concept in OCPR, we shall develop methods in both the magnitude and intensity senses.

Now, one last practical point. The optimal filter value at a given frequency is a function of *every other frequency's filter value*. Optimizing a filter is rather like getting a good haircut. There is not an ideal length for a hair considered alone – the optimal length of each hair depends on the lengths of the other hairs. Similarly the optimal value of a frequency's filter value depends simultaneously on the values it can take on and the values that are realized at every other frequency. Fortunately, we can express the effect of the whole set of filter values in a small set of parameters, and these parameters become a vastly reduced set of independent variables. The parameter values are not known a priori, so they become items to search through during the filter optimization. The ramifications of this point are more difficult to accommodate for physically realistic filter SLMs than for idealized SLM models, and this accounts for the development of ad hoc methods of filter optimization for OCPRs to use before moving on to the analytic methods of this chapter. A positive feature of the algorithms in this chapter is the size of the parameter set which is very small compared with the number of frequencies in the filter, so the search space is exceedingly small compared with the number of filter frequencies. (Consider that the parameter set size is typically one complex scalar per reference image, compared with tens of thousands of frequencies at which filter values are to be computed.) Nicely, the parameter values are computable functions of the filter values once the search has produced the optimal parameter set, so we have a consistency check built in. The parameter set implies the filter that implies the parameter set; if a candidate parameter set does not reproduce itself through the filter, the filter is not optimal.

8.2 History, formulas, and philosophy

A few points on nomenclature: The *filter* is the whole set of complex values implemented on the filter SLM, and the *domain* of the filter is the possible set of values for the filter SLM. The term *spectral* means "as a detailed function of frequency," and we will employ the term *global* to connote quantities that have been assembled from spectral components. Some concepts, such as SNR, that are ordinarily thought of in their global connotations also have spectral variants. The *filter drive* is the control value (often voltage) that we apply to the SLM in order to evoke the filter. The digital precursor of the actual voltage sent to the SLM is also referred to as a filter drive, and the calibration of the SLM is most often done in terms of complex response to a digital control. We particularly eschew the loose terminology in which the digital control values would be called a filter.

For historical reasons we distinguish between *magnitude-based* and *measurement-based* quantities. Much of the published OCPR work (including our own) operates with the expected value and the variance of the correlator's electromagnetic magnitude. Consistent with our increasing interest in applying SPR theory to optical correlation, we have come to favor an approach in which measurements – the observable outputs, or intensities – from the correlator are considered rather than the correlation magnitude. But because of the considerable body of magnitude-based literature, we include discussion of magnitude-based computations and optimizations.

In relation to linear systems analysis, the optical device is actually a convolver, but the result can be cast in the form of a correlation. In earlier chapters, we discussed convolution and correlations, but we assumed real signals and images in this discussion. In this chapter, we allow the input signals and images to be complex-valued. As a reminder, here are the definitions of correlation and convolution in their complex-form.

Convolution of x and h, both functions of t:

$$(x * h)(t) = \int_{\tau=-\infty}^{\infty} x(\tau)h(t-\tau)d\tau \tag{8.1}$$

Correlation of x and r, both functions of t:

$$(x \otimes r)(t) = \int_{\tau=-\infty}^{\infty} x(\tau)r^*(t+\tau)d\tau \tag{8.2}$$

From these definitions we see at a glance that the impulse response function $h(\cdot)$ and the correlation function $r(\cdot)$ are the reversed and conjugated versions of each other. This being true, and since we know that we alter the transform of the reference object to build the filter (whitening, altering phase, scaling, converting to realizable optical values), we expect that changes are made to the reference image in the original coordinate system. It is instructive to take the inverse Fourier transform of the optimal realizable filter and see how different an object we are actually correlating with, compared to the one we went in with. Generally it will be a complex object, so a special effort has to be made to envision the phase of the complex object. One method uses color imagery as follows. Consider the hue–saturation–intensity (HSI) coordinate system (see any image-processing text [84] or program). One maps the magnitude of the inverse transform to the intensity, and the phase of the complex object to hue, with a constant saturation. The frequency content of the modified reference image is altered, being of lower intensity where the input noise is higher. The modified reference object is a result of the entire suite of actions befalling the original reference image on its way to becoming an optimal filter.

For OCPR, we define the SPR metrics in terms of intensities and probability density functions, measured at the center of the correlation plane (for the centered appearance of the reference object). C is a class (Λ denotes the accept class, Ψ the reject class). Function $\langle\cdot\rangle_C$ is the mean of a quantity over class C.

We begin the process with an input $\upsilon(x)$ that might be a video signal. It is encoded onto light by the complex function $e(\upsilon)$, transformed optically, filtered, re-transformed, and detected. As has become traditional in the OCPR community we use a one-dimensional scalar (x or f, in this case) to represent the two-dimensional vector quantity.

8.2.1 Nomenclature

υ is the level (e.g., video voltage) of drive value applied to the (input or filter) SLM.

$e(\cdot)$ is the complex encoding of the (input or filter) SLM.

$s(x) = e(\upsilon(x))$ is the optical signal when υ is the input drive.

$H(\upsilon(\cdot)) = e(\upsilon(\cdot)) = M(\upsilon(\cdot))\exp(j\theta(\upsilon(\cdot)))$ is the filter's phasor when υ is the filter drive.

$S(f) = \Im\{s(x)\} = A(f)\exp(j\phi(f))$ is the optically transformed signal.

$H(f) = M(f)\exp(j\theta(f))$ is the filter.

$n(x)$ is the noise that we presume to be additive to $s(x)$, stationary, and known only as far as its power spectral density $P_n(f)$.

$P_n(f) = E\{|\Im\{n(x)\}|^2\}$ is the additive noise power spectral density.

$D = B\exp(j\beta) = \sum_k H_k S_k$ is the central correlation field.

$\Upsilon = \sigma^2_{\text{mag},n} = \sum_k |H_k|^2 P_{nk} = \sum_k M_k^2 P_{nk}$ is the filtered-noise-induced variance in the magnitude of the central correlation field.

$I = B^2$ is the square of the central correlation magnitude with no noise, subscripted where necessary by the index number of a reference signal.

u is the measurement in its native units (e.g., volts, counts after digitizing, etc.). Note that I is a digital term for computations, u is an optical term for measurements.

R is the instrumental responsivity.

$u = IR$ is the connection between digital and optical quantities.

$p_C(u)$ is the probability density function of measurements, given that the class is C.

P_C is the a-priori probability of class C. Note that this is distinct from the noise power spectral density, $P_n(f)$.

$\sigma^2_{CT} = \sigma^2_C + \langle\sigma_n^2\rangle_C + \sigma_{Id}^2$ is the total intensity variance over class C, modeled as having three sources described below.

$\sigma^2_C = \langle(I - \langle I\rangle_C)^2\rangle_C$ is the component of variance arising in different correlation intensities for the objects within training class C.

$\langle\sigma_n^2\rangle_C$ is the variance owing to additive input noise that passes through the filter.

$\sigma^2_{\text{mag},n}$ is the magnitude variance owing to noise in the correlation plane detection; that is, unresponsive to the filter.

$\sigma^2_{I,\,d}$ is the intensity variance owing to noise in the correlation plane detection; that is, unresponsive to the filter. 'I' is included in the subscript to distinguish the intensity variance from the magnitude variance.

For some of the above quantities we may also use vector expressions.

$\mathbf{A} = [A_k \exp j\varphi_k]$ is the transform of a reference object.
$\mathbf{P}_n = [P_{nk}]$ is the power spectral density of input noise following the input SLM.
$\mathbf{H} = [M_k \exp(j\theta_k)]$ is the filter.
$B \exp(j\beta) = \mathbf{H}^{\mathsf{T}}\mathbf{A}$ is the expectation of the central correlation field from $\mathbf{A}$.
$\mathbf{M}^2 = [|H_k|^2]$ is the vector of filter power.
$\Upsilon = (\mathbf{M}^2)^{\mathsf{T}}\mathbf{P}_n$ is the magnitude variance.

Probability density function of measurements u for class C: $p_C(u) = \dfrac{1}{N_C}\dfrac{\mathrm{d}N_C}{\mathrm{d}u}(u)$, in which $\mathrm{d}N$ is the number of measurement values between u and $u + \mathrm{d}u$.

General form for magnitude-based metrics, to detect objects i or discriminate them from objects j (the us are somewhat arbitrary non-negative weights):

$$J(\mathbf{H}, \mu_1, \mu_2, \mu_3) = \frac{\sum_i B_i^2}{\mu_1 \sum_j B_j^2 + \mu_2 \sum_k |S_{jk}H_k|^2 + \mu_3\Upsilon + \left\langle \sigma^2_{\mathrm{mag},d} \right\rangle} \tag{8.3}$$

Probability density function of measurements u for class C:

$$p_C(u) = \frac{1}{N_C}\frac{\mathrm{d}N_C}{\mathrm{d}u}(u) \tag{8.4}$$

in which $\mathrm{d}N$ is the number of measurement values between u and $u + \mathrm{d}u$.

Chernoff and Bhattacharyya upper bounds [29]:

$$E_{\mathrm{Bayes}} \leq E_{\mathrm{ub}}(s) = P_\Lambda^s P_\Psi^{1-s} \int_{\mathrm{all}\,u} \rho_\Lambda^s(u)\rho_\Psi^{1-s}(u)\,\mathrm{d}u$$

$$\leq \sqrt{P_\Lambda P_\Psi} \int_{\mathrm{all}\,u} \sqrt{\rho_\Lambda(u)\rho_\Psi(u)}\,\mathrm{d}u \tag{8.5}$$

in which $0 \leq s \leq 1$ that minimizes the error's upper bound gives the Chernoff upper bound (the first equality), and $s = 0.5$ expresses the Bhattacharyya upper bound that follows the second inequality.

The relative entropy between two population densities ρ_Λ and ρ_Ψ, also known as their Kullback–Liebler distance [83], is denoted $\mathfrak{D}(\rho_\Lambda \| \rho_\Psi)$.

$$\mathfrak{D}(\rho_\Lambda \parallel \rho_\Psi) = \int\limits_{\text{all } u} \rho_\Lambda(u) \log\left(\frac{\rho_\Lambda(u)}{\rho_\Psi(u)}\right) du \qquad (8.6)$$

8.2.2 Getting specific to optical correlation

There are different sources of measurement variance for the optical implementation of correlation pattern recognition than for the digital, so the noise model is different. In this section we shall describe our measurement variance model for OCPR.

The within-class variance, σ^2_C, is the scatter in correlation intensity values resulting from different responses among the members of the training set. Expanding the definition,

$$\sigma^2_C = \frac{1}{N_C - 1} \sum_{i \in C} \left(I_i - \langle I \rangle_C\right)^2 = \frac{1}{N_c - 1} \sum_{i \in C} \left(I_i - \frac{1}{N_C} \sum_{\ell \in C} I_\ell\right)^2 \qquad (8.7)$$

We model the intensity's noise variance as resulting from a process in which the noise's electromagnetic field – a zero-mean circular Gaussian RV – is added to the correlation field prior to intensity detection. Under the usual assumptions that (1) we have no information about the phase of clutter objects'

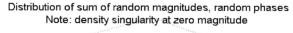

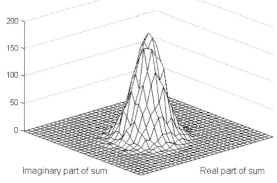

Figure 8.1 Probability density function of a sum of complex numbers. This is the density of 8192 sums having 1024 randomly drawn complex numbers per sum. The original distribution is described by the magnitudes and the phases being uniformly distributed. That original distribution is peaked with a singularity at zero

frequencies, so that (2) the clutter's individual frequencies pass through the filter and add as random uncorrelated complex numbers, (3) that there is independence of clutter and signal, and (4) that there are a large number of such frequencies, the electromagnetic field resulting from clutter then has a circular normal density as plotted in the complex plane. This is because of the Central Limit Theorem – that regardless of the distribution from which x is drawn, in the limit of large numbers of samples the sum of the samples has Gaussian distribution. Figure 8.1, Figure 8.2, and Figure 8.3 demonstrate this effect. In these

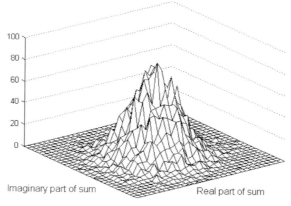

Figure 8.2 The same as Figure 8.1, except that the underlying probability density function is uniform over the unit disk

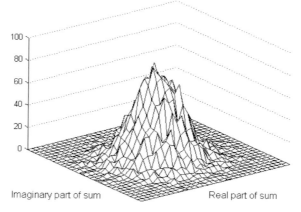

Figure 8.3 The same as Figures 8.1 and 8.2, except that the underlying distribution is uniform along the unit circle

figures we have 8192 samples of averages of 1024 complex numbers drawn from varying distributions. In Figure 8.1, the magnitude is distributed uniformly in [0, 1], and the phase is distributed uniformly in [0, 2π], with phase and magnitude statistically independent of each other. This results in the density of the complex numbers being summed having a singularity at the origin. In Figure 8.2 the complex numbers being summed are distributed uniformly in the unit disk. In Figure 8.3 the complex numbers being summed are distributed uniformly around the unit circle.

In spite of the great differences in the original distributions the sums are seen to approximate a normal distribution. This fact is used in creating a model of the noise-modified distribution of correlation measurements. The filtered signal produces an offset in the complex correlation's electromagnetic field, and the noise adds randomly, and before detection, as a sum described by a zero-mean normal. In Figure 8.4, this situation is indicated; the field is the sum of a static offset and a zero-mean, circular normal RV. The field values in the differential annulus are detected to yield a measurement in the differential range shown. The statistics of the measurement process are derived in the appendix of a paper by Juday [52], and produce the results that are quoted, but not developed, here.

In Figure 8.4 we detect the intensity of the noisy correlation field in which the filtered noise is distributed as a circular Gaussian RV. Integrating through the noise's density produces the following distribution:

$$p_I(u/R) = \frac{1}{\sigma_{\text{mag}}^2} \exp\left(-\frac{\left(B - \sqrt{u/R}\right)^2}{\sigma_{\text{mag}}^2}\right) \exp\left(-\frac{2B\sqrt{u/R}}{\sigma_{\text{mag}}^2}\right) I_0\left(\frac{2B\sqrt{u/R}}{\sigma_{\text{mag}}^2}\right)$$

(8.8)

in which $I_0(\cdot)$ is the modified Bessel function of the first kind, order zero, given by

$$I_0(x) = \frac{1}{2\pi} \int_0^{2\pi} \exp(x\cos\theta)\mathrm{d}\theta.$$

(8.9)

The class average noise-induced intensity variance is

$$\langle\sigma_n^2\rangle_C = \Upsilon^2 + 2\Upsilon\langle B^2\rangle_C$$

(8.10)

Here, B^2 and Υ are both explicitly functions of the filter, and so therefore is $\langle\sigma_n^2\rangle_C$.

The items defined above are digital representations of what happens in the correlator and are what we manipulate in computationally optimizing a filter.

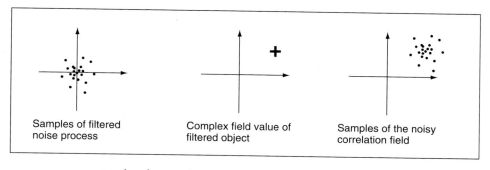

| Samples of filtered noise process | Complex field value of filtered object | Samples of the noisy correlation field |

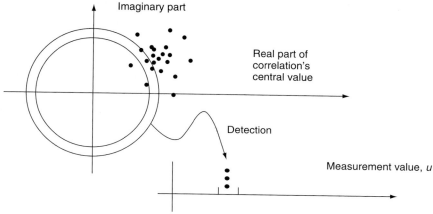

Figure 8.4 An offset normal distribution of complex sums is detected. The resulting probability density function of measurements is described by a Bessel function

The intensity detection variance σ_{Id}^2 is actually an optical quantity, however, and we must have a separate discussion on relating computational quantities to their optical equivalents. If the correlator is operated in a linear regime (i.e., no saturation), its responsivity R can be converted between the digital quantities and optical ones. The responsivity is found from:

$$u = RI = RB^2 \tag{8.11}$$

with u being the observed optical intensity at the center of the correlation plane for a noise-free correlation signal with digital value $I = B^2$. In practice σ_{Id}^2 comes from noise in the detection process, from light scattered by the optics, from light that passes through parts of the filter SLM that are not modulating ("dead" areas), etc. To optimize a filter in computations we need a digital version of the σ_{Id}^2 term that suitably represents the optically observed effect.

The digital computations' value for σ_{Id}^2 is obtained as follows. For a strongly correlatable signal s_{bw}, the ratio of B_{bw}^2 to I_{bw} (computed central correlation intensity to the observation) gives us the conversion units to change the observed intensity variance into σ_{Id}^2. For s_{bw} we construct a high-energy object (e.g., a binary object with a coin-flip determining whether a given pixel is black or white). By techniques presented in the literature [41] we compute the maximal intensity filter for that high-energy object. That filter sweeps as much of the light into the correlation spot as it can so we take the ratio of large numbers rather than small ones. We use the computed value B_{bw}^2 for the digital correlation intensity, and we measure the corresponding physical correlation intensity I_{bw} in its native units (it might be volts on a screen trace through the correlation peak, or it might be counts in a grabbed frame) at the observed peak. Then $\sigma_{optical}^2$, the variance in the detected light (and detection noise that is indistinguishable from such light) scattered around at other locations in the correlation plane, is what we build the value of σ_{Id}^2 from. Assuming proportionality between both pairs of related digital and optical items, we then have:

$$\sigma_{Id}^2 = \left(\frac{B_{bw}^2}{I_{bw}}\right)^2 \sigma_{optical}^2 = R^2 \sigma_{optical}^2 \tag{8.12}$$

as the digital version of the detection contribution to intensity variance. Having described this quantitative connection between digital and optical quantities we shall largely let the conversion between them be implicit during the remainder of this chapter.

8.3 Physical view of the OCPR process

In this section we shall take a very physical view of OCPR and how a correlation spot is formed. The ideas in this section hinge on recognizing that the light in the correlation plane is nothing more than a reorganization of a coherent lightwave that begins at the input plane. This viewpoint is important for several reasons.

- Firstly, in understanding correlation as a linear systems process.
- Secondly, in visualizing the Fourier components in a signal and how phase is important in understanding some aspects of filter optimization.
- Thirdly, in understanding how limited modulation range affects filters, thus making best use of what complex action one has.
- Fourthly, in understanding that a well designed correlation filter will lead to a correlation that the peak may not be exactly at the desired location, but that it will be close.

- Finally, in knowing the limits on how sharp a correlation spot may be produced. We shall not, for now, consider noise at input and at detection in this process.

OCPR is a subset of Fourier optics. Fourier optics differs from conventional optics in that the coherence of the light in the system becomes important in the system analysis. For an example of Fourier optics, the imaging properties of a convex lens can be shown to be the result of two successive Fourier transforms that are expressed in a re-statement of the diffraction integrals for coherent light. In OCPR, an image is encoded into a coherent lightwave, and the physics of the light propagation is arranged so that the light pattern is converted to a simpler form – ordinarily a bright spot – if the information that the system is looking for is present in the input image.

Although the previous paragraph is technically accurate, let's see if we can draw some more penetrable descriptions of what happens. The really fundamental tools are a laser, some lenses, some modulators driven by computer control, an input image source (often a television image), and an imager with which to read the processed light.

8.3.1 Mathematical representation of SLM action

Driving the input SLM with the scene's electronic signal encodes that scene onto the wavefront as a complex field. It is important to know the complex representation of the input scene and to set up that encoding in favor of our pattern recognition process. The SLM's attenuation and retardation are given as phasors, and the action of the SLM is assumed to be uniform over its face. Attenuation is given in the magnitude, and retardation in the phase, of the phasor.

The magnitude linearization of the input SLM – that is, a conversion from υ to $|e|$ – is an interesting item. Suppose for the moment that the input SLM is analog and real (i.e., continuously variable and its phase does not change with the applied drive). This is a condition often sought in setting up a correlator and approximated in modeling it. Typically a correlator has an adjustable hardware conversion from input drive value to the control parameter actually applied to the SLM's pixels. Here is the intuitive but naïve version of how to map υ onto e: υ is delivered from the image sensor so that υ is proportional to the intensity of the viewed scene; hence I want the correlator's output intensity to be linear with υ. That is, if I were to set the filter to a constant value (this act causes the input image to be reconstructed at the correlation plane), I want the intensity $|e|^2$ of the captured correlation plane image to faithfully follow the gray-level presentation υ of the viewed scene. This would have $|e|^2$ being proportional to υ.

The logical version is different, however. Here is the logical version: I want to use the optical correlator to do linear system processing (filtering) of the input signal, followed by intensity detection. I note that in the quantum mechanical sense it is a signal's (complex) amplitude, not intensity, that adds during the internal processing before detection. Thus, to achieve the linear processing that is central to all of the correlation theory we have developed, I want the *magnitude* of the input SLM's encoding, not its *intensity*, to be linear with the input drive value. This would have $|e|$ being proportional to υ, as is desired if we want to correlate $\upsilon(\cdot)$ rather than $\sqrt{\upsilon(\cdot)}$. To linearize the input SLM for intensity rather than magnitude is to do square-root emphasis of the low-magnitude harmonics over the high-magnitude harmonics in the correlation process. Clearly this would be an unwanted emphasis, so I reject the intensity linearization.

The mathematics we use depends on linearity in system response. To remind the reader of the linearity concept introduced in Chapter 3, if input drive value υ_1 produces an encoded value e_1 (expressed as $\upsilon_1 \rightarrow e_1$), and if input υ_2 is encoded on the light as e_2, then for arbitrary (but real, in the present case) scalars α and β, linearity of the input process demands that: $(\alpha \upsilon_1 + \beta \upsilon_2) \rightarrow (\alpha e_1 + \beta e_2)$. If there is curvature in the complex plane of input encoding values, or if the straight line of the input encoding domain does not pass through the origin, then no pre-warping, $e(\upsilon)$, of the input is possible to produce general linearity. The matter of linearity is significant because we need a filter constructed for an input signal $\upsilon(x)$ to work well for real multiples of the input drive. (For example if the signal being recognized arises in a CCD imager, and the scene illumination changes, or if the CCD has automatic gain control and something bright appears in part of the scene and reduces the signal from the object we wish to see. This magnitude variation has received less attention in the literature on filter optimization than other variations such as size or aspect changes. We suppose this is because the linearity of input encoding has not been perceived as a problem.) Let us illustrate these principles.

In Figure 8.5 we show three versions of an input coding range. Figure 8.5(a) is representative of a coupled modulator that has been set up for its minimum phase change. Figure 8.5(b) is a straight line that does not pass through the origin. Figure 8.5(c) is a straight line through the origin.

We see that Figure 8.5(a) does not permit linear encoding, since the two dots on the curve (representing encoding e) joined according to $(\alpha e_1 + \beta e_2)$ fall on the straight line joining the points, and in the presence of the curvature these points are not in the range. So a drive value $(\alpha \upsilon_1 + \beta \upsilon_2)$ cannot produce the encoding $(\alpha e_1 + \beta e_2)$ and linearity fails.

Since the encoding curve in Figure 8.5(b) is straight, the exact argument just advanced for a curved encoding domain does not apply. However, if the line

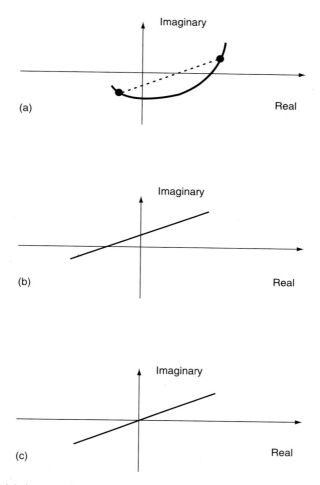

Figure 8.5 (a) A curved operating curve. (b) An offset, but straight, operating curve. (c) A straight line through the origin

does not pass through the origin, linearity nevertheless fails. (One giveaway reason is that a linear system includes zero.)

We are not necessarily home free in Figure 8.5(c). If the magnitude of the encoded signal is not in proportion to the drive, we need a pre-warping operation to achieve linearity. In Figure 8.6, we plot a sigmoidal magnitude as a function of drive, as might result from breakaway starting and saturation at higher drive levels (this is reasonably characteristic of many SLMs). If the input drive v is put through a pre-warping amplifier whose gain characteristic is the inverse of the encoding, then the s-to-magnitude relationship is linearized, and Figure 8.5(c) finally results in a linear encoding.

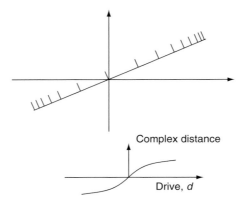

Figure 8.6 The SLM's action occurs on a straight line through the origin, but the change in action is a variable function of the applied drive

The literature reports work in which an input image is encoded onto the phase of the lightwave. The modes of SLM operation in which only phase varies typically have large magnitude, so there is potentially more light energy present in a phase-encoded image than in a magnitude-encoded one. As is evident from the preceding discussion, this is at the expense of linearity in the system response, the more strongly so if a larger phase gain (radians per drive unit) is used. A quasi-linearity ensues from a small phase gain when the DC term in the transform of the input object is blocked, in which case one might just as well have set up the correlator with minimum phase variation anyway.

8.3.2 Bumpy-lens analogy

Examine the form of the central correlation intensity $B^2 = \left| \sum_k M_k \exp(j\phi_k) S_k \right|^2$. The correlation process takes the sinusoids composing $s(t)$, scales each by magnitude M_k, retards each by phase ϕ_k, and detects the intensity of their sum. Correlation tends to line up the sinusoids so that their maxima align at the center before summing, thus producing the bright spot. The physical way in which the sinusoids are lined up is by delaying the ones that are "ahead of the pack" compared with the shifts that build the bright correlation spot. In this section we'll draw an analogy to adaptive optics to describe how that is done. In adaptive optics a phase change is discovered, and implemented, over the aperture of an optical system in order to flatten a wavefront. Our analogy for the OCPR process is a bumpy lens.

We start with a conventional lens. A conventional lens focuses collimated light – the sort of lightwave that originates in a small, bright, far-off source.

It puts concave spherical curvature into the arriving flat wavefront and, since waves propagate normally to their wavefronts, the lightwave collapses to a point. As the planar wave arrives from different directions the location of the bright spot moves, but it remains focused.

Now, if instead of a small bright far-off source we have light coming from holes in a plate, the waveform that forms is more complicated than the plane wave. If the apertures emit lightwaves that oscillate together because the light falling on them is coherent, the waveform is stable (and in principle computable).

The conventional lens does not convert that more detailed wavefront into a spherically converging shape. However, we can envision an equally highly figured glass that will retard the wavefront so that it exits the glass with the wavefront straightened out and flat.

If we add a converging lens behind the bumpy lens (or grind the back side of the lens with an appropriate curvature), we have a combination lens that focuses the aperture plate's wave into a sharp spot. If the aperture plate is a great distance from the bumpy lens (or if we put its virtual image at infinity by placing the plate at the focal plane of another lens), then as the plate moves there is only a linear phase ramp added to the phase pattern that the bumpy lens straightens out.

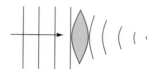

Figure 8.7 Conventional lens focusing light from a plane wave

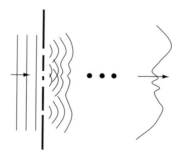

Figure 8.8 Apertures in a plate passing waves that combine to a complicated wavefront. The three dots indicate propagation over a great distance – or equivalently, passing through a lens one focal length from the aperture plate. In the far field, the field is the Fourier transform of the pattern of apertures

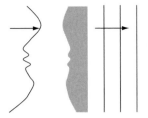

Figure 8.9 Lightwave and bumpy lens conjugating the wavefront

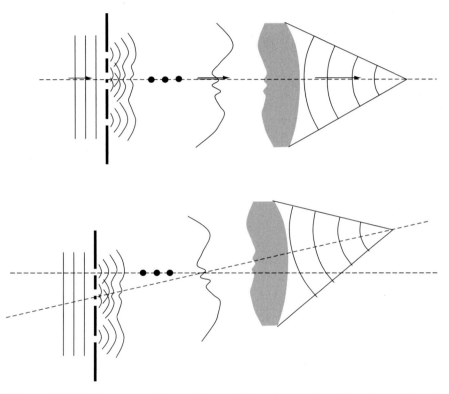

Figure 8.10 (a) Aperture plate at focus of lens; formation of conjugated and focused wave. (b) From linear systems theory, a translation at the aperture plate adds a phase ramp in the transform plane. The result is a movement of ultimate focus corresponding to the shift at the aperture plate

The linear ramp survives passage through the bumpy lens, and the location of the bright spot moves around but the spot stays bright. The spot tracks the motion of the hole pattern in the aperture plate. We have built an optical correlator.

That exquisitely bumpy lens focuses to a point the light originating in the full detail of its reference object, as embossed into the lightwave upon passing the first modulator. The implications and applications for this technology arise in the speed and parallelism with which this bumpy lens focuses light. It works in nanoseconds, and in doing so it does not mind how complicated (i.e., how many pixels) the input image is. A complicated part of the operation is computing exactly how to set up the second modulator, the one on which the "bumpy lens" is implemented. A complication is that, to complete the analogy, we model the material from which the bumpy lens is created as being absorptive in some relation to the thickness. It is as though we were building an actual lens from smoky glass, so that there is a relationship between the phase adjustments and the overall local transmittance of the lens. (Such an SLM is described as having coupled phase and magnitude characteristics.) We would seek a compromise between having a thick convex lens that created exactly spherical converging wavefronts at a reduced intensity, and brighter converging wavefronts that were not exactly spherical. In the latter case the light at the focal plane would be spread further out, but nevertheless would be brighter. We shall develop the mathematics for building optimal filters on the coupled SLM. In addition to allowing us to make the brightest correlation spots using the smoky-glass lens, this mathematics will allow us to make the lens less transmissive to frequencies that we wish *not* to contribute to the correlation plane.

The bright spot in the output plane of the correlator is a measure of how closely the input scene resembles the reference object the correlator is looking for. An attractive feature is that the correlation is a whole-scene-at-a-time operation; you are not fixating (unless you want to) on corners, or lines, or other individual locations on the input scene. Thus, if part of the object is obscured, the correlation pushes right along and shows you how strongly the remaining part continues to resemble the reference object. We artfully choose or construct reference objects so that their resemblance to the actual scene conveys information for the task at hand. Most other vision methods have a great deal of trouble with signal obscuration. Similarly, other methods are confused by background clutter; but if the background does not resemble the reference object, the correlation method is clearly superior to methods that must, for example, segment the image, find all the edges, etc., in order to understand what object they are viewing.

8.4 Model, including circular Gaussian noise

8.4.1 Variance in magnitude, intensity, and measurement

To reiterate the philosophy: in OCPR we make our judgements on things we can observe. We do not directly observe the complex field amplitude in the correlation plane; we observe its squared modulus, the intensity, through the instrument's responsivity R and in the presence of detection noise. If there is no detection noise, the measurement u resulting from a noiseless digital computational intensity I is

$$u = RI \qquad\qquad (8.13)$$

The physical measurement is inevitably corrupted by noise, and we shall take it to be additive and of zero mean. We shall try to be consistent in separating the digitally achieved quantities and the optical ones, and the responsivity R converts between the computer-modeled quantities and the things one measures on the optical bench. We will not always carefully distinguish between the digital intensity and the optical measurement, as they are linearly related. (There is a slight problem in that following square-law detection, a zero-mean complex Gaussian noise biases the measurement to a higher value than in Eq. (8.13), but this statement is heuristically close. The correct expression is Eq. (8.20).) A point often overlooked in the literature is that when we compute a filter to implement in the correlator, we need to know the relationships between the digital elements in the computation and their optical equivalents on the bench. Only then can we reliably compute the drive values to apply to the filter SLM, in order to gain the optimal system performance. (Which drive one puts onto the input SLM is less of a free variable, ordinarily being an input video signal with all its pixels put through a common transfer function.) Two relationships are of principal importance: (1) the complex actions that result from drive values applied to the SLMs, and (2) the measurable optical effect in the correlation plane as it relates to its digital simulation. The first we measure during the *characterization* of the SLMs. The second is the system *responsivity*, and it too is determined by laboratory measurement. One critical use for the responsivity is the correct representation of instrument noise in the filter computation, and another is knowing, for example, what digitally computed values may drive the detected correlation into saturation. We shall have more to say about these points later, but they are an essential part of the philosophy.

The entirety of this chapter is aimed at reducing the ill effect of measurement variance as induced from its various sources. To minimize the effect we must first model it.

The foundation of OCPR and the optimization of SNR have been found in a classical model for statistical pattern recognition [29]. In that model a continuous-time one-dimensional real voltage signal is received, having been corrupted to some degree with additive noise, and our job is to build a filter to help us decide whether the target we seek is present amongst the noise. The radar systems on which statistical pattern recognition was based detected the received electromagnetic energy turning it into voltage before detection was done, and the pattern recognition was performed on that voltage.

The voltages that we deal with in OCPR follow from the detection of intensity. In much of the OCPR in the literature, you will find that the expected signal is assumed to be the squared value of the expected magnitude. In fact, however, the expectation of the squared magnitude is not the squared expectation of the magnitude. This becomes significant when noise processes are considered; we often model the noise as zero mean and additive in complex amplitude. Even with a zero-mean additive noise, the expectation of squared (noise plus magnitude) is not the same as noise variance plus the squared expectation of the magnitude. Similar comments apply in computing measurement variance. The noise-induced distribution of the measurement is described in part, but not entirely, by the variance of the magnitude. Recently we have begun to model the variance in the measurement as arising in an intensity variance, not in a magnitude variance.

Let us elucidate the distinctions. First we develop what we shall call the "magnitude model" and then the "measurement model."

In the *magnitude model*, for input signal s, input noise n, and frequency-plane filter H, the expectation of the correlation electromagnetic field at the center of the correlation plane is the DFT of the product of filter and transformed signal, evaluated at the center.

$$D = B\exp(j\beta) = \sum_k H_k S_k \qquad (8.14)$$

with the sum being taken over all frequencies, indexed by k. The term B is the magnitude, and β the phase, of the sum. The expected intensity with zero input noise (in the measurement model we shall see that the noise contributes to the expected intensity) is

$$\langle I \rangle = |D|^2 \qquad (8.15)$$

and the variance of the field magnitude is

$$\begin{aligned}
\text{var}|D| &= \sigma^2_{\text{mag},n} + \sigma^2_{\text{mag,det}} \\
&= \sum_k |H_k|^2 P_{n,k} + \sigma^2_{\text{mag,det}} \qquad (8.16) \\
&= \Upsilon + \sigma^2_{\text{mag,det}}
\end{aligned}$$

with P_n being the power spectral density of the noise and $\sigma_{\text{mag,det}}$ being the detector noise contribution to magnitude variance which is independent of the filter. Equation (8.16) defines the effect of filtered input noise on magnitude variance, $\Upsilon = \sigma^2_{\text{mag},n}$. The magnitude version of SNR has been optimized in the OCPR literature and in Chapter 5:

$$\text{SNR}_{\text{mag}} = \frac{|D|^2}{\text{var } D} \tag{8.17}$$

This model has had success, but it is not completely in tune with the philosophy stated in the Introduction. As stated previously, the squared value of the expectation of the correlation magnitude is not the same as the expectation of the squared noisy magnitude. The latter is what is observed and should be used in metrics of the correlator performance. Ergo, we continue now to the intensity (or measurement) model.

In the *measurement model*, for the SNR to be a dimensionless quantity based on the optical measurement, we would take either

$$SNR_{\text{meas},1} = \frac{\langle I \rangle}{\sqrt{\text{var } I}} \tag{8.18}$$

or equivalently

$$\text{SNR}_{\text{meas},2} = \frac{\langle I \rangle^2}{\text{var } I} \tag{8.19}$$

(The responsivity R is a common factor in numerator and denominator and hence cancels out.) In both cases we need to determine the statistics of the noise- and clutter-influenced observed intensity. These statistics are dependent on some of the quantities in the magnitude model.

In the unidimensional real-valued signal case, we often assume that the noise is additive and Gaussian, with a power spectral density that is somehow known (e.g., from examples of clutter scenes), and that we model the noise as additive. Without presenting justification here, we shall assume that the input noise and clutter are additive, circular Gaussian, and of known power spectral density P_n. In the absence of any other information about the additive input noise, we shall assume that the variance of I and its statistics have uniform behavior over the correlation plane, that the input noise's electromagnetic field is a complex quantity of zero mean and magnitude variance Υ, and that the field has isotropic Gaussian distribution. In statistical optics this is one example of circular complex Gaussian RVs; see Goodman [32] for details. This model suffices to convert the variance of the complex correlation electromagnetic field into variance in the observed intensity, after the expectation of the correlation electromagnetic field has been computed. If the reader has information about their correlator that is different from this model, such information can be used.

Figure 8.4 indicates the model we use for determining intensity variance. It can be shown [52] that the effect of (filtered) noise is to increase the expected intensity from $|D|^2$ to

$$\langle I \rangle = |D|^2 + \Upsilon \tag{8.20}$$

with the magnitude variance defined in Eq. (8.16). Under the stated assumptions the measurement's variance, in terms of Υ, is

$$\sigma^2_{\text{meas}} = \Upsilon^2 + 2\Upsilon|D|^2 + \sigma^2_{\text{det}} \tag{8.21}$$

That is, the measurement's expectation value and its variance are different from those that enter the OCPR magnitude formulations in Eqs. (8.16) or (8.20).

8.4.2 Known problems with the DFT representation

Using the DFT to represent the diffraction integral assumes that all points in the aperture have even illumination. Using the complex signal $s(x) = e(\upsilon(x))$ assumes that the SLM's encoding produces a lightwave in proportion to e at all points. In fact e is a complex factor that multiplies the incoming lightwave. So if the illumination has complex amplitude $a(x)$ over the aperture, the actual light signal entering the system is $a(x)\,e(\upsilon(x))$, and the effect at the transform plane is the convolution of $\Im(a(x))$ with $S(f)$. Shift invariance is lost, since the eccentric appearance of the reference object is accompanied by a change in its appearance as it moves into a different illumination situation. Practical correlators use a compromise between sending all of the source's light to the input SLM (this results in variations of $a(\cdot)$ over the aperture) and overfilling the aperture (this gives more nearly uniform illumination but is less efficient in light use).

Another physical problem with this representation is scattered light from the signal. Optics is not a perfect conveyor of information, and in at least two ways, imperfections cause a difference between the DFT representation and what physically happens. One is that light scatters from unmodeled characteristics of the SLM (namely the sub-pixel structure of the pixels), and another is the unmodelable scattering from dust and other irregularities in the optics. The DFT assumes a discrete point-like character of the pixels, as a two-dimensional array of impulses. To obtain a more nearly correct expression we would convolve that array of impulses with the detailed spatial structure of an individual pixel, which we know to be expressed as the product of transforms at the next position in the optical transform chain. Limiting the light to a single diffraction order moderates this effect but does not eliminate it, since the transform of a pixel's internal structure can have significant variation within the single passed order.

8.5 Metrics and metric potential

In order to optimize the performance of an optical correlator we must quantify its performance. The metrics we develop in this section are of various types. We might judge the correlator by its central correlation magnitude or the central correlation intensity. We might ask the correlator to detect a single reference object in a noisy background, or to detect and/or discriminate several reference objects. The metrics are in some cases ad hoc, in others they are grounded in statistical pattern recognition. We currently have to judge a filter on the basis of its performance at the center of the correlation plane (if the input object is centered), but we hope to develop metrics that regard the whole of the correlation plane and also provide an analytic means to optimize the filter.

Table 8.1 describes what sort of filter is applicable in several noise environments and to several detection and discrimination tasks. The less demanding applications are up and left within this table.

A general form of the tradeoff metrics was given in Eq. (8.3).

8.5.1 *The statistical pattern recognition metrics*

We will hardly ever take the luxury of using an optical correlator to detect only a single object, or a single appearance of the object if there is indeed only one object (we won't distinguish between multiple objects and multiple views of a single object any more). The point of an optical correlator is that a set of filters

Table 8.1. *Filter applicability* (filter type in **bold**)

Task		Noise environment		
		Low noise	Detection noise, white input noise	Colored input noise
Detect and/or track	Single object	**Intensity**		**Magnitude SNR**
	Any of a class	Statistical pattern recognition (**Bayes error, A_{ROC}**, etc.)		
Discriminate	Between pairs	**Rayleigh quotient**		
	Between classes	Statistical pattern recognition (**Bayes error, A_{ROC}**, etc.)		

can rapidly be run against a single input image. Fast SLMs and correlation detection mean that OCPR can be a rapid process. One method of running a correlator is to have a single filter per object. However, the correlator can run even faster, functionally speaking, if a filter is designed so that it responds to several objects. The optical filters described in this section are designed to promote detection of any of a set of desired objects. Using the measurement basis, we shall introduce the optical versions of several standard SPR metrics and show how to optimize each one. In increasing degrees of competence, we shall examine the Rayleigh quotient, the Fisher ratio, the area under the receiver operating characteristic (ROC) curve, the average information, and the Bayes error.

8.5.2 Rayleigh quotient

The Rayleigh quotient's digital optimization was shown in Chapter 2 by using Lagrange multipliers and eigenvalue–eigenvector arguments. In the notation of Chapter 5, the digital Rayleigh quotient for a single pair of filtered images is:

$$J(\mathbf{h}) = \frac{\mathbf{h}^+ \mathbf{A} \mathbf{h}}{\mathbf{h}^+ \mathbf{B} \mathbf{h}} \tag{8.22}$$

In the digital form, the value of $\mathbf{h}$ that maximizes J is the eigenvector for the dominant eigenvalue of $(\mathbf{B}^{-1} \mathbf{A})$, *presuming that* $\mathbf{B}$ *has an inverse*

Let's extend the Rayleigh quotient to apply to optical correlation. The optical Rayleigh quotient is the ratio of two intensities – one for the reference object we want to accept, and the other for the one we want to reject. Added to the correlation intensity of the object we wish to reject will be filtered noise and detector noise. The (weighted) inner product formed in Eq. (8.22) is the same as the correlation evaluated at the center, or

$$J_{\text{Rayleigh}} = \frac{B_1^2 + \left\langle \sigma_{n,1}^2 \right\rangle}{B_2^2 + \left\langle \sigma_{n,2}^2 \right\rangle} \tag{8.23}$$

The expected correlation magnitudes and the input-noise-induced variances are functions of the filter and so, too, is the Rayleigh quotient.

8.5.3 Fisher ratio

A statistical pattern recognition metric of long standing is the Fisher ratio, J_{Fisher}. It is the squared difference between class means divided by the average of the classes' variances. In our optical terms,

$$J_{\text{Fisher}} = \frac{\left(\langle I \rangle_\Lambda - \langle I \rangle_\Psi\right)^2}{\sigma_{\Lambda T}^2 + \sigma_{\Psi T}^2} \tag{8.24}$$

The intent is that the members of the classes to be distinguished shall be well separated, when the length of the yardstick used in reporting their separation is proportional to the average width of the classes' distributions. The means and variances of measurements are functions of the filter, and so too is the Fisher ratio, which can be maximized by choice of filter. However well intentioned, the Fisher ratio does not generally give us a direct measure of the classification error.

8.5.4 Area under ROC curve

The "receiver operating characteristic curve" (ROC curve) is historically based on electrical processing of radar signals. The area under the ROC curve (called the *power of the test* in statistical decision theory), A_{ROC}, is quite a useful gauge of a filter's practical utility.

We let Ψ represent the class of sheep and Λ represent the class of wolves, where we wish to set the correlator's threshold at T to detect wolves but be blind to sheep. The probability density function of correlation intensities for class C we denote by p_C. Then the false alarm probability P_f is given by:

$$P_f(T) = 1 - \int_{-\infty}^{T} p_\Psi(m)\mathrm{d}m = \int_{T}^{\infty} p_\Psi(m)\mathrm{d}m \tag{8.25}$$

and the probability of detection p_d is

$$P_d(T) = \int_{T}^{\infty} p_\Lambda(m)\mathrm{d}m \tag{8.26}$$

Given these definitions, it is straightforward that the area under the ROC curve is:

$$\text{area} = \int_{T=-\infty}^{\infty} \int_{m=T}^{\infty} p_\Lambda(m) p_\Psi(T)\mathrm{d}m\mathrm{d}T \tag{8.27}$$

Maximizing the power of the test by choice of filter is far more complicated than, say, creating the brightest correlation spot for given wolf and a filter domain.

8.5.5 Expected information

Suppose a measurement returns a value m and that we must choose between deciding Λ and Ψ on the basis of the measurement. The a-posteriori likelihood ratio of Λ and Ψ is:

$$\text{ratio}(m) = \frac{P_\Lambda\, p_\Lambda(m)}{P_\Psi\, p_\Psi(m)} \tag{8.28}$$

The definition of information arises from the fact that the probability of independent outcomes is given by the product of the individual events. But we would like information to be a quantity that adds, not multiplies, as independent events are considered. For this reason "information" is taken as the logarithm of likelihood ratios. Then the information that the value m provides in favor of deciding that we have class Λ instead of Ψ is $\log((P_\Lambda\, p_\Lambda(m))/(P_\Psi p_\Psi(m)))$. For example, if the likelihood ratio is unity, we have no information resulting from the appearance of m as the measurement, and taking the logarithm of the likelihood ratio indeed returns zero.

This is for a single value that m might take on. But note that not all measurement values are equally likely. A value of m that is very informative but almost never occurs will, in the long run, return less information on average than a much more-likely-to-occur value of m even it if yields less information per occurrence. We can extend the concept of information at a given value of m to all values of m by taking the expectation of information, using the PDF of the considered class to weight the information. The result tells us how well the populations separate when all values of the measurement are considered (given that we have equal priors). In a slight generalization of the Kullback–Liebler distance (the generalization includes unequal a-priori probabilities), the expected information that the process will provide is:

$$\langle \mathfrak{D}_\Lambda \rangle = \int_{\text{all } m} p_\Lambda(m) \log\left(\frac{P_\Lambda p_\Lambda(m)}{P_\Psi p_\Psi(m)}\right) dm \tag{8.29}$$

Since the measurement densities are functions of the filter, so too is the expected information. We can analytically maximize $\{\mathfrak{D}_\Lambda\}$ by choice of filter, although, as may be apparent, it is another messy problem.

8.5.6 Bayes error

Suppose in the following discussion that we have a filter in hand, and can thus compute the measurement PDFs for the two classes we wish to separate.

Given a measurement value m, Eq. (8.28) gives the a-posteriori likelihood ratio that the measurement arose in either of the two classes.

If the *ratio* is greater than unity the a-posteriori likelihood is greater that the class is Λ, and vice versa. If we minimize "Bayes error" by choosing the more likely class, then for measurement m the likelihood of error is:

$$P(\text{error})|_m = \min\{P_\Lambda p_\Lambda(m), P_\Psi p_\Psi(m)\} \tag{8.30}$$

We find the entire likelihood of making an error by integrating over all values of m:

$$E_{\text{Bayes}} = \int_{m=-\infty}^{\infty} \min\{P_\Lambda p_\Lambda(m), P_\Psi p_\Psi(m)\} \mathrm{d}m \tag{8.31}$$

By letting $\mathbb{R}_C$ be the set of values of m assigned to class C, we can also express the Bayes error as follows:

$$E_{\text{Bayes}} = P_\Psi \int_{\mathbb{R}\Lambda} \rho_\Psi(m)\mathrm{d}m + P_\Lambda \int_{\mathbb{R}\Psi} \rho_\Lambda(m)\mathrm{d}m \tag{8.32}$$

Bayes error is a function of the measurement densities and, hence, of the filter. Again, in this case we can analytically optimize the Bayes error by choice of filter.

8.5.7 *Nonlinearities in filter design and in metrics*

The MED projection process that is fundamental to our optimization scheme induces nonlinearities in filter computation. Our form is evoked by an algorithm based on a detection and decision scheme. In addition to including only physically observable quantities, a metric should represent the utility of the observable in the application for the correlator. Consider the case in which we have made so bright a correlation spot that it saturates the imaging detector; the utility of that spot clearly includes an essential nonlinearity in the composite response to a frequency's filter value. For another example, it does little good to carry a correlation response much further beyond the detection threshold even if the measurement does remain linear. One should also keep in mind that the digital version of the correlation intensity (e.g., B^2) may or may not have a linear relationship with the optically measured intensity; we have referred to this relationship as the instrument's responsivity. Even if, as usual, the imager collecting light at the correlation plane is linear over a great range, it can still be driven into saturation, which is prototypical nonlinear behavior.

8.6 Gradient concepts

We typically discuss filters in terms of their magnitude and phase (M, θ), rather than their real and imaginary parts (x, y). The reasons should be apparent in the forms of the metrics; you do not see the real part of a filter expressed in the metrics, but you do see its magnitude. Also, absolute phase does not matter in the non-interferometric measurement of intensity in the correlation plane. Phase affects the value of the real and imaginary parts of an electromagnetic field, but it does not affect the magnitude.

In this section we'll find how J varies as a function of an arbitrary value of H_m, and use that information to optimize J by a suitable choice of a realizable filter. Increment $\Delta J(H_m)$ is the contribution to J that H_m would make if realized, and it is built up from gradient expressions $\nabla J(H_m)$. We will use magnitude (radial) and phase (azimuthal) components of ∇J.

Now we describe our method of optimizing filters for analytic metrics. Strictly speaking the complex derivative of J does not exist. (Note that some of the components of J are squared magnitudes of complex quantities that would be given as z^*z, and z^* does not have a derivative.) However, J certainly is a function of H_m, and we can plot the contours of J as we move H_m around in the plane of its complex values. To optimize the metric we use the realizable value of H_m that is at the highest (or lowest, as appropriate) contour. Rather than implement exactly that process, however, it is easier to find the gradient of J in the plane of H_m and use some of the gradient's properties. Although the background just developed is adequate for our purposes, the matter of taking derivatives of scalar functions of complex variables is developed in Kreyszig [19] or Therrien [22].

The metric J is a function of the filter **h**. Suppose **h** is optimal; then no allowed small adjustment of **h** can improve J. (By "allowed adjustment" we mean a change to another member of the limited set of values the filter SLM can take on.) Presuming that the elements of the optimal **h** can be independently chosen, then either (1) the partial derivative of the metric with respect to the allowed variation of each element is zero, or (2) the optimal value of the mth filter value is at an end or discrete point of the operating curve. (Suppose to the contrary that we are not at an extremum, and the mth partial derivative is not zero. Then we can slightly readjust H_m and improve the value of the metric, so we did not begin with the optimizing **h**.) From that necessary condition of optimality we can infer how to construct an optimal filter that is restricted to the filter SLM's range. To do this we take the partial derivative of J with respect to the magnitude of the mth frequency's filter value, infer the whole gradient of the metric and, finally, from the set of realizable values, deduce where the optimal filter value lies. The partial derivative method is more thoroughly developed by Juday [41, 52].

The quality of the metric that is critically important is that it is a continuous and differentiable function of the spectral filter values. Then putting a metric to an extremum is as conceptually simple as maximizing a quadratic. If we have the coefficients of the powers of x in $y = ax^2 + bx + c$, we know that y is an extreme value when we use $x_0 = -b/(2a)$. We find x_0 by setting $\mathrm{d}y/\mathrm{d}x = 0$ and examining the consequences. It is possible to optimize the very much more complicated forms of the optical pattern recognition metrics similarly.

In this section we shall assume that for all $k \neq m$ the filter components are at their optimal values. Although we know J is a function of the set of all frequencies' filter values, this perspective allows us efficiently to optimize the whole filter, one frequency at a time.

To be strictly correct we would say we are computing the contribution $(\Delta J)_m$ to the metric from the mth frequency's filter value H_m. Instead, we shall use the looser terminology, $J = J(H_m)$, since the partial derivative expression of optimality must hold at each frequency. To make the equations more easily readable we shall occasionally use "Num" for the numerator of the metric, and "Den" for its denominator. We shall be examining several differential equations to infer an optimal filter value, and we shall get an integration constant for the differential equations from assuming that a zero filter value contributes nothing to the metric; thus $J(0) = 0$.

Let's lay some groundwork. Suppose we have a real scalar field $F(z)$, with z being a complex number, and that F is continuous and differentiable – at least in the sense that there is a derivative of F in any selected direction. Then it has a gradient in the plane of z, and F can be obtained from the gradient and a boundary condition. Our boundary condition will be that a zero filter value contributes nothing to the metric. The partial derivative of F radially in the θ direction is:

$$\frac{\partial F}{\partial M} = \frac{\nabla F \cdot [\mathrm{d}M \, \exp(j\theta)]}{\mathrm{d}M} \tag{8.33}$$

with the dot product for complex numbers being defined as:

$$w \cdot u \equiv |w||u| \cos(\arg w - \arg u). \tag{8.34}$$

As a particular example that will be useful later, suppose we find

$$\frac{\partial F}{\partial M} = -k_1 M + k_2 \cos \theta + k_3 \sin \theta \tag{8.35}$$

with the k terms being constant. The term k_1 will usually be associated with an input noise power spectral density term. The composite of k_2 and k_3 will usually be associated with the signal's complex spectrum. Using the function atan2 $(\cdot,\cdot)$ as the two-argument arctangent satisfying $\phi = \text{atan2}(a\sin\phi, a\cos\phi)$ for any $a > 0$ and for $0 \leq \phi < 2\pi$, we see Eq. (8.35) as the form for the gradient's being:

$$\nabla F = -k_1 M + \sqrt{k_2^2 + k_3^2}\, \exp[j\,\text{atan2}(k_3, k_2)] \tag{8.36}$$

and for F's being:

$$F(z) = -k_1 \frac{|z|^2}{2} + \left(\sqrt{k_2^2 + k_3^2}\right) z \cdot \exp[j\,\text{atan2}(k_3, k_2)] \tag{8.37}$$

after we apply the boundary condition $F(0) = 0$. The first term on the right in Eq. (8.37) is a quadratic circular component of F centered at the origin, and the other term on the right is a planar ramp rising in the direction of $\phi_0 = \text{atan2}(k_3, k_2)$. The net behavior is that F has a maximum in the direction ϕ_0 from the origin, and there is circular behavior about that maximum. The location of the maximum depends on the balance among the k terms, including that F increases without bound with just the ramp behavior if $k_1 = 0$, but not if both k_2 and k_3 are zero. Otherwise, if $k_1 \neq 0$ the maximum is found at a distance r_{max} from the origin, where

$$r_{max} = \frac{\sqrt{k_2^2 + k_3^2}}{k_1} \tag{8.38}$$

so the maximum of F is found at

$$z_{max} = r_{max} \exp(j\phi_0) \tag{8.39}$$

If we are searching in the z plane for the best filter value (i.e., to maximize F) and not all locations z are available, then we select the one closest to z_{max}. The behavior of F resulting from the quality of its radial partial derivative stated in Eq. (8.35) will be found time and again as we look at the metrics optimized in this chapter. The values of the k terms are functions of the selected metric, the spectral values of the signal(s), the noise in the input, and the noise added during the correlation detection. The partial derivative statement – that the change in metric for differential allowed change of filter value is zero – is consonant with this groundwork; the direction of allowed change is parallel to the isometric contours, and since those contours are concentric circles we choose the closest realizable filter value to the location of the maximum. That is the necessary groundwork.

There are some partial derivatives that will come up repeatedly as we compute metric gradients. Two in particular are the derivative of the expected value of central correlation intensity, and that of Υ, with respect to the mth filter's magnitude M_m. We develop them now. Firstly we obtain the partial derivative of the correlation intensity:

$$
\begin{aligned}
I = B^2 = |B\exp(j\beta)|^2 &= \left|\sum_k H_k S_k\right|^2 = \left|\sum_k M_k \exp(j\theta_k) A_k \exp(j\phi_k)\right|^2 \\
&= \left(\sum_k M_k A_k \exp(+j[\theta_k + \phi_k])\right)\left(\sum_k M_k A_k \exp(-j[\theta_k + \phi_k])\right)
\end{aligned}
\tag{8.40}
$$

From this follows:

$$
\begin{aligned}
\frac{\partial I}{\partial M_m} &= \left(\sum_k M_k A_k \exp[j(\theta_k + \phi_k)]\right)\frac{\partial}{\partial M_m}\left(\sum_k M_k A_k \exp[-j(\theta_k + \phi_k)]\right) \\
&\quad + \left(\sum_k M_k A_k \exp[-j(\theta_k + \phi_k)]\right)\frac{\partial}{\partial M_m}\left(\sum_k M_k A_k \exp[j(\theta_k + \phi_k)]\right) \\
&= B\exp(j\beta)A_m \exp[-j(\theta_k + \phi_k)] + B\exp(-j\beta)A_m \exp[j(\theta_k + \phi_k)] \\
&= 2BA_m \cos[\beta - (\theta_m + \phi_m)]
\end{aligned}
\tag{8.41}
$$

The partial derivative of Υ is easier:

$$
\frac{\partial\Upsilon}{\partial M_m} = \frac{\partial}{\partial M_m}\sum_k |H_k|^2 P_{n,k} = \frac{\partial}{\partial M_m}\sum_k M_k^2 P_{n,k} = 2M_m P_{n,m}
\tag{8.42}
$$

8.7 Optimization of the metrics

In some cases we shall see that the ad hoc methods (e.g., matching phase on a POSLM) were optimal even though an analytic basis was not available when they were developed; in other cases we shall see that they were not quite right (e.g., matching phase on a coupled SLM).

Now we present the differential equation from which we shall derive the optimal realizable filters for a variety of optical correlator metrics. If J is the metric, and υ_m is the control value applied to the mth frequency's pixel (e.g., the voltage across a liquid crystal cell), then the necessary condition of optimality is either that

$$\frac{\partial J}{\partial v_m} = 0 \qquad (8.43)$$

or that we are at an appropriate extreme value of the SLM's operating curve. The condition is justified by consideing that if at some frequency the partial derivative is *not* zero, at that frequency the filter can be adjusted to beneficial effect (improving the value of the metric), and the proffered filter was not optimal.

We remark that this formula as written applies to univariate control (as having one voltage to apply to each pixel). This is the case for many SLMs, such as those using liquid crystals or deformable structures. If *two* controls are available for a pixel, then two partial derivative statements identical to Eq. (8.43) apply, with little consequence to the analysis that follows. Spatial light modulator architectures that use two control values have been proposed, such as passing light through two SLMs in series so that their effects multiply [85], or by spatially averaging the effect of multiple sub-pixels [86], or by splitting and recombining a beam so that the effects of two SLMs add in the quantum mechanical sense [87].

In this section we optimize the filter's performance while explicitly taking into account the limitations of the correlator. We find the gradient of the metric in the plane of complex values for the mth frequency, and that gradient tells us which realizable filter value to use. In a comparatively unguided approach such as simulated annealing, applied to filter values at the entire ensemble of frequencies, the dimensionality of the search is as large as the number of frequencies. We can reduce that huge dimensionality of the optimizing search to a very few parameters. We work out two examples – intensity and SNR. Of those examples, intensity is a maximal-projection metric (as is the Rayleigh quotient), and SNR requires taking the closest value to a finite ideal filter value (as do the rest of the entire suite of SPR metrics).

8.7.1 Intensity

$$J(\mathbf{H}) = \frac{B^2}{\langle \sigma^2 \rangle} \qquad (8.44)$$

Equivalently, $J = B^2$ is to be maximized. Eq. (8.41) is seen to be the inner product of $BA_m \exp j(\beta - \phi_m)$ and $\exp j\theta_m$, indicating that J has a uniform gradient in the direction of $\exp j(\beta - \phi_m)$. So we choose for H_m the realizable value with the largest projection in the direction of $\exp j(\beta - \phi_m)$. We do not know β as we begin; it is a parameter to be searched over and tested for consistency. (Note that in this case, the parameter suite to be searched over reduces to a single real quantity.) Some value, $\beta^\diamond$, produces the optimal filter, $\mathbf{H}^\diamond$. Via the definition of β, the optimal value of $\mathbf{H}$ will produce the same value $\beta^\diamond$ as was used to compute $\mathbf{H}^\diamond$, but this is not true of values of β and $\mathbf{H}$ for which the necessary condition does not hold, that the partial derivative of J be zero for all frequencies. The value β is a constant for all m, of course.

The intensity metric points up the difference in the optimal filters under the digital filtration situation, when all values in the complex plane are available, and the optical situation, in which only a limited subset within the unit disk is accessible. In the latter case the algorithm directs us to take the filter value with the largest realizable projection, reflecting a physical limitation on the achievable intensity.

8.7.2 Magnitude SNR

$$J = \frac{B^2}{\Upsilon + \langle \sigma^2 \rangle} \tag{8.45}$$

At the best possible filter value (whether realizable or not), the total gradient is zero. First we find those locations where the azimuthal component is zero:

$$\left(\frac{\partial J}{\partial \theta_m} = 0 \right) \Rightarrow (\Upsilon + \langle \sigma^2 \rangle) \frac{\partial B^2}{\partial \theta_m} = B^2 \frac{\partial \Upsilon}{\partial \theta_m} = 0 \tag{8.46}$$

with the final equality since Υ does not depend on θ_m. Thus,

$$\frac{\partial B^2}{\partial \theta_m} = 2BM_m A_m \sin(\beta - \phi_m - \theta_m) \tag{8.47}$$

which is to be zero by choice of θ_m (the $M_m = 0$ solution is uninteresting). Setting $\theta_m^I = \beta - \phi_m$ will assure the equation. In looking for where the total gradient is zero, we next set $\partial J / \partial M_m = 0$, and after some manipulation our choice of θ_m^I produces:

$$M_m^I = \left(\frac{\Upsilon + \langle \sigma^2 \rangle}{B} \right) \frac{A_m}{P_{n,m}} \tag{8.48}$$

So the location where the total gradient of J is zero is:

$$H_m^I = \Gamma \exp(j\beta) \frac{A_m}{P_{n,m}} \exp(-j\phi_m) \tag{8.49}$$

in which the filter gain Γ is defined by reference to Eq. (8.48), and correlation phase β by reference to Eq. (8.14). The optimal realizable value of H_m is the one closest by Euclidean measure [41, 52] to H_m^I. In the optimization process, the complex quantity $\zeta = \Gamma \exp(j\beta)$ is to be searched over, and the algorithm produces $J(\zeta)$. We have one reference image in the filtering objective, and we have one complex number to search over in the parameter set. The value of ζ that maximizes J will be found to satisfy the consistency requirements constituted by its definitions.

8.7.3 *Statistical pattern recognition metrics*

This is going to seem anticlimactic, but a generalization of the basic algorithm given in Eq. (8.49) optimizes the Fisher ratio, A_{ROC}, Bayes error, etc. The generalization is:

$$H_m^I = \sum_i \Gamma_i \exp(j\beta_i) \frac{A_{i,m}}{P_{n,m}} \exp(-j\phi_{i,m}) \tag{8.50}$$

where i indexes the objects in the training set and, as usual,

$$H_m = \mathrm{MED}\{H_m^I\} \tag{8.51}$$

Here MED refers to the minimum Euclidean distance mapping which maps a point to the nearest point in the operating region of the SLM. The set of complex parameters $\{\Gamma_i, \beta_i\}$ is the search space for optimizing the filter. Experimentally we find that the optimizing parameters are different among the SPR metrics, and so, of course, are the filters. The demonstration of the optimization is roughly the same as the process for optimizing magnitude SNR, but it is too tedious to include here. See Juday [52] for the details. Some remarks are in order. There is not a royal road to ascertaining the optimizing values of the parameters, and, hence, of the metric-optimizing filter. The reader must numerically evaluate the metric according to its definition, and they must do so for each trial set of search parameters. Optimizing the SPR filters is substantially more complicated for an optical filter and its restricted values than for digital filtering, but at least we are assured that we can optimize the optical filter on the basis of the SPR metrics.

8.7.4 *Peak-to-correlation energy (PCE)*

This metric is defined as:

$$\mathrm{PCE} = \frac{\left|\sum_k H_k S_k\right|^2}{\sum_k |H_k S_k|^2} \tag{8.52}$$

Similar manipulations as those for optimizing SNR produce:

$$H_m^I = \Gamma \exp(j\beta) = \Gamma \exp(j\beta) \frac{1}{S_m} \qquad (8.53)$$

which is an inverse filter. The parameters Γ and β are to be searched over.

8.8 SLMs and their limited range

The metric-optimizing algorithms depend on knowledge of the values that the SLM can take on, so we shall discuss some realistic and some ideal characteristics of SMLs.

A modulator is the tool with which we control the light. A simple example is the liquid crystal display (LCD) on your wrist; it modulates the light landing on it so that the pattern is readable by eye. The modulators in OCPR are hardly more complicated than those LCDs, although they are much faster and finely detailed. The individual elements in the modulator are about 5 to 40 µm on a side, which is about 25 to 200 mm^{-1} – smaller than you can see without a microscope. A wristwatch ordinarily changes its display once or twice per second; the fastest 2-D analog modulators in common use can change some thousands of times per second. A significant difference between a modulator's use in a watch and in OCPR is that OCPR productively uses them to delay the light in addition to changing its intensity. The necessary delay is not large: being able to retard the red light of a helium–neon laser by a wavelength – just more than half a millionth of a meter – is quite sufficient for our purposes. This ability to delay the light lets Fourier optics do some far more powerful things than the LCD on your wrist is used for. It lets us process light by the "bumpy lens" analogy developed earlier.

We have seen that (1) the metric-optimizing algorithms specify an ideal filter value and (2) its optimal representation is the closest realizable value on the SLM (by Euclidean measure – hence the "MED," or minimum Euclidean distance, filter value). In the remainder of this section we shall look at some SLM operating curves that are mathematically tractable, in that the conversion from an ideal filter value to the MED representation is simple. Then we will show a lookup table method for working with an SLM that is uniform across its face, but does not have a simple analytic expression for the closest realizable value to the ideal value. Finally, we shall give a method of working with an SLM that is non-uniform across its face, in addition to having no easy analytic MED conversion.

8.8.1 Continuous mode

The most tractable and capable of SLMs would express any filter value within the unit disk. It would be able to express any phase in $0 \leq \theta < 2\pi$, and independently it would express any magnitude in $0 \leq M \leq 1$. If the ideal filter H^I has magnitude M^I and phase θ^I, the MED value of H^I is:

$$
H_{\text{MED}} = \begin{cases} H^I, & M^I \leq 1 \\ \exp(j\theta^I), & M^I > 1 \end{cases} \tag{8.54}
$$

No single SLM is currently available that can express the unit disk. There are costs associated with obtaining full regional coverage (whether all of the unit disk or a subset area). If you average the effects of adjacent pixels [86, 88], you reduce the count of independent pixels and thus perhaps the information-bearing capacity of the SLM. If you get the areal coverage by affecting the light sequentially with two SLMs [85, 89], your correlator architecture becomes longer and the drive circuitry becomes more complicated. One can arrange beamsplitters and SLMs so that the virtual image of one SLM is at the physical location of the other, or so that their virtual locations are together. Either of these conditions we will call putting the SLMs into virtual register. Further, we will want the path lengths to and from the position of register to be well within the coherence length of the processing light beam, and for the relative longitudinal positions to be stable well within a wavelength of light. If you split and recombine a light beam so that it is simultaneously affected by two SLMs in virtual register [87], the drive circuitry must be provided for the second SLM and the SLM alignment is very exacting. If you give a pixel a random value so that its statistical expectation is the desired value, the filter you create is never exactly the one you want, but some interesting results have been shown [90]. This is a little like having one hand in hot water and the other in an ice bath, and then claiming that on average you are comfortable, or going to Las Vegas and betting that a rolled die will come up 3.5, since that is the average value of

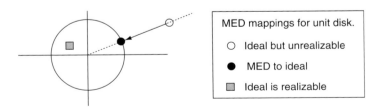

Figure 8.11 MED mappings for unit disk

the number of spots on the sides of the die. A correlator uses light to process information and does it so fast that its response is to the instantaneous value on the SLM, not to its average value.

When optimized on the unit-disk SLM, the maximum-projection filters (i.e., those that maximize correlation intensity) reduce to unit-magnitude filters with matched phase. It is the filters that do noise rejection (e.g., optimized for SNR, or tradeoff involving the noise PSD) that make use of interior points of the unit-disk SLM.

Phase-only (1-DOF)

In the common parlance a phase-only (SLM POSLM) has unit (or constant) magnitude and (usually) any phase in $0 \leq \theta < 2\pi$. On the POSLM, phase matching is equivalent to MED mapping of both varieties (H^I is finite or infinite). The ideal filter value $H^I = M^I \exp(j\theta^I)$ is represented by a unit vector with its same phase:

$$H_{\text{MED}} = \exp(j\theta^I) \qquad (8.55)$$

A filter written on a POSLM is an all-pass filter; each frequency arrives in the correlation plane with its original magnitude but with its phase possibly shifted. If the POSLM is not able to express the full range of 2π radians, then Eq. (8.55) is not correct, but the filter for the incomplete POSLM is nevertheless an MED filter.

Magnitude-only (1-DOF)

Again in the common parlance, a magnitude-only SLM; in this chapter we reserve "amplitude" for the possibly complex version of magnitude, but historically this SLM has been described as "amplitude-only" and we leave the "A" as the first letter. Another reason is that a type of SLM known as a magneto-optic SLM has primacy rights to "MOSLM." (An AOSLM) can produce a magnitude anywhere in $0 \leq M \leq 1$ at a constant phase (taken as zero). The MED mapping is simple: the real part of H^I is truncated to the interval from zero to one:

$$H_{\text{MED}} = \begin{cases} 0, & \text{Re}(H^I) < 0 \\ \text{Re}(H^I), & 0 \leq \text{Re}(H^I) \leq 1 \\ 1, & 1 < \text{Re}(H^I) \end{cases} \qquad (8.56)$$

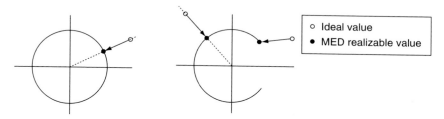

Figure 8.12 Phase-only operating curves

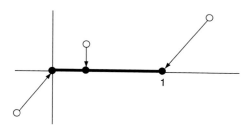

Figure 8.13 MED mapping for amplitude only SLM (AOSLM)

We remark that although there is a strong urge to map the magnitude of H^I into the magnitude of the AOSLM, this is *not* an MED mapping and it gives sub-optimal filters. Since maintaining the phase of H^I in mapping it onto a POSLM is so successful one wonders why maintaining the magnitude of H^I in mapping it onto an AOSLM is not equally successful. The answer is that maintaining phase is an MED process, whereas maintaining magnitude is not. Compare the two mappings in Figure 8.14, a considerably greater average distance between H^I and its mapped value is evoked by maintaining magnitude than by the mapping specified in Eq. (8.56).

Coupled SLMs

Even though we may find analytic expressions for the optimal mapping from ideal to the best realizable filter, using the lookup-table method for the operating curves in this sub-section would probably be preferable for operational purposes.

None of the SLMs we have looked at in this section have yet had coupled behavior. In a 1-DOF SLM, the operating curve typically runs along neither constant-magnitude circles nor exactly along constant-phase lines. There is

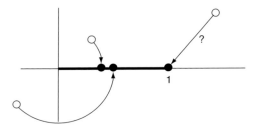

Figure 8.14 Magnitude-preserving mapping for AOSLM. Note that an ideal filter value with magnitude greater than unity is not well specified in its realizable mapping. Preserving magnitude is not an MED process

coupling between the magnitude and phase. Such an SLM has more complicated algorithms for finding the optimal realizable value than the SLMs we
have looked at so far.

We remark that although a strongly coupled SLM is somewhat more
complicated to compute the optimal realizable filter for than the AOSLM or
POSLM, its character (in our opinion) can be better suited to correlation
pattern recognition than the AOSLM or POSLM. The reason is that many
(often most) frequencies have less signal energy than the noise/clutter objects'
spectral energy. Whitening the transformed and conjugated training object
then causes a large population of ideal filter values near zero magnitude.
Having a set of available filters valued near the origin permits the SLM to
come close to realizing these small magnitudes, so the noise in these frequencies is suppressed.

The details of a coupled SLM's realizable range are highly variable with
the optical environment in which it is operated, so we don't give an equation
for the values as is possible for a true AOSLM or POSLM. Figure 8.15 gives
an example typical of the realizable range for a nematic liquid crystal SLM.
The 8-bit drive value runs from 0 to 255 as the filter value moves along the
spiral curve.

Since a coupled SLM usually does not admit an analytic expression for its
values, then it also does not admit an analytic expression for the MED value
for arbitrary H^I, either. Our approach is to make a two-dimensional tabulation
of the MED value. In Figure 8.16 we have plotted a filter drive value for each
point in the complex plane of ideal filter values. The drive value is that of the
closest point on the operating curve. The axes and the curve itself have been set
off in contrast so that they can be seen. An interesting feature is that near the
center of curvature of the operating curve, there is a strong variation in the
MED value.

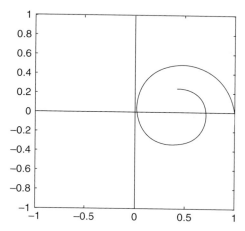

Figure 8.15 A spiral operating curve. The complex value begins at (1,0) with zero drive, passes near the origin at mid-level drive, and continues to wind through the unit disk's first and fourth quadrants as drive increases to its maximum

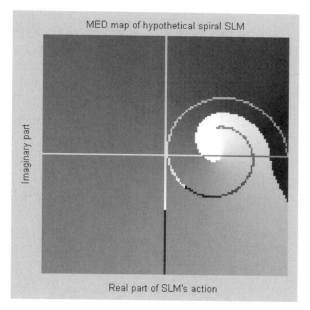

Figure 8.16 MED filter drive values for the spiral operating curve in Figure 8.15. Lightness corresponds to the drive value of the closest point on the operating curve, but the axes and the curve itself are set in contrast to make them visible. Black is zero drive value, white is the largest drive value, gray is in-between

8.8.2 *Discrete mode*

Certain SLMs operate in discrete modes. These include the ferro-electric SLM and the magneto-electric SLM (MOSLM), both having speed of operation as a positive consideration in their use. They are essentially binary in their operation and "snap" from one state to another. The MOSLM has been caused to operate in a ternary fashion as well, with the third state resulting from an interdigitation of the other two states within the pixels of the SLM. The discrete SLMs are often operated in states of polarization that cause balanced values of the two states (referred to as +1 and −1, although in fact their light efficiency is much closer to zero than to unity). They can also be set to have one polarization blocked and the other transmitted, in which case the SLMs' states are referred to as 0 and 1. The ternary SLMs are said to operate at +1, 0, and −1. If we properly normalize the noise levels in the filter computations we can just as well use the unity-sized transmittances, and so we shall do just that. For the balanced binary SLM (Figure 8.17), the MED mapping is given by:

$$H_{\text{MED}} = \begin{cases} -1, & \text{Re}(H^I) < 0 \\ +1, & \text{Re}(H^I) \geq 0 \end{cases} \tag{8.57}$$

This mapping is shown in Figure 8.17. For the ternary SLM,

$$H_{\text{MED}} = \begin{cases} -1, & -0.5 < \text{Re}(H^I) \\ 0, & -0.5 \leq \text{Re}(H^I) \leq +0.5 \\ +1, & +0.5 < \text{Re}(H^I) \end{cases} \tag{8.58}$$

as is shown in Figure 8.18.

In many filtering operations it is advantageous to be able to write an opaque value at selected spatial frequencies. We think it is almost without exception

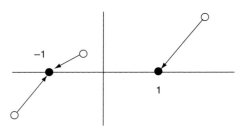

Figure 8.17 Binary phase-only operating curve

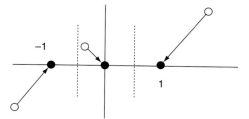

Figure 8.18 Ternary operating curve

desirable to block the DC component of a filtered scene, for one example. In cases where a set of frequencies is always found to contribute more to noise than it contributes to the construction of the correlation peak, it would be preferable to put black paint on the SLM at those frequencies. A filter's region of support (ROS) is that set of frequencies at which the value of the filter may be non-zero, and the black paint is the complement of the ROS. One quantifies the advantage of painting the SLM by adding zero as a value that the filter computations may choose. For the POSLM this would be expressed as:

$$H_{\text{MED}} = \begin{cases} 0, & |H^I| < 0.5 \\ \exp(j\theta^I), & \text{otherwise} \end{cases} \qquad (8.59)$$

If the computations suggest that a significant gain will result from adding a complementary ROS to the SLM, then those places where H_{MED} is zero indicate where to paint the filter SLM.

8.8.3 *Unit disk*

Of all passive filter SLMs, one that can realize any location in the unit disk is the most useful. Rather than simply being a tautology, this statement is the basis for quantifying the performance of an SLM. Choose a pattern recognition problem, a candidate SLM, a noise environment, and a metric by which to evaluate a filter. In general, when the metric is optimized the result will be poorer for the candidate SLM than for the unit-disk SLM. The ratio of metric performance quantifies how well the candidate SLM works in the stated conditions. Since the unit-disk SLM is the most competent we shall use it as the reference for even the less detailed evaluation of an SLM. The more like a unit-disk SLM our candidate SLM looks, the better it is, the benefit being that it can reach any point in the unit-disk. We have seen that minimizing distance in the complex plane is a good thing to do, so we could judge the candidate by the average of how closely it approaches points in the unit disk.

This is a modification of the "coverage" defined by Zeile and Lüder [91]. For any point ζ in the unit disk, the SLM has a closest realizable point, $z_0(\zeta)$. They integrate $|\zeta - z_0|^2$ over the unit disk and obtain the mean-squared distance to the closest realizable point, calling the result the SLM's coverage of the unit disk. We propose a more informative method of quantifying the property that Zeile and Lüder intended their coverage to show; it conforms to the exact problem to which we are applying the SLM. For any given pattern recognition problem (i.e., the set of reference objects, the specification of the noise environment, and a choice of metric) we can compare the performance of a candidate SLM with that of the unit-disk SLM. The value of the metric achieved by the candidate SLM, normalized to the performance of the unit-disk SLM, is a direct and meaningful measure of the candidate SLM's performance.

$$J_{\text{norm}} = \frac{J(\text{actual SLM})}{J(\text{unit disk SLM})} \tag{8.60}$$

Perhaps unfortunately, J_{norm} varies from problem to problem with a single SLM, whereas Zeile and Lüder's coverage by the SLM is an absolute measure of its performance. Nevertheless, the ratiometric method is more directly an answer to the long-standing question, "How good is this SLM for my job?" For example, the ratiometric method accounts for the nonlinear effects of the MED mapping, which is not the case with Zeile and Lüder's coverage.

8.9 Algorithm for optical correlation filter design

The filters described in this chapter refer to the performance of the correlator as measured at the center of the correlation plane. We can refer to them as "inner product filters," since the central correlation performance is principally determined by the inner product of the reference object (as limited in its representation by the filter SLM's operating curve) and the input object. In summary of the development in this chapter, the general algorithm for computing such optically implemented filters is the following.

- Select initial complex scalars, the search parameters.
- Weight the training images with the search parameters.
- Whiten the weighted sum of the transformed and conjugated training images by normalizing to the power spectral density of noise.

We now have an ideal filter value H_m^I for each frequency.

- Having these H_m^I in hand as being the locations where the gradient of the metric is zero, realize each if you can; if you cannot, then use the realizable value that is closest by Euclidean measure in the complex plane.

- Repeat this process for each spectral component, indexed by m.
- Evaluate J. It is a function of the search parameters.
- Do a gradient search in the space of those parameters, plotting the value of the metric J as a function of the parameters and moving through the parameter space so as to take J to its extremum.

In our experience there does not seem to be much difficulty arising from hanging J up on local extrema, but the reader should remain alert to the possibility. For all of the metrics and SLMs mentioned in this chapter there is a maximum of one complex scalar factor per training object – fewer, in degenerate conditions such as the POSLM, in which only phase is considered during the filter optimization. In all of this chapter's metrics and filter SLMs, the complex scalars that optimize a metric are not computable before starting the optimization, but there exist confirming formulas that a putatively optimizing set of scalars must meet.

Since we are dealing with optics rather than digits here, the gain Γ and phase α have physical significance. Indeed they are crucial factors in optimizing a filter on a physical modulator. They are not arbitrary; the optimal filter will produce values for them that are physically consonant with their definitions. Consider now the following equation:

$$H \overset{?}{=} \frac{S^*}{P_n} \tag{8.61}$$

The signal S in an optical processor is an electric field with dimensions of $[V/m]$, and P_n is a noise power spectral density. But the filter simply converts an incoming electric field into an outgoing one, so it is dimensionless, although probably complex. On the very face of it, this supposed equation is physically incorrect since the dimensions do not match. That this sort of dimensional mismatch permeates the OCPR literature indicates that the physical considerations we use have not been widely taken into account. The factors such as the Γ in this section provide the necessary dimensional matching. In addition they bolster confidence that the optimal filter has been computed, by exhibiting a loop-closure for Γ and β from their definitions that follow from the gradient expression. Consider Eq. (8.48), repeated here:

$$M_m^I = \left(\frac{\Upsilon + \langle \sigma^2 \rangle}{B} \right) \frac{A_m}{P_{n,m}} \tag{8.62}$$

All of the terms within the parentheses are functions of the filter, and when the optimizing value of Γ is found in the search, the indicated mathematical operations within the parentheses will reproduce Γ. Similarly, the optimizing value of β from the search parameters will replicate as the phase of the expression $\sum_k H_k S_k$.

8.10 Some practical points

8.10.1 Storage and representation of filters

We may artificially decrease the domain of filter values for practical reasons. The prime example is that we can store more filter drives in a given amount of memory if we use fewer bits per frequency. A lookup table is very often used in optical correlators to convert the index values for a smaller set of discrete filter values into the control actually applied to a pixel. More control is generally better, but there is a diminishing additional return for every bit of added control. Intuitively, we know that there will be a limit to the return on investment as we add bits. It is our experience that 16 to 32 well chosen filter values give nearly as good a performance as the common 8 bits (256 levels) of voltage resolution in addressing liquid crystal SLMs. The term J_{norm} is a measure of performance loss in response to fewer realizable filter values, as well as being a guide to which values to drop and which to retain.

8.10.2 Specifying lot uniformity

Let us consider how to write specifications for intra-lot variability of an SLM. If we are to build a substantial number of correlators and wish to have interchangeable SLMs, then we have to specify how nearly identical their performance is. If we are unreasonably tight in the specification we needlessly run up the cost, but if we are not tight enough we will not achieve functional interchangeability. SLM performance tolerance should be tied to system performance. We propose two tools to specify the SLM's performance tolerance, although we have not put them into practice.

The first tolerance tool, NEΔH, is more complicated than J_{norm}. It is similar to the characterization of a thermal radiometer in its noise-equivalent random temperature change (NEΔT). To compute NEΔT for a radiometer, one observes the noise level in the output and determines, by knowing the calibration of the instrument, the level of random temperature fluctuation that would itself produce the same level of variation in the output. Such a level of temperature fluctuation is then called the noise-equivalent change in temperature, NEΔT. The dynamic range of the radiometer is the total range of measurable temperatures divided by NEΔT. We can similarly compute the effect of a noise-equivalent random change of an SLM's operating curve, NEΔH. The method is straightforward and gives a direct system-based specification on SLM performance tolerance, but it would be tedious in practice.

Characterizing the noise figure of a radiometric thermometer begins with observing the noise-induced variance, σ_n^2, in the output, and knowing the instrumental responsivity, R, in terms of, for example, (change in volts)/(change in temperature) at the temperature of interest. Then the noise-equivalent change in temperature is:

$$\text{NE}\Delta\text{T} = \frac{\sqrt{\sigma_n^2}}{R} \qquad (8.63)$$

Physically we can think of NEΔT as the random variation in temperature which, if viewed by a noiseless but otherwise identical system, would have the observed output variance. The temperature change NEΔT is a measure of the smallest temperature difference the radiometer can distinguish.

We can adapt the idea of NEΔT to our situation. The simulation and analysis method for NEΔH is rather more complicated than finding NEΔT. Below we lay out the method for determining NEΔH, which is then a system-based and reasonable way to specify the uniformity of behavior for a lot of SLMs.

1. Construct an optimal filter under the circumstances appropriate to the correlator's task.
2. Note the peak correlation intensity predicted by the digital model of the correlator; denote this C_{digital}.
3. Run the filter and image in your optical correlator. Note the peak intensity and denote this C_{optical}.
4. Observe the noise variance in the correlation peak and denote this $\sigma_{\text{optical}}^2$; compute the optical signal to noise ratio $\xi_{\text{optical}}^2 = C_{\text{optical}}^2 / \sigma_{\text{optical}}^2$.
5. Return to the digital simulation of the correlator. Disturb the list of the SLM's values as follows, noting that all filter frequencies using a common filter value have that value changed by the same amount. Construct a set of complex disturbances $\{\Delta H\}$ to add to the optimal realized filter values, each filter value being displaced by its own ΔH. Their phase should be distributed uniformly in $[0, 2\pi)$, and independently of the magnitude. It is a mere detail of implementation whether they have a common magnitude d, or are drawn from a Gaussian population with standard deviation d.
6. Ascertain the value of C_{digital} with the disturbed optimal filter.
7. Repeat Steps 5 and 6 a sufficient number of times, using a new set $\{\Delta H\}$ each time, to estimate the variance $\sigma_{\text{digital}}^2$ in C_{digital}. Compute $\xi_{\text{digital}}^2 = C_{\text{digital}}^2 / \sigma_{\text{digital}}^2$, and note that $\xi_{\text{digital}}^2 = \xi_{\text{digital}}^2(d)$.
8. Adjust d to find the value NEΔH such that $\xi_{\text{optical}}^2 = \xi_{\text{digital}}^2(\text{NE}\Delta\text{H})$.

8.10.3 *Choosing a finite subset of filter SLM values*

Now we have NEΔH in hand, fully particularized to the situation in which the SLM has been designated to work. This quantity is a direct measure of how distant two SLM values must be in order to be functionally distinct from each other. We can use that separation to count the distinct number of values on the SLM and to choose a finite subset to work with. We count how many disks of radius NEΔH will just fit along the SLM's operating curve, if we require that the center of the disk must occupy an operating value and that the disks cannot overlap. The operation is like stringing beads onto a necklace, and the bead count is how many distinctly different values the SLM can make, given the whole environment of the correlator. In the example shown in Figure 8.19, we reckon we can make about 15 usefully distinct values of the filter, so we would probably choose 16 values, as a close power of 2, and store filter drives with a 4-bit representation.

NEΔυ is a related concept that is one step more particularized to a given system than NEΔH. We use noise-equivalent change in the filter drive (measured in bits) to ascertain the useful number of filter drive bits to employ.

8.10.4 *MED maps*

The algorithms for optimizing a filter on a limited set of realizable values call for repeatedly finding the spectral filter value that either is the closest to a

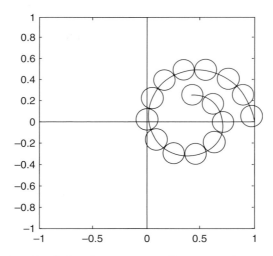

Figure 8.19 A bunch of "beads" corresponding to NEΔH, perching along a string of realizable values

computed filter value, or has the greatest projection in a computed direction. We have found it efficient to compute a map from locations in the complex plane to their closest realizable filter values. These are minimum Euclidean distance (MED) maps, so named because they find the realizable value that is the minimum Euclidean distance from the arbitrary filter value; using the map is efficient because it needs to be computed only once. We treat the two cases of a spatially uniform SLM, and a non-uniform one. The user can determine the advantage of using the more complicated non-uniform method by using computer simulation to check the benefits of the more accurate description. In Chapter 7 we described methods of characterizing the SLMs and we assume here that those data are in hand.

8.10.5 *MED map for a uniform SLM*

We wish to build a lookup table – a map – of an SLM's closest realizable value to an arbitrary H^I. Our practical method as described in text is shown in Figures 8.20 and 8.21. The easiest condition arises if the SLM is sufficiently nearly uniform over its surface such that all points may be treated as having the same operating characteristics. Under the pressures of commerce and driven by the human eye's sensitivity to local abnormalities, a great deal of effort has gone towards exactly this point in the case of SLMs that are adapted from the display industry. In particular, the magnitude characteristics of the displays are often uniform in such cases, although there is less incentive for phase uniformity since phase is not ordinarily important to visual display.

We begin with a list of the realizable values $R = \{z_k\}$, and the drive values $V = \{v_k\}$ that produce them. Three statistics are most important to the MED map. First is the largest magnitude $d_{\max} = \max_k |z_k|$, second is the average spacing $d_{\text{avg}} = \langle |z_k - z_{k+1}| \rangle$ between successive members of R, and third is the minimum spacing $d_{\min}$ between members of R. Two quantities determine the map – its expanse in the complex plane and the resolution with which we view that area. We let the map extend through $\pm E$ in both real and imaginary axes, dividing the space with N marks per axis. (We like N be odd so that there is a mark at zero.) Two rules of thumb are that E is at least twice the maximum magnitude, and that the resolution $2E/(N-1)$ is no larger than the average spacing in R. Nothing is gained by making the resolution smaller than the minimum spacing in R.

Having sized this array, we proceed to fill it. Let $0 \leq n < N$ index the real values, and $0 \leq m < N$ index the imaginary values. For each element in the array we compute its complex value $z = x + jy$ from setting the real part to

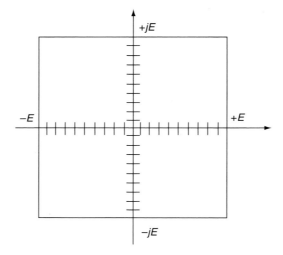

Figure 8.20 Coordinates for the MED map. The magnitude runs from $-E$ to $+E$ on each axis, and there are (in this example) $N = 21$ marks per axis

$$x = E \, \frac{2n - N + 1}{N + 1} \tag{8.64}$$

and the imaginary part to

$$y = E \, \frac{2m - N + 1}{N + 1} \tag{8.65}$$

These invert to

$$n = \left\lfloor \frac{\dfrac{x}{E}(N + 1) + N}{2} \right\rfloor \tag{8.66}$$

and a similar equation for m, in which the symbol $\lfloor \cdot \rfloor$ indicates rounding down (towards negative infinity) to the next integer. For every member of the array we compute $z = x + jy$ and find the member of R that is closest to it, designated MED$\{z\}$. Whether this is a drive value, or the realizable filter value itself, depends on the application. What we store in the array itself depends on how we will later use the information. If we are to create a set of drive values to apply to the SLM, we would store the drive value of that closest member of R; but if we are to simulate the effect of a computed filter, we might store the complex value of that closest member. We use the map by converting a computed ideal filter value to indices m and n and reading the stored information from the array.

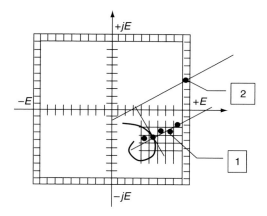

Figure 8.21 (1) Filling cells in the plane of H_m with MED values from the operating curve. (2) Filling a ring of H_m's phase values with maximum-projection values from the operating curve. The indicated value on the SLM operating curve is found along the perpendicular to the curve, and in the peripheral cell in the direction of that perpendicular from the origin

We also have the question of how to deal with computed ideal filter values that lie outside the $(\pm E, \pm jE)$ region of the complex plane. This becomes important especially when maximum-projection filters are being built, such as to optimize intensity or in some forms of the Rayleigh quotient. It is convenient to reserve the outer ring of cells in the square array for these values. In that outer ring of map cells we place the representation of the realizable filter value having the largest projection in the direction of each cell. It is for this reason that we recommend E to be at least twice as large as the largest realizable magnitude. A computed ideal filter value lying outside the $(\pm E, \pm jE)$ region is brought back along a ray towards the origin and assigned to the boundary cell it passes through. If E is sufficiently large, there is little difference between the largest projected value as stated, and a strictly computed MED representation for a filter value that falls outside the map.

8.10.6 MED maps for a spatially variant SLM

The previous section assumes that the SLM is uniform throughout. The calibration methods described in Chapter 7 include a way of characterizing the SLM for each pixel's operation. If the SLM's action varies over its surface, we are faced with having a different MED map for each pixel. This is a difficult situation with no really good way out, but it is a situation to be lived with. The set of all pixels' operating curves is usually too much data to use in raw form when building a filter. One economy is to cluster the pixels according to similarity in their operating curves, and then use one MED map per cluster. This requires that the cluster's members show behavior that is "close enough" to each other. We let $\rho_{i,j}$ be the distance measure between the ith and jth operating curves, computed by:

$$\rho_{i,j} = \sqrt{\sum_d |z_i(\upsilon) - z_j(\upsilon)|^2} \qquad (8.67)$$

where υ is the drive value indexing the set of complex operating values. (In practice we do not bother with the square root, and instead threshold on ρ^2.) Then, we cluster the set of operating curves by setting a threshold value ρ_t, picking a first pixel's curve as a prototype, and finding all of the pixels whose operating curves fall within ρ_t of the prototype. If not all pixels are assigned to clusters yet, we choose an unassigned pixel's operating curve as the next prototype, find its neighbors, and assign those neighbors to the cluster. When all pixels have been assigned, we have a set of clusters that is dependent on the size of ρ_t. If there are inoperably many clusters we must adjust ρ_t, making it larger to admit more members per cluster, but at the expense of reduced accuracy in representing an operating curve by its cluster.

We can employ another economy: using transformations under which the MED mapping is invariant. On average this will put more pixels in each cluster, meaning that fewer MED maps can handle the whole SLM. Cartesian translations, scale changes, and rotations do not alter the MED relationship between an SLM's operating curve and an off-the-curve point z. In Figure 8.22 we illustrate an operating curve, and an MED mapping that is unchanged, if the curve and a point we wish to represent on it are both modified by the same complex scale factor and complex translation. Letting **b** be a bias vector, and **k** a complex scalar that expresses the rotation and scale, we now compute the distance between the ith cluster and the jth candidate pixel's operating curve according to:

$$\rho_{i,j} = \min_{\mathbf{k},\mathbf{b}} \left\{ \sqrt{\sum_d |z_i(\upsilon) - [\mathbf{k}z_j(\upsilon) - \mathbf{b}]|^2} \right\}. \qquad (8.68)$$

When assigning the jth pixel to the ith cluster, we make note of not only the cluster index i, but also the minimizing values $\mathbf{b}_j$ and $\mathbf{k}_j$, for later use.

At this point we build an MED map for each prototype as described in the preceding section. When building a filter drive, at the jth pixel we are given an ideal filter value $H^I(j)$ to represent some point on the pixel's operating curve. We choose

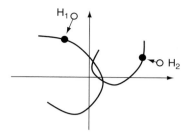

Figure 8.22 Drive $v(H_1)$ represents H_1 onto the operating curve OC_1. If the operating curve is modified to $OC_2 = \mathbf{k}OC_1 + \mathbf{b}$, then $v(H_1)$ also represents $H_2 = \mathbf{k}H_1 + \mathbf{b}$

$$z = \frac{\mathbf{b}_j + \mathrm{MED}_i\{\mathbf{k}_j H^I - \mathbf{b}_j\}}{\mathbf{k}_j} \qquad (8.69)$$

as the representative of H^I, where the subscript i indicates we are using the MED map for the ith cluster. The equation indicates that we begin with H^I; convert it by the translation, scale, and rotation that best fitted the pixel to its (the ith) prototype curve; find the closest prototype value; and rotate, scale, and bias back to where it came from. This is not exact; if the drive value version of the MED process is used, we can use the operating curve of the jth pixel to tell us what filter value will actually be implemented. However, that would imply storing all the operating curves rather than just those of the prototypes. With a sufficiently small value of ρ_t we can have as much precision as we wish, but this is at the expense of enlarging the number of prototypes and their MED maps.

8.11 Some heuristic filters

Except to compare the effect of different methods of filtering, one will not ordinarily implement a filter that does not use all of the ability of the SLM. We would neither use a suboptimal algorithm for the full SLM behavior, nor restrict ourselves to less than the full domain of the SLM. We have presented algorithms that make best use of a filter SLM if we have an analytic form for a metric to optimize, and know the list of realizable filter values. In Section 8.8 we described some mathematically tractable SLMs and showed how to create an optimal filter on them strictly from analysis. We also showed how to deal with departures from analytically tractable SLMs. In this section we look at some algorithms of historical interest that in some cases are seen to be optimal, and in other cases, not.

We mention three heuristic algorithms that have been used in mapping from digitally computed filter values to the values realizable by the SLM, and then discuss their shortcomings.

- The first is applying phase matching to coupled SLMs,
- the second is mapping a computed filter value's magnitude onto the magnitude of an AOSLM, and
- the third is binarizing a filter.

As noted previously, the phase-only filter has been a considerable success, although for a long time there was not a body of theory to show why it performed so well. We suggest that the MED principle is one such theory; it shows that when the SLM's operating curve is the unit circle, phase matching is optimal. Further, those metrics demanding a large magnitude of the filter will cause only the unit circle to be occupied for the most capable of filter domains, the unit disk. The metrics that demand large magnitude will cause POSLM behavior from a unit disk filter. However, one should keep in mind that for the maximum-projection metrics, magnitude is the dominant consideration, not the maintenance or continuity of phase. In Figure 8.23, we see that if offered three points that have large magnitude and are appropriately disposed around a smaller-magnitude complete circle, the maximum-projection algorithms will choose from the triad of distinct points, in preference to the inner circle having all phase values.

The success of the matched-phase filter on a POSLM led to some overuse, in our opinion. As one example, suppose that an SLM has a nearly complete circle of phase-only operating values. The MED principle would have the wedge of missing phase split down the middle and ideal filter values in its halves mapped to one of the end points. See Figure 8.24.

The MED value is obtained as indicated and is optimal – as opposed, say, to uniformly compacting the 2π phase range to fit on the available SLM domain. The success of matching phase on a POSLM should not dissuade us from exploring other operating curves that may be available on the SLM. (Varying the input and analyzer polarizations around a liquid crystal SLM produces such variable operating curves; see Chapter 7.) The more powerful metrics,

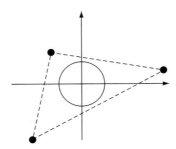

Figure 8.23 A set of three points set around a phase-only circle so that their convex hull falls outside the circle

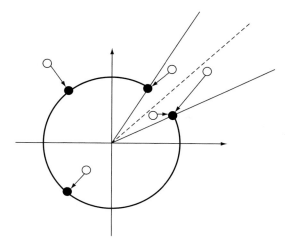

Figure 8.24 Incomplete ring of PO values. Lines show the missing wedge, including the dashed one that splits it. Ideal points in the wedge are MED-mapped to end points. Other points are mapped so as to maintain phase

such as SNR, that take into account the power spectral density (PSD) of additive incoming noise and seek finite values of the filter, will respond well to an operating curve that passes close to the origin. If the PSD of input noise is known and can be used in the filter optimization, then coupled SLMs should certainly be considered, rather than setting up the SLM to operate in a phase-only mode. We discussed metrics of SLMs per se without regard to the total environment they are in, and we examined a couple of ways of measuring the utility of the SLM. In the case where we are performing noise rejection, as for optimizing the SPR metrics, the POSLM is overshadowed by a coupled SLM.

Since mapping the phase of an ideal complex filter H^I onto the phase of a POSLM produces good results, how about mapping the magnitude of H^I onto the magnitude of an AOSLM? Well, in short, bad things happen. This point is made clearly in Chapter 5. New images were constructed by "crossing" the spectra, and the output images far more strongly resembled the images whose phase they bore. These results are replicated here, using MED projections onto the real axis rather than using the magnitude of the transform.

We attribute the predominance of phase fundamentally to its being a leveraging quantity; that is, for larger magnitudes a phase difference has a larger effect than a magnitude difference, since $|\Delta z| = 2M \sin(\Delta\phi/2)$ for a purely phase difference, whereas $|\Delta z| = \Delta M$ for a purely magnitude difference, with z being the complex number being altered. Thus, roughly speaking, at large-magnitude portions of the spectrum the effect of phase is dominant over that of magnitude.

Figure 8.25 Appliance image

To illustrate this point, in Figures 8.26 and 8.27, we disturb Figure 8.25 by exactly the same set of $|\Delta z|$ at each frequency, only having the disturbance aligned in the magnitude direction in Figure 8.26, and in the phase direction in Figure 8.27 so that the phase's leveraging effect is eliminated. The images were created by adding a zero-mean RV to the complex value of the transform of the image of appliances, then inverse transforming, and displaying the magnitude as intensity. If $z = x + jy$, and r is an RV distributed uniformly in $[-b, +b]$, then for the magnitude-disturbed image we compute $z_{\text{disturbed}} = z + rx/|z| + jry/|z|$, and for the phase-disturbed image, $z_{\text{disturbed}} = z + ry/|z| + jrx/|z|$, for every pixel in the transform of the original image. (Parameter b was chosen to display the effect clearly but without entirely destroying the image.) Phase error and magnitude error are seen to have distinctly different kinds of impact, but overall the Figure 8.26 and Figure 8.27 images are visually degraded to approximately the same degree from Figure 8.25.

Next, we show the effects of combining the magnitude, MED mapping onto magnitude, random, and constant magnitudes; and phase, random phase, constant phase, from one image and another. For this illustration, we use the register and appliances images in Figure. 8.28. In Figure. 8.29, we show the various reconstructions. The degree of resemblance to the image at the head of

Figure 8.26 Magnitude-disturbed appliance image

Figure 8.27 Phase-disturbed appliance image

Fig. 8.28 Images of register (left) and appliances (right)

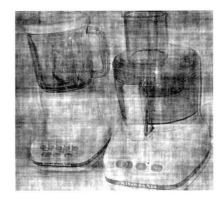

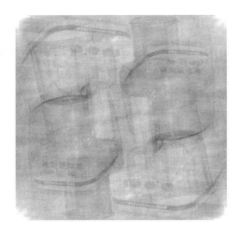

Fig. 8.29 Images reconstructed from various combinations of magnitudes and phases of the Fourier transforms of the two images in Figure 8.28

the columns decreases from top to bottom. First place goes to the phase using the alternate magnitude, so the phase is seen to carry the most weight, as shown by Oppenheim and Lim [43]. Second place goes to a constant phase with an MED-mapped complex amplitude, a mapping they did not examine. Third place goes to the phase with a constant magnitude, and fourth place to the phase with a random magnitude. We are hard pressed to see the image by the time we get to fifth place, the phase being accompanied by the MED-mapped complex amplitude of the alternate image; we take this near-invisibility of images as showing the near-equivalence of the phase and the MED-mapped complex amplitude for information-bearing capacity. The final three mappings have essentially no visible information on the leading image.

Let's think about binarization as a filtering ploy on a real-valued AOSLM. Say the AOSLM operates on the range $[-1, +1]$ with the continuum of values available. Any of the maximum-projection metrics will pick out only the two end points of that range. In this case, a binary SLM is just as capable as the bipolar real-valued SLM we posited. The interior values of the bipolar real SLM are not used, analogous to how the interior SLM values on the unit disk are not used by the maximum-projection algorithms. We conclude that for the maximum-projection metrics (intensity, for example), a binary SLM with values ± 1 is just as useful as an SLM with the continuum of values available in $[-1, +1]$. The finite-value metrics (SNR, for example) will make use of the interior values on the continuum SLM and produce a greater value of the metric than the binary SLM.

Our point is this. If one has optimizing filter code that knows the available filter values, it will select a subset of the available operating curve if that is appropriate. One is ill advised to restrict the values available to expression on the SLM without considering the metric to be optimized, and the effect of the diminution.

8.12 Chapter summary

- No physical device can fully represent complex filters. Therefore we must develop some method of using only those filter values that our filter SLM can realize. Implicit in this statement is that the SLMs must be characterized so that their behavior is known in mathematical terms.
- We recognize two main sorts of filter computation methods. The first is ad hoc, well-intentioned methods such as phase-matching onto coupled SLMs. The second is to choose a metric that expresses beneficial effect, and then to determine analytically what realizable value to select at each frequency, so as to optimize the metric.

- The metrics we divide into two categories. The first category (based on correlation magnitude statistics) was developed earlier than the second (based on measurement statistics). Intensity and magnitude SNR are examples of the first category, and the statistical pattern recognition metrics such as Bayes error and the area under the receiver operating characteristic curve exemplify the latter.

- A general algorithm is shown to optimize many of the metrics. Reference objects are transformed, whitened, conjugated, multiplied by complex search parameters, summed, and projected onto the closest realizable filter value. Different values of the search parameters optimize different metrics. No general rule is available for selecting the parameters without searching.

- The chapter shows several practical methods for using filter SLMs that might have no convenient analytical expression for the closest realizable value, and for using filter SLMs that might be spatially variant over their surface.

9

Application of correlation filters

The preceding chapters have described the techniques for designing correlation filters and their underlying mathematical foundations. The objective of this chapter is to provide a better understanding of the end-to-end process of designing and applying correlation filters to solve a pattern recognition problem. To facilitate this process we discuss two examples. The first is geared towards the recognition of targets in synthetic aperture radar (SAR) imagery. For this purpose, we use the public MSTAR SAR data set [3]. Details of a sample pseudo-code needed to construct and apply the filters to this data set are also provided. The second example discusses applications of correlation filters for face verification. Face recognition is just one example of a growing research field called biometric recognition [92] which includes other biometrics such as fingerprint, iris, etc. Correlation filters should prove useful in all such image recognition endeavors.

Correlation filters can be used to recognize patterns in images generated by many different types of sensors. Once the sensed information is converted into image pixels, the correlation pattern recognition algorithms can be applied in a fairly universal fashion. Thus correlation filters can find uses in all areas of automation including industrial inspection, security, robot vision, space applications, and defense. For instance, systems have been developed for fingerprint recognition [93]. Although the choice of sensor depends on the phenomenology associated with the pattern recognition/machine vision problem of interest, the approach for designing correlation filters generally remains the same.

9.1 Recognition of targets in SAR imagery

For many years, one of the focus problems for the image-understanding research community was the automatic recognition of objects in SAR imagery for targeting and surveillance applications. The problem of recognizing objects in SAR images is characterized by strong variations induced by changes in orientation of the object, and the squint and grazing angles of the radar. The benefits of SAR imaging include the fact that it is an active sensor that produces predictable object signatures, and that it provides images at long range in almost all weather conditions. Synthetic aperture radar imaging is fundamentally different from electro-optical imaging (range, along-track instead of elevation, azimuth image). We are not used to looking at it.

The application to SAR imagery is chosen for inclusion because of the availability of a common public database [3], and we do not intend to suggest that correlation filters are useful only for SAR images or vehicular target recognition. In the next section, we will discuss the face verification application, once again because a large database, known as the Pose, Illumination, and Expression (PIE) database [4] is available. In this section, we discuss the fundamentals of designing maximum average correlation height (MACH) filters, and distance classifier correlation filters (DCCFs), and discuss their performance using the MSTAR database. The techniques are illustrated for a simplified two-class automatic target recognition (ATR) problem. This section also assesses the performance of these algorithms on target and clutter data, and characterizes the performance of the ATR system in terms of probabilities of correct recognition and false alarm. The benefits of using a specialized form of the MACH filter for reducing false alarms are also discussed.

9.1.1 SAR ATR using MACH and DCCF filters

The public MSTAR database provides a wealth of information that can be used for algorithm development and evaluation. The 1-foot resolution SAR data is well calibrated and ground truth is known precisely. In fact, it is an ideal database for comparing different algorithms with respect to performance and throughput efficiency. In this chapter, we restrict our attention to two of the target sets in the public database. Specifically, we formulate a two-class problem to distinguish between the BTR70, an armored personnel carrier (APC) and the tank T72. Photographs of each type of vehicle and examples of these targets as seen in SAR images are shown in Figures 9.1 and 9.2, respectively. Note the speckle and the shadows in the SAR images. It should be noted that there are a variety of serial numbers and configurations of these vehicles, and that the photographs may not show the exact same models of the targets as in the SAR imagery.

(a)   (b)

Figure 9.1 Sample images of the two targets (a) T72 and (b) BTR70

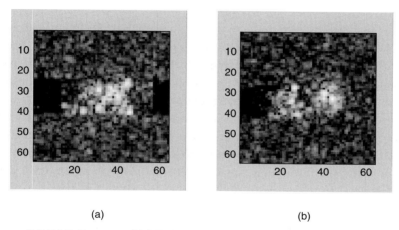

(a)   (b)

Figure 9.2 SAR Images of (a) T72 and (b) BTR70

9.1.2 MACH filter design and performance analysis

The database provides the images of each class of target at various elevations.
The 17° elevation imagery was considered to be the training set, whereas the
test set is at an elevation of 15°. Each set contains several hundred images of
the objects covering many different viewing angles. There are images of several
serial numbers of the tank in the database of which the T72 SN_S7 was used
for these experiments. The number of available training images is 228 for the
tank, and 233 for the BTR70.

There is a performance tradeoff between distortion tolerance and discrimination. Therefore, a filter that generalizes over a wide range of angles is unable
to discriminate effectively between the classes. To mitigate this problem, the
data can be partitioned into several smaller sets, to build filters that generalize

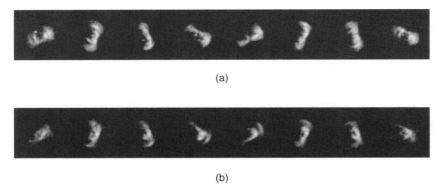

(a)

(b)

Figure 9.3 Space domain mean images of the eight clusters for (a) the BTR70 and (b) the T72

over limited angular ranges. Depending on the application, these partitions may be chosen as natural clusters or arbitrarily demarcated, often into uniform angular bins or clusters. For illustrative purposes, we partition the data set by orientation into eight uniform clusters, each spanning a range of 45°. The images are cropped to size 64×64 and manually registered so that the object is at the center. Registration is not necessary during the testing stage (because of the shift-invariance of correlation filters), but is an important step for synthesizing the filters. The log-magnitude of the complex SAR images is used for both training and testing, because the dynamic range of the SAR image magnitudes is too large to fit into the usual 8-bit representations. Some of the background pixels are removed by applying a simple threshold and such background removal helps to make the performance more robust.

The first step in synthesizing either the MACH filter[1] or the DCCF, is to estimate the mean $M(k, l)$, the power spectrum $D(k, l)$, and the spectral variance $S(k, l)$ for both classes in the frequency domain. It is easier to visualize the space domain version of the mean images as shown in Figure 9.3. It is also insightful to visualize the spectral variance in the frequency domain. The log-magnitude of $S(k, l)$ of the eight clusters for both classes is shown in Figure 9.4. The frequencies where $S(k, l)$ has a large magnitude are unattractive for in-class distortion tolerance since these are the frequencies at which the target class exhibits a large variance. Therefore, the inverse of $S(k, l)$ tends to attenuate frequencies where the training patterns vary most, which results in the MACH filter exhibiting the necessary distortion tolerance. The inverse of $D(k, l)$ tends to whiten the data and helps to produce sharp correlation peaks. The filters

---

[1] The optimal tradeoff version of the MACH filter is used.

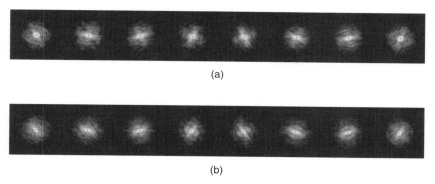

(a)

(b)

Figure 9.4 Spectral variance images of the eight clusters for (a) the BTR70, and (b) tank T72

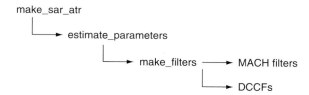

Figure 9.5 Algorithm flow diagram for synthesizing correlation filters

were synthesized using Eq. (6.75) with $\alpha = 0.2$, $\beta = 0.75$, a white noise model with $\gamma = 0.5$.

We now describe the main steps in designing and testing correlation filters using a software package such as MATLAB. In what follows, we use the syntax of MATLAB, but it may be possible to use other similar software packages. Typically, the code is a set of scripts and function calls which implement the process shown in Figure 9.5. We will now review the pseudo-code for each of these modules.

At the highest level, the code `make_sar_atr.m` synthesizes a two-class SAR ATR system using the MSTAR data set. For convenience, the SAR images may be cropped to 64×64 size with *d1* and *d2* as the image size parameters. Also, the number of aspect angle clusters in our example is 8, but can be chosen to be any other number. The following lines accomplish these settings:

```
d1=64; d2=64; d=d1*d2;
max_cluster=8;
```

The MSTAR data will probably be in directories specific to the computer system. The next few lines are an example of how to set up the path to each of the classes and to estimate the necessary statistics. The names of the data files for each class may be read from a list. The program `estimate_parameters.m` goes through each image and computes the mean images *M1* and *M2*, and the variance images *S1* and *S2* for the two classes.

```
name = 'btr70c71. lst'
loc = 'E:\TARGETS\TRAIN\17_DEG\btr70\sn_c71\'
estimate_parameters
M1 = M;
S1 = S;
name = 't72sn_s7.lst'
loc = 'E:\TARGETS\TRAIN\17_DEG\t72\sn_s7\'
estimate_parameters
M2 = M;
S2 = S;
```

Finally, these class statistics are used by the module, `make_filters`, to design the MACH and DCCF filters which are saved to files for use by the test code.

```
nclass = 2;
M = [M1 M2];
S = [S1 S2];
make_filters
save filters Hmach H
```

Note that the statistics of all the classes are combined into common variables **M** and **S** prior to filter synthesis.

Let us now discuss how the module, `estimate_parameters.m`, may be formulated. Firstly the variable spaces for accumulating the mean and the variance images are initialized using:

```
M = zeros (d,max_cluster);
S = zeros (d,max_cluster);
count = zeros (max_cluster,1);
```

The variable `count` will be used to keep track of the number of images that fall into each cluster. Note that these matrices have as many rows as total number of pixels in the images (i.e. $d = 4096$). Further, the number of columns is equal to the number of aspect angle clusters so that the estimates for all the clusters can be stored in the same variable. We also initialize a rectangular image of all 1s called `boxwin` which represents a support region roughly the same size as the targets. This is used to mask (remove) the background that is present in the MSTAR training imagery.

```
boxwin = ones (20,45); boxwin = pad (boxwin,64,64);
```

The main body of the code which estimates the mean and the variance simply loops through reading all the images, obtains the logarithm of the magnitude of the complex SAR data, masks out the background, performs the FFT and computes the necessary averaging calculations as follows:

```
for i=start:finish

[x,header]=rd_mstr(filename);      %Get input data
aspect=header (1,1);               %Check aspect from header

id=determine_cluster(aspect)       %Determine which cluster
count(id)=count(id)+1;             %Update cluster count

x=abs(x);                          %preprocess image
x=log10(crop(x,d1,d2));
boxwinmask=imrotate(boxwin,aspect,'crop');
x=x.*boxwinmask;

figure(1); imagep(x); pause(0.01);
X=fft2 (x);                        %Compute 2-D FFT
M(:,id)=M(:,id)+X(:);              %Compute mean
S(:,id)=S(:,id)+X(:).*conj(X(:));  %& power spectrum
count(id)=count(id)+1;
end

for i=1:max_cluster                %Accumulate average
M(:,i)=M(:,i)/count(i);
S(:,i)=S(:,i)/count(i);
end
```

At the end of this process, the matrices **M** and **S** contain the frequency domain mean and power spectrum of all clusters of one class of targets. Of course, this routine is repeated for all classes of targets as shown in the script make_sar_atr.

We are now ready to make the filters. Recall that make_sar_atr combines the parameters for all classes as blocks of the matrices **M** and **S**. Therefore, the code make_filters may include a simple set of lines such as:

```
D=S;                               %The Power spectrum ACE
S=S-M.*conj(M);                    %The Variance term for ASM

Hmach=M;                           %initialize filter with means

%Compute MACH filters for all classes and clusters

for i=1: max_cluster*nclass
   %Optimal Tradeoff MACH Filter expression
   Hmach(:,i)=Hmach(:,i)./(0.2*D(:,i)+0.7*S(:,i)+0.1);
end
```

In this example, the optimal tradeoff parameter values for average correlation energy (ACE), average similarity measure (ASM), and output noise variance (ONV) parameters are 0.2, 0.7, and 0.1 respectively. Any other suitable choice of these numbers may be made depending on the application. The key MATLAB commands that may be used to synthesize the DCCFs are as follows. Most of the commands are self-explanatory, with functions of these commands indicated by the comments on the right.

```
for ii=1: max_cluster              %For each cluster
    for j=1:nclass                 %index the class parameters
        k=(j-1)*max_cluster+ii;
        CM(:,j)=M(:,k);
        SM(:,j)=S(:,k);
    end

Mmean=mean(CM')';                  %Overall mean

Smean=mean(SM')';                  %Overall variance

H=zeros(d+1,3)                     %initialize filter space
h=CM(:,1)-CM(:,2))./Smean;         %The DCCF transform
H(1:d,1)=h;

for i=1:nclass
buf=conj(h).*CM(:,i);              %The transformed class mean
b=buf'*buf/d;                      %The class constant
buf=h.*buf;                        %The auxiliary filter
H(:,i+1)=[buf(:);b];               %Storing filters
end
Hdccf=[Hdccf H];                   %Store DCCF and repeat for all
                                     clusters

end
```

It should be noted that all of the filters are synthesized and stored in the frequency domain.

For well-designed filters, the desired class should yield sharp and tall correlation peaks, and for the clutter class, we should not see sharp and tall peaks. One way to characterize the peak sharpness is through the peak-to-sidelobe ratio (PSR) defined as follows:

$$PSR = \frac{(\text{peak} - \text{mean})}{\text{std}} \qquad (9.1)$$

Where the "peak" is the largest correlation value, and the "mean" and the "std" are the mean and the standard deviation of the correlation output in an annular region centered at the peak. Figure 9.6 illustrates how the PSR is estimated. Firstly, the peak is located (shown as the bright pixel in the center of the figure). The mean and the standard deviation of the 20×20 sidelobe region (excluding a 5×5 central mask) centered at the peak are computed. The PSR is

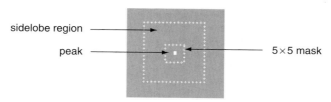

Figure 9.6 This figure shows how the peak-to-sidelobe ratio (PSR) is estimated

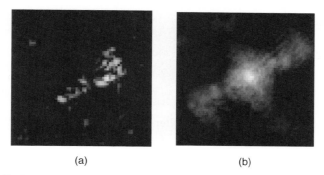

Figure 9.7 Example of BTR70 test image and the correlation surface produced by the MACH filter trained to recognize BTR70 targets at this orientation

the ratio of (peak − mean) to standard deviation as shown in Eq. (9.1). The mask sizes should be chosen based on careful experimentation.

For testing, the 15° elevation images of both classes were correlated with all 16 filters (8 filters for each class). In each case, the highest PSR was used for deciding the class of the object (i.e., the class with the highest PSR over all aspect angle bins). It was also required that the maximum PSR should exceed a threshold of 6.0 for the object to be classified, or else the decision was made to reject the input as "unknown."

Now consider the performance of the algorithm on a typical test image. Examples of a BTR70 test image, and the output of the MACH filter trained to recognize the target at this orientation are shown in Figure 9.7. The PSR of the correlation peak was 6.3, which is above the required threshold. The correlation surface is also shown as a 3-D surface plot in Figure 9.8. It is evident that the sidelobes are substantially smaller than the main peak. Figure 9.9 shows the PSR of the peak outputs of all 16 MACH filters in response to the test image in Figure 9.7(a). Clearly, the highest PSR above threshold is produced

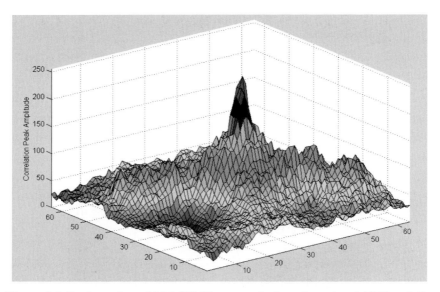

Figure 9.8 Surface plot of MACH filter correlation plane (peak PSR = 6.3)

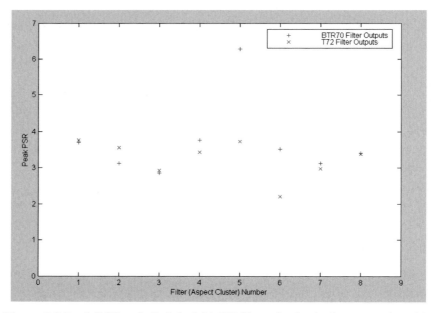

Figure 9.9 Peak PSRs of all eight MACH filters for both classes produced in response to the test image in Figure 9.7(a)

Table 9.1 *Confusion matrix for two-class ATR using MACH Filters*

	BTR70	T72	Unknown
BTR70	189	7	0
T72	6	185	0

by the fifth BTR70 filter, indicating that this is the class (and the orientation bin) of the input image.

The results of processing all test images of both classes are summarized in the form of a *confusion matrix* in Table 9.1. The entries in the confusion matrix are counts of the various decisions made by the ATR system. The diagonal numbers represent correct decisions, while off-diagonal numbers are error counts. Of the 196 BTR70 test images, 189 were correctly classified and only 7 were mistaken to be a tank. For the T72, there were 185 correct classifications while 6 images were erroneously classified as BTR70. On average, over the set of 387 test images, the ATR achieved a correct recognition rate of approximately $P_c = 96.6\%$, (i.e., an error rate of $P_e = 3.4\%$).

9.1.3 *Performance improvements using DCCFs*

Clearly, it is possible to achieve reasonable performance using MACH filters. DCCFs, however, may be used to correct some of the errors made by the MACH filters and further improve the performance of the ATR system. For instance, Figure 9.10 shows one of the T72 images incorrectly classified as a BTR70 by the MACH filters. In this instance, the highest PSR of 6.9 was produced by a BTR70 filter. The T72 filter for the correct cluster yielded a somewhat lower PSR of 6.3. Although the similar PSR values for the two classes suggest a borderline case, the decision must be made to classify the image as a BTR70 since it has the higher PSR. Such errors can be corrected if the image is further processed by the DCCFs.

A simplified architecture for a two-class ATR system using MACH filters and DCCFs is shown in Figure 9.11. The DCCFs were synthesized for each of the eight clusters. The DCCF of the cluster with the highest MACH filter PSR is invoked to compute the distances to both classes. The structural symmetry of the targets gives rise to a 180° orientation ambiguity in SAR images. Thus it is not uncommon for the MACH filters of diametrically opposing clusters also to yield high PSRs (e.g., filters trained over front views may correlate well with rear views). It is therefore prudent to interrogate the DCCF at the most likely

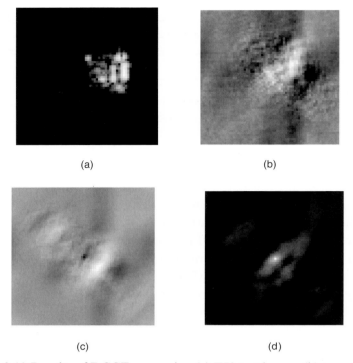

(a)   (b)

(c)   (d)

Figure 9.10 Results of DCCF processing (a) T72 test image, (b) transformed test pattern, (c) ideal transformed reference for class 1, (d) ideal transformed reference for class 2

orientation as well as for the diagonally opposite cluster. Of the two DCCFs, we must eventually decide on only one for the purpose of making a decision. A *distance ratio* is computed for each DCCF as the ratio of the smaller distance to the larger one. Smaller ratios indicate better matches (ideally zero if there is an exact match with one of the classes), while larger ratios imply greater ambiguity (the ratio is 1.0 when the distances to both classes are equal). The DCCF class yielding the smallest distance ratio is selected for making a decision. Finally, the PSR and the distance ratio are tested against thresholds. In the present case it is required that the PSR be at least 6.0, and the distance ratio be smaller than 0.90. If these conditions are met, the decision is made in favor of the class with the smaller distance.

Recall that the T72 image in Figure 9.10 had an acceptable PSR above threshold, but was misclassified on the basis of a higher PSR with the other class. On the other hand, the DCCF corresponding to the cluster with the highest PSR yielded a distance 2.92 to class 1 (BTR70, wrong class) while the distance to class 2 (T72, correct class) was 2.17. It is comforting to note that

Table 9.2 *Improved results of two-class ATR using DCCFs*

	BTR70	T72	Unknown
BTR70	196	0	0
T72	2	187	2

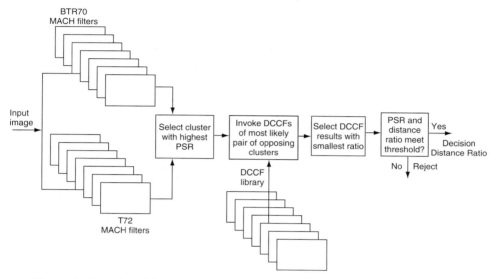

Figure 9.11 A simplified two-class SAR ATR architecture

the input is *closer* to the correct class. The ratio of the two distances is 0.74, which is also acceptable. Accordingly, the image is correctly classified as a T72 based on the DCCF measurements. Figure 9.10 shows the ideal transformed reference patterns for classes 1 and 2, and the pattern produced by the test image after transformation by the DCCF. Subjectively, the transformed test pattern in Figure 9.10(b) has more regions (dark and bright areas) that match with the ideal pattern for class 2 in Figure 9.10(d), than the pattern for class 1 in Figure 9.10(c). It should be noted that this test image is one of the more challenging cases that could be easily confused as the wrong class but is correctly recognized by the DCCF.

The results of processing all 387 images of both classes using the MACH filters and DCCFs are summarized in Table 9.2. It is interesting to note that all errors associated with class 1 are corrected by the DCCFs and all BTR70 images are recognized correctly. The performance on class 2 also improves with only 2 remaining errors and 187 correct decisions. Two T72 images are rejected by the

DCCFs as ambiguous (i.e., the distance ratio was not sufficiently small) rather than incorrectly classified. In many instances, the decision to reject may be preferable to making an error. The overall performance rates are $P_c = 99.0\%$, $P_e = 0.5\%$, and P_r (i.e., probability of reject) $= 0.5\%$.

We now discuss the pseudo-code for implementing the algorithm for testing the SAR ATR. A MATLAB script `test_sar_atr.m` for testing the two-class SAR ATR may be set up as follows. We specify parameters such as the size of the images, the filter file names, the number of aspect clusters, and a space to accumulate the confusion matrix using the following lines:

```
d1=64; d2=64;
load filters
nclass=2;
max_cluster=8;
conf_mtrx=zeros (nclass, nclass+1);
```

The data from each class at $15°$ depression is tested using code that points to the location of the appropriate files and a list of their names:

```
name= 'btr70c71.tst'
true_class=1;
loc= 'E:\TARGETS\TEST\15_DEG\btr70\sn_c71\'
run_mach_dccf

name= 't72sn_s7.tst'
true_class=2;
loc= 'E:\TARGETS\TEST\15_DEG\t72\sn_s7\'
run_mach_dccf
```

The script `run_mach_dccf.m` is the main portion of the test code. Some arrays where results will be stored are initialized as:

```
count=0;                %Initialize variables
aspects=[ ];
distances=[ ];
PSRs=[ ];
```

The main loop that processes each image is given below. The images are read in one at a time, cropped to the desired size, converted to log-magnitude form and then Fourier transformed.

```
for i=start:finish
count=count+1;
    file=[loc name_list(k1:k2)]

    [x,header]=rd_mstr;           %Read Image data
    true_aspect=header(1,1);
```

```
mask = x == 0;
x = x + mean(x(:))*mask;
x = crop(x,128,128);                    %Crop and convert to logmag
x = log10(x(33:96,33:96));

figure(1); imagep(x); pause(.1);
X = fft2(x);                            %Fourier transform
```

The following lines are used to run all of the MACH filters. Each filter is retrieved from the storage array and reshaped into image format. Recall that all the filters are stored in the frequency domain. The correlation is implemented as usual as a product of frequency domain terms which is inverse transformed and shifted by a quadrant to move the DC term to the center of the image. The location and PSR of the peak is computed and accumulated for each correlation.

```
R = [ ]; C = [ ]; PSR = [ ];
nfilt = max_cluster*nclass;

%Retrieve filter and correlate:

for i = 1:nfilt
        F = reshape(Hmach(:,i),d1,d2);
        g = real(fftshift(ifft2(X.*conj(F))));
        [peak,col] = max(max(g)); [peak,row] = max(max(g'));
        mu = mean2(g); stdg = std(g(:)); psr = (peak-mu)/stdg;
        PSR = [PSR; psr];
        R = [R; row]; C = [C; col];
end
```

The section of the code that computed the distances is shown below. Here, a pointer is used to first extract the set of DCCFs for each aspect cluster:

```
distances = [ ]; ratio = [ ];

for j = 1:max_cluster
    dccf_ptr = (j-1)*(nclass+1)+1;          %Point to a cluster
    Hdccf = H(1:d,dccf_ptr);                 %Retrieve DCCF set
    H1 = reshape(H(1:d,dccf_ptr+1),d1,d2);
    H2 = reshape(H(1:d,dccf_ptr+2),d1,d2);
    b1 = H(d+1,dccf_ptr+1);
    b2 = H(d+1,dccf_ptr+2);
```

Once the DCCF filters have been retrieved from storage, the distances to the two classes are calculated as follows. The power term is directly calculated in the frequency domain, but divided by the number of pixels to obtain the space domain equivalent. The correlations $g1$ and $g2$ are computed to search for the best match (the smallest distance) over all possible shifts of the image. Of course, the best (the smallest) distances are then obtained where $g1$ and $g2$ exhibit their maximum values. The distances and a ratio (defined as the smaller distance divided by the larger one) are then stored for each cluster.

```
p=X(:).*conj(Hdccf);              %Compute power term
p=p'*p/d;                          %space domain eq.
g1=real(ifft2(X.*conj(H1)));       %test class 1 shifts
g2=real(ifft2(X.*conj(H2)));       %test class 2 shifts
dist1=p+b1-2*max(g1(:));           %best class 1 match
dist2=p+b2-2*max(g2(:));           %best class 2 match

distances=[distances;[dist1 dist2]];%Store distances

ratio=[ratio;min(dist1,dist2)/max(dist1,dist2)];
end
```

The next few lines are book keeping to determine whether the maximum PSR occurs amongst the class 1 or class 2 MACH filters. Depending on the outcome, the results of the corresponding DCCF will be examined to make the final decision.

```
[psr,k]=max(PSR);
if k>max_cluster                   %Which DCCF ?
    psr_des=2;
    k=k-max_cluster;
else
    psr_des=1;
end
```

Since targets in SAR images tend to be symmetric with the $180°$ (opposite) angle, it is a good idea to check DCCF results at both orientations. The following lines accomplish this:

```
if k>max_cluster/2                 %Determine opposite angle
    opp_k=k-max_cluster/2;
else
    opp_k=k+max_cluster/2;
end

[r,kk]=min([ratio(k),             %select the smaller ratio
ratio(opp_k)]);                    between opposite angles

if kk==1
    kk=k;
else
    kk=opp_k;
end
```

The final decision is made as follows. For the target to be classified in favor of the class with the smaller distance, the PSR must be larger than a threshold of 6.0, and the distance ratio must be less than 0.9. If these conditions are not met, the image is rejected as unrecognizable. The confusion matrix is updated to reflect the results of the decision.

```
if (psr > 6.0) & (r < 0.90)
    dist1=distances(kk,1);dist2=distances(kk,2);
    if dist1 < dist2
    des=1;
else
    des=2;
end
else
    des=3;
end
[psr r des]
conf_mtrx(true_class,des)=conf_mtrx(true_class,des)+1
end
```

9.1.4 Clutter tests of the MACH/DCCF algorithms for SAR ATR

It is not sufficient to test an ATR's performance on target images alone. Almost invariably, large areas must be searched for targets, which gives rise to the opportunity for false alarms. While the two-class ATR in the previous section demonstrated an extremely low error rate for targets, it is important to ascertain how its probability of false alarm (P_{fa}) varies with the probability of correct detection (P_d). Of course, this is quantified by the *receiver operating characteristic* (ROC) curve introduced in Chapter 4.

Fortunately, the public release MSTAR data set contains SAR images of clutter which can be used to estimate the false alarm rate. For illustrative purposes, we processed 50 clutter images using the proposed ATR system. Each file contains a 1478×1784 1-foot resolution SAR image of natural and urban areas such as the sample image in Figure 9.12. Each image was divided into 64×64 chips for a total of 31 050 clutter samples, which were all processed by the ATR at the same PSR and distance ratio threshold settings as the target images. As the PSR is varied, however, there is a tradeoff (represented by an ROC curve) between the number of targets classified by the ATR (i.e., not rejected as unknowns) and the number of false alarms caused by clutter. The PSR threshold was then varied from 4.0 to 9.0 in increments of 0.5. The false alarm rate is estimated as the percentage of clutter chips that are called targets, while the probability of target detection is the percentage of targets that pass the PSR threshold and are correctly classified by the ATR system.

Figure 9.12 Sample clutter scene from MSTAR public data set

The ROC curve in Figure 9.13 shows that the simplified two-class ATR can achieve a P_d of around 90%, while rejecting clutter at a modest false alarm rate of approximately 10%. If, however, a lower false alarm rate is desired, a specialized *screener* stage must be used (Figure 9.14). The requirements on the screener are simply that it should use inexpensive computations to eliminate most of the clutter from consideration by the ATR system. Several different types of screeners are in existence. Mahalanobis and Singh [94] have proposed a variant of a correlation filter as a screener to further improve the performance of the correlation-based ATR system. Other screeners based on the use of wavelets and the use of principal component analysis (PCA) have been proposed.

We now discuss the design of a correlation-based screener. The idea is that a small two-dimensional finite impulse response (FIR) filter (say of size 11×11 or smaller) may be designed to enhance target-like textures while suppressing other textures associated with natural terrain or background clutter. The technique for designing such a texture-discrimination filter is discussed elsewhere [94]. Essentially, the correlation energy produced by the filter **h** in response to a texture to be enhanced (say **x**) can be expressed as:

$$E_x = \mathbf{h}^\mathrm{T} \mathbf{R}_x \mathbf{h} \tag{9.2}$$

where $\mathbf{R}_x$ is the autocorrelation matrix for the texture image estimated over a window the same size as the filter **h**. Thus, if **h** is a 5×5 filter, $\mathbf{R}_x$ is a block

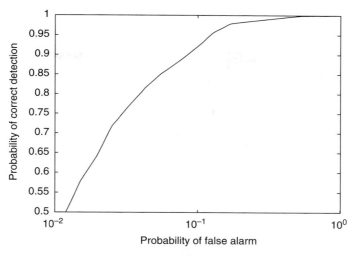

Figure 9.13 Example of receiver operating characteristic (ROC) curve showing performance of a simple two-class SAR ATR system using MACH filters and DCCFs

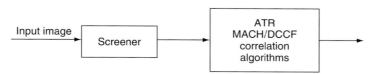

Figure 9.14 Using a screener for the purposes of lowering false alarm rate

Toeplitz matrix of size 25×25. Similarly, the correlation energy for the texture to be suppressed (say $\mathbf{y}$) can be expressed as:

$$E_y = \mathbf{h}^{\mathrm{T}}\mathbf{R}_y\mathbf{h} \qquad (9.3)$$

where $\mathbf{R}_y$ is the autocorrelation matrix for $\mathbf{y}$. To enhance $\mathbf{x}$ while attenuating $\mathbf{y}$, we find the filter $\mathbf{h}$ to maximize the Rayleigh quotient given below:

$$J(\mathbf{h}) = \frac{\mathbf{h}^{\mathrm{T}}\mathbf{R}_x\mathbf{h}}{\mathbf{h}^{\mathrm{T}}\mathbf{R}_y\mathbf{h}} \qquad (9.4)$$

Of course, the optimum filter is the dominant eigenvector of $\mathbf{R}_y^{-1}\mathbf{R}_x$. It has been shown that the resulting filter can discriminate between textures in various types of imagery, including SAR imagery. To design a screener for targets, we design $\mathbf{h}$ to enhance vehicle textures while suppressing clutter textures. Thus, we use the target training images to estimate $\mathbf{R}_x$ (instead of using them directly to synthesize a conventional filter), and estimate $\mathbf{R}_y$ from samples of clutter.

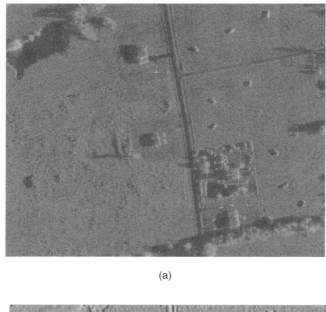

(a)

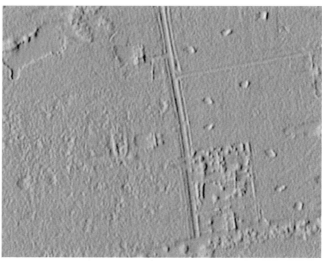

(b)

Figure 9.15 Target texture enhancing filter for screening (a) clutter scene with nine targets, and (b) results of filtering showing enhanced pixels on targets

The performance of the screener is shown in Figure 9.15 [95]. Targets were inserted in a clutter scene to create the image in Figure 9.15(a). Importantly, the pixels on the targets are not the largest values in this image, and several clutter objects exist which are stronger than the targets. The results of filtering

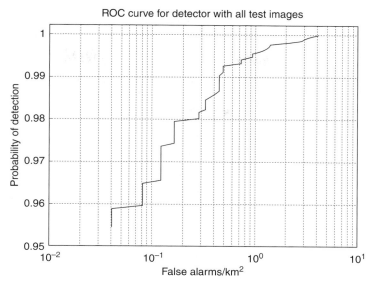

Figure 9.16 ROC curve of a correlation-based SAR ATR system with screener stage

this image are shown in Figure 9.15(b). The nine targets now possess the largest pixel values so that a simple threshold can isolate their locations and reject all the clutter.

Using this approach, Mahalanobis *et al.* [95] obtained false alarm rates as low as 0.1 false alarms per square kilometer over the entire clutter set, in the public release version of the MSTAR data set, while maintaining a P_d of about 96%. The ROC curve in Figure 9.16 characterizes the performance of an ATR system using a screener.

In summary, the MACH/DCCF algorithms work well on targets. Given a target chip to be classified the ATR algorithms can classify the targets very accurately. A simple screener (a small FIR filter) can be efficiently implemented to eliminate large areas of clutter (non-target type textures) and enhance *regions of interest* for the ATR system to consider. The combination of the two stages achieves the objective of having a low false alarm rate while maintaining a high probability for recognizing targets.

9.2 Face verification using correlation filters

Face verification is an important tool for authentication of an individual, and it can be of significant value in secure access and e-commerce applications. Most current methods for access control (e.g., automatic teller machines, to secure

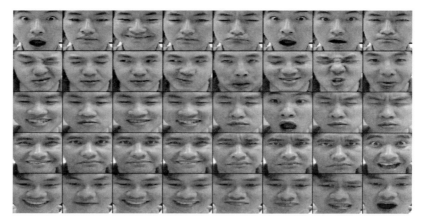

Figure 9.17 Sample images from the Carnegie Mellon University Advanced Multimedia Processing Lab's facial expression database

facilities) rely on passwords or personal identification numbers (PINs). Passwords and PINs can be forgotten or stolen, compromising the security. A better method would be to use biometrics that are unique to an individual. Examples of biometrics are face images, fingerprints, iris images, etc. In this section, we will show a simple example of applying correlation filters for face verification. Of course, correlation filters can also be applied to other biometrics such as fingerprints and iris images.

Various techniques have been popularized for face recognition including the well-known eigenface method [96]. In this section, we will illustrate the use of correlation filters for recognizing faces. Specifically, the minimum average correlation energy (MACE) filter outlined in Chapter 6 is used for face verification.

The computer simulations described in this section utilize a facial expression database collected at the Advanced Multimedia Processing Lab [97] at the Electrical and Computer Engineering Department of Carnegie Mellon University. The database consists of 13 subjects, whose facial images were captured with varying expressions. Each subject in the database has 75 images of varying facial expressions. The faces were captured in a video sequence where a face tracker [98] tracked the movement of the user's head based upon an eye localization routine and extracted registered face images of size 64×64. Example images are shown in Figure 9.17.

This facial expression database was used to evaluate the performance of the MACE filter for face verification. The computer simulation proceeded as follows. A single MACE filter was synthesized for each of the 13 subjects using a variable number of training images from that person. In the test stage, for each filter, cross-correlations were performed with all of the face images from all of the

(a) Sample Correlation Plane for Test Image from true-class: PSR = 37.0

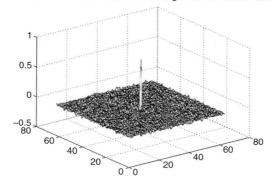

(b) Sample Correlation Plane for a false-class image: PSR = 3.8

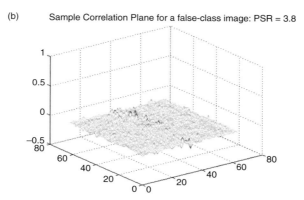

Figure 9.18 Correlation outputs when using a MACE filter designed for Person A. (a): Input is a face image belonging to Person A. (b): Input is a face image not belonging to Person A

people (i.e., $13 \times 75 = 975$ images). For authentics, the correlation output should be sharply peaked (i.e., PSR should be large); the correlation output should not exhibit such strong peaks for impostors (i.e., PSR should be small).

Figure 9.18(a) shows a typical correlation output for an authentic face image. Note the sharp correlation peak resulting in a large PSR value of 37. The correlation output in Figure 9.18(b) shows a typical response to an impostor face image, exhibiting low PSRs (<10).

Initially, only three training images were used for the synthesis of each person's MACE filter. These three images were used to capture some of the expression variations in the data set (e.g., images 1, 21, and 41). To evaluate the performance of each person's MACE filter, cross-correlations of all the images in the data set were computed using that person's MACE filter resulting in $13 \times 75 = 975$ correlation outputs (corresponding to 75 true-class images and the 900 false-class images), and the corresponding PSRs were measured and recorded.

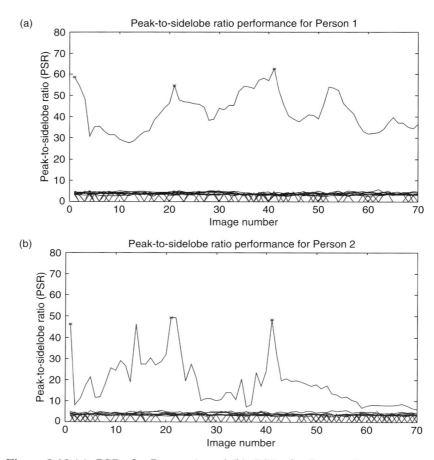

Figure 9.19 (a): PSRs for Person 1, and (b): PSRs for Person 2

Figure 9.19 shows the best MACE filter PSR performance (plot (a), Person 1), and the worst PSR performance (plot (b), Person 2) as a function of image index. Peak-to-sidelobe ratios of authentics are shown using a solid line (upper curve) and those of impostors using dotted lines (lower curves).

One very important observation from all 13 PSR plots (we have shown only two PSR plots in Fig. 9.19) is that all of the false class images ($12 \times 75 = 900$) yielded PSR values consistently smaller than 10 (these are the dotted lines at the bottom of the plot) for all 13 subjects. The three '$\times$' symbols indicate the PSRs for the three training images used to synthesize the MACE filter for that person, and as expected they yield high PSR values. The plot for Person 2, whose filter yields the worst performance (exhibiting the smallest margin of separation between the authentic and the impostor PSR values), suggests that the expected distortions in the test set were not adequately captured by the training set, and

Table 9.3 *Error percentages for all 13 MACE filters synthesized using only three training images*

Person	1	2	3	4	5	6	7	8	9	10	11	12	13
FAR, FRR = 0	0	1.3	0	0	1	0	0	0	0	0	0	0	0
EER	0	0.9	0	0	1	0	0	0	0	0	0	0	0
FRR, FAR = 0	0	0.2	0	0	2.6	0	0	0	0	0	0	0	0

Table 9.4 *Error percentages for all 13 MACE filters synthesized using the first five training images*

Person	1	2	3	4	5	6	7	8	9	10	11	12	13
FAR, FRR = 0	0	2.4	0	0	0	0	0	0	0	0	0	0	0
EER	0	1.3	0	0	0	0	0	0	0	0	0	0	0
FRR, FAR = 0	0	2.6	0	0	0	0	0	0	0	0	0	0	0

indeed a close look at the data set shows that Person 2 exhibits significantly more variation in facial expressions than others. Thus more training images may be needed to improve authentication of face images belonging to Person 2. Nevertheless even Person 2's filter designed using only three training images performed reasonably well, yielding a 99.1% verification performance.

Table 9.3 shows the error rates achieved using MACE filters designed from only three training images. The abbreviations FAR, FRR, and EER, in this table refer to false acceptance rate, false rejection rate, and equal error rate, respectively. The setting is EER when FRR equals FAR. Table 9.3 shows that the overall EER (13 filters each tested on 975 images) is only 0.15% from MACE filters designed from only three training images per person.

Another numerical experiment was performed by using the first five training images from each person in the data set to design that person's filter. These five images exhibit a different range of variability and have been placed there out of sequence. Table 9.4 summarizes the results of using the first five training images of each person. There is some improvement in that Person 5 is now 100% correctly classified. However, class 2 gives 1.3% EER for an overall EER of 0.1%.

Another simulation was performed where the training data set size was increased to 25 face images per person sampled at regular intervals of the 75-image video sequence. Figure 9.20 shows the PSR plots for Person 1 and Person 2 for the MACE filter synthesized from 25 training images. This figure

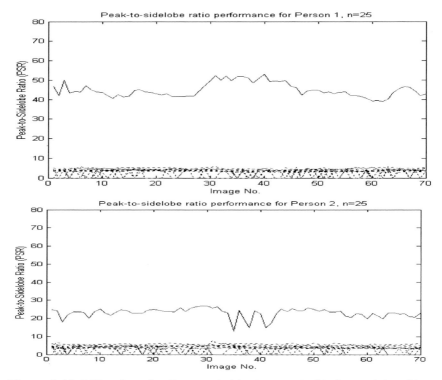

Figure 9.20 PSR plots for Person 1 (a), Person 2 (b) for MACE filters designed from 25 training images

shows a larger margin of separation than in the previous cases. In this case, the 13 MACE filters yielded 100% verification for all people.

Although we have shown that the verification accuracy of the MACE filters increases as more training images are used for filter synthesis, it is attractive that this method can work well with as few as three training images per class for this database.

9.3 Chapter summary

In this chapter, we have provided insights into the application of correlation filters for automatic target recognition and face verification. The former example focused on the use of the MACH filter and DCCF algorithms for processing large regions in SAR imagery, whereas the latter focused mainly on how MACE correlation filters can be used to verify a person's face image. These examples should inspire the reader to try applying correlation filters to other image pattern recognition problems.

References

[1] R. O. Duda, P. E. Hart and D. G. Stork, *Pattern Classification*, 2nd edn., New York, John Wiley, 2001.

[2] K. Fukunaga, *Introduction to Statistical Pattern Recognition*, 2nd edn., New York, Academic Press, 1990.

[3] T. D. Ross and J. C. Mossing, "The MSTAR evaluation methodology," in *Algorithms for Synthetic Aperture Radar Imagery VI* (ed. E. G. Zelnio), Photo-Optical Instrumentation Engineering, 3721, 1999, 703–713.

[4] T. Sim, S. Baker, and M. Bsat, *The CMU Pose, Illumination, and Expression (PIE) Database of Human Faces*, Tech. Report CMU-RI-TR-01-02, Robotics Institute, Carnegie Mellon University, January 2001.

[5] A. VanderLugt, "Signal detection by complex spatial filtering," *IEEE Transactions Information Theory*, **10**, 1964, 139–145.

[6] D. O. North, "An analysis of the factors which determine signal/noise discrimination in pulsed carrier communication systems," *Proceedings of the IEEE*, **51**, 1963, 1016–1027.

[7] C. Hester and D. Casasent, "Multivariant technique for multiclass pattern recognition," *Applied Optics*, **19**, 1980, 1758–1761.

[8] B. V. K. Vijaya Kumar, "Tutorial survey of composite filter designs for optical correlators," *Applied Optics*, **31**, 1992, 4773–4801.

[9] B. V. K. Vijaya Kumar, M. Savvides, C. Xie, K. Venkataramani, J. Thornton, and A. Mahalanobis, "Biometric verification by correlation filters," *Applied Optics*, **43**, 2004, 391–402.

[10] *Applied Optics* – Information Processing Division, Optical Society of America, Washington, DC.

[11] *Optical Engineering*, SPIE, Bellingham, WA.

[12] Society of Photo-Optical Instrumentation Engineers (SPIE) Conferences on *Automatic Target Recognition* and *Optical Pattern Recognition*, Orlando, FL.

[13] C. R. Rao, *Linear Statistical Inference and its Applications*, New York, John Wiley, 1973.

[14] E. Oja, *Subspace Methods of Pattern Recognition*, New York, John Wiley, 1983.

[15] H. Murakami and B. V. K. Vijaya Kumar, "Efficient calculation of primary images from a set of images," *IEEE Transactions on Pattern Analysis and Machine Intelligence*, **4**, 1982, 511–515.

[16] G. Strang, *Linear Algebra and Its Applications*, 3rd edn., San Diego, Harcourt Brace Johanovich, 1976.

[17] G. W. Stewart, *Introduction to Matrix Computations*, New York, Academic Press, 1973.

[18] G. H. Golub and C. F. Van Loan, *Matrix Computations*, 2nd edn., Baltimore, MD, Johns Hopkins University Press, 1989.

[19] E. Kreyszig, *Advanced Engineering Mathematics*, New York, Wiley, 1999.

[20] A. Papoulis, *Probability Theory, Random Variables and Stochastic Processes*, New York, McGraw Hill, 1965.

[21] H. Stark and J. Wood, *Probability and Random Processes with Applications to Signal Processing*, 3rd edn., Upper Saddle River, NJ, Prentice Hall, 2002.

[22] C. W. Therrien, *Discrete Random Signals and Statistical Signal Processing*, Englewood Cliffs, NJ, Prentice Hall, 1992.

[23] D. Casasent and D. Psaltis, "Scale invariant optical transform," *Optical Engineering*, **15**, 1976, 258–261.

[24] H. Araujo and J. M. Dias, "An introduction to the log-polar mapping [image sampling]," *Proceedings of Second Workshop on Cybernetic Vision*, 1996, 139–144.

[25] J. W. Cooley and J. W. Tukey, "An algorithm for machine computation of complex Fourier series," *Mathematics of Computation*, **19**, 1965, 297–301.

[26] A. V. Oppenheim, R. W. Schafer, and J. R. Buck, *Discrete-Time Signal Processing*, 2nd edn., Upper Saddle River, NJ, Prentice-Hall, 1999.

[27] S. K. Mitra, *Digital Signal Processing: A Computer-Based Approach*, New York, McGraw-Hill, 1998.

[28] C. S. Burrus and T. W. Parks, *DFT/FFT and Convolution Algorithms: Theory and Implementation*, New York, John Wiley, 1985.

[29] H. L. Van Trees, *Detection, Estimation and Modulation Theory: Part I*, New York, John Wiley, 1968.

[30] S. M. Kay, *Fundamentals of Signal Processing: Estimation Theory*, Upper Saddle River, NJ, Prentice Hall, 1993.

[31] C. R. Rao, "Information and accuracy attainable in the estimation of statistical parameters," *Bulletin of the Calcutta Mathematical Society*, **37**, 1945, 81–91.

[32] J. W. Goodman, *Introduction to Fourier Optics*, New York, McGraw-Hill, 1968.

[33] D. A. Jared and K. M. Johnson, "Ferroelectric liquid crystal spatial light modulators," in *Spatial Light Modulators and Applications III, Critical Reviews of Science and Technology*, ed. U. Effron, Proceedings of SPIE, Bellingham, WA, SPIE, **1150**, 1989, 46–60.

[34] U. Effron *et al.*, "Silicon liquid crystal light valve: status and issues," *Optical Engineering*, **22**, 1983, 682.

[35] C. S. Weaver and J. W. Goodman, "A technique for optically convolving two functions," *Applied Optics*, **5**, 1966, 1248–1249.

[36] J. E. Rau, "Detection of differences in real distributions," *Journal of the Optical Society of America*, **56**, 1966, 1490–1494.

[37] R. A. Sprague, "A review of acousto-optic signal correlators," *Optical Engineering*, **16**, 1977, 467–474.

[38] P. A. Molley and B. A. Kast, "Automatic target recognition and tracking using an acousto-optic image correlator," *Optical Engineering*, **31**, 1992, 956–962.

[39] B. V. K. Vijaya Kumar, F. M. Dickey, and J. DeLaurentis, "Correlation filters minimizing peak location errors," *Journal of the Optical Society of America A.*, **9**, 1992, 678–682.

[40] P. Réfrégier, "Filter design for optical pattern recognition: multi-criterial optimization approach," *Optics Letters*, **15**, 1990, 854.

[41] Richard D. Juday, "Optimal realizable filters and the minimum Euclidean distance principle," *Applied Optics* **32**, 1993, 5100–5111.

[42] J. L. Horner and P. D. Gianino, "Phase-only matched filtering," *Applied Optics* **23**, 1984, 812–816.

[43] A. V. Oppenheim and J. S. Lim, "The importance of phase in signals," *Proceedings of the IEEE* **69**, 1981, 529–541.

[44] B. V. K. Vijaya Kumar and Z. Bahri, "Phase-only filters with improved signal-to-noise ratio," *Applied Optics*, **28**, 1989, 250–257.

[45] R. Bracewell, *Fourier Transform and its Applications*, New York, McGraw-Hill, 1978.

[46] D. L. Flannery, J. S. Loomis, and M. E. Milkovich, "Transform-ratio ternary phase-amplitude formulation for improved correlation discrimination," *Applied Optics*, **27**, 1988, 4079–4083.

[47] J. A. Davis and J. M. Waas, "Current status of magneto-optic spatial light modulator," in *Spatial Light Modulators and Applications III, Critical Reviews of Science and Technology*, ed., U. Effron, Proceedings of SPIE, Bellingham, WA, SPIE, **1150**, 1989, 27–43.

[48] F. M. Dickey and B. D. Hansche, "Quad-phase correlation filters for pattern recognition," *Applied Optics*, **28**, 1989, 1611–13.

[49] F. M. Dickey, B. V. K. Vijaya Kumar, L. Romero, and J. M. Connelly, "Complex ternary matched filters yielding high signal-to-noise ratios," *Optical Engineering*, **29**, 1990, 994–1001.

[50] B. V. K. Vijaya Kumar and J. M. Connelly, "Effects of Quantizing the Phase in Correlation Filters," *Proceedings of SPIE*, Bellingham, WA, SPIE, **1151**, 1989, 166–173.

[51] B. V. K. Vijaya Kumar, Richard D. Juday, and P. K. Rajan, "Saturated filters," *Journal of the Optical Society of America A.*, **9**, 1992, 405–412.

[52] Richard Juday, "Generality of matched filtering and minimum Euclidean distance projection for optical pattern recognition," *Journal of the Optical Society of America A.*, **18**, 2001, 1882–1896.

[53] J. M. Florence, "Design considerations for phase-only filters," *Proceedings of SPIE*, Bellingham, WA, SPIE, **1151**, 1989, 195–202.

[54] B. Javidi, P. Réfrégier, and P. Willett, "Optimum receiver design for pattern recognition with spatially disjoint target and scene noise," *Proceedings of SPIE*, Bellingham, WA, SPIE, **2026**, 1993, 29–33.

[55] P. Réfrégier, B. Javidi, and G. Zhang, "Minimum mean-square-error filter for pattern recognition with spatially disjoint signal and scene noise," *Optics Letters*, **18**, 1993, 1453–1455.

[56] C. Chesnaud, P. Réfrégier, and V. Boulet, "Statistical region snake-based segmentation adapted to different physical noise models," *IEEE Transactions on Pattern Analysis and Machine Intelligence*, **21**, 1999, 1145–1157.

[57] A. Furman and D. Casasent, "Sources of correlation degradation," *Applied Optics*, **16**, 1977, 1652–1661.

[58] B. V. K. Vijaya Kumar, A. Mahalanobis, and A. Takessian, "Optimal tradeoff circular harmonic function correlation filter methods providing controlled in-plane rotation response," *IEEE Transactions on Image Processing*, **9**, 2000, 1025–1034.

[59] R. Wu and H. Stark, "Rotation-invariant pattern recognition using a vector reference," *Applied Optics*, **23**, 1984, 838–843.

[60] Y. Hsu and H. H. Arsenault, "Optical pattern recognition using circular harmonic expansion," *Applied Optics*, **21**, 1982, 4016–4022.

[61] R. D. Juday and B. Bourgeois, "Convolution-controlled rotation and scale invariance in optical correlation," *Proceedings of SPIE*, Bellingham, WA, SPIE, **938**, 1988, 198–205.

[62] A. Mahalanobis, B. V. K. Vijaya Kumar, D. Casasent, "Minimum average correlation energy filters," *Applied Optics*, **26**, 1987, 3633–3640.

[63] B. V. K. Vijaya Kumar, "Minimum variance synthetic discriminant functions," *Journal of the Optical Society of America A.*, **3**, 1986, 1579–84.

[64] J. Figue and P. Réfrégier, "Optimality of trade-off filters," *Applied Optics*, **32**, 1993, 1933–1935.

[65] A. Mahalanobis, B. V. K. Vijaya Kumar, S. Song, S. R. F. Sims, and J. F. Epperson, "Unconstrained correlation filters," *Applied Optics*, **33**, 1994, 3751–3759.

[66] A. Mahalanobis and B. V. K. Vijaya Kumar, "On the optimality of the MACH filter for detection of targets in noise," *Optical Engineering*, **36**, 1997, 2642–2648.

[67] G. F. Schils and D. W. Sweeney, "Rotationally invariant correlation filtering," *Journal of the Optical Society of America A.*, **2**, 1985, 1411–1418.

[68] G. F. Schils and D. W. Sweeney, "Optical processor for recognition of three-dimensional targets viewed from any direction," *Journal of the Optical Society of America A.*, **5**, 1988, 1309–1321.

[69] A. Mahalanobis, B. V. K. Vijaya Kumar, and S. R. F. Sims, "Distance classifier correlation filters for multi-class automatic target recognition," *Applied Optics*, **35**, 1996, 3127–3133.

[70] A. Mahalanobis and B. V. K. Vijaya Kumar, "Polynomial filters for higher order correlation and multi-input information fusion," *Euro American Workshop on Optoelectronic Information Processing*, Stiges, SPIE, 1997, 221–231.

[71] G. Ravichandran and D. Casasent, "Advanced in-plane rotation-invariant correlation filters," *IEEE Transactions on Pattern Analysis and Machine Intelligence*, **16**, 1994, 415–420.

[72] J. Garcia, J. Campos, and C. Ferreira, "Circular-harmonic minimum average correlation energy filter for color pattern recognition," *Applied Optics*, **33**, 1994, 2180–2187.

[73] J. Fisher and J. Principe, "Recent advances to nonlinear MACE filters," *Optical Engineering*, **36**, 1997, 2697–2709.

[74] O. Gualdrón, J. Nicolás, J. Campos, and M. J. Yzuel, "Rotation invariant color pattern recognition by use of a three-dimensional Fourier transform," *Applied Optics*, **42**, 2003, 1434–1440.

[75] L. Hassebrook, B. V. K. Vijaya Kumar, and L. Hostetler, "Linear phase coefficient composite filter banks for distortion-invariant optical pattern recognition," *Optical Engineering*, **29**, 1990, 1033–1043.

[76] G. Ravichandran, and D. Casasent, "Minimum noise and correlation energy optical correlation filter," *Applied Optics*, **31**, 1992, 1823–1833.

[77] J. W. Goodman, *Statistical Optics*, New York, John Wiley, 1985.

[78] E. L. O'Neill, *Introduction to Statistical Optics*, New York, Dover, 1992.

[79] R. M. A. Azzam and N. M. Bashara, *Ellipsometry and Polarized Light*, New York, North Holland, 1987.

[80] E. Wolf, *Proceedings of the Royal Society of London A* **230**, 1955, 246.

[81] L. Mandel and E. Wolf, *Optical Coherence and Quantum Optics*, New York, Cambridge University Press, 1995.

[82] R. Juday *et al.*, "Full-face full-complex characterization of a reflective SLM," in *Optical Pattern Recognition* XI, ed. D. P. Casasent and T. H. Chao, Proceedings of SPIE, Bellingham WA, SPIE, **4043**, 2000, 80–89.

[83] T. Cover and J. Thomas, *Elements of Information Theory*, New York, Wiley, 1991.

[84] A. Jain, *Fundamentals of Digital Image Processing*, Englewood Cliffs, NJ, Prentice-Hall,1989.

[85] D. A. Gregory, J. C. Kirsch, and E. C. Tam, "Full complex modulation using liquid crystal televisions," *Applied Optics* **31**, 1992, 163–165.

[86] J. M. Florence, *Spatial Light Modulator with Full Complex Capability*, US Patent Specification 5148157, September 15, 1992.

[87] R. D. Juday, Full Complex Modulation using Two One-Parameter Spatial Light Modulators, US Patent Specification 5416618, May 16, 1995.

[88] J. M. Florence and R. D. Juday, "Full complex spatial filtering with a phase mostly DMD," *Proceedings of SPIE*, Bellingham, WA, SPIE, **1558**, 1991.

[89] R. D. Juday and J. M. Florence, "Full complex modulation with two one-parameter SLMs," *Proceedings of SPIE*, Bellingham, WA, SPIE, **1558**, 1991.

[90] R. W. Cohn, "Pseudo-random encoding of complex-valued functions onto amplitude-phase coupled modulators," *Journal of the Optical Society of America A.*, **15**, 1998 868–883.

[91] C. Zeile and E. Lüder, "Complex transmission of liquid crystal spatial light modulators in optical signal processing applications," **1911**, 1993, 195–206.

[92] *Biometrics: Personal Identification in Networked Society*, ed., A. Jain, R. Bolle and S. Pankanti, Norwell, MA, Kluwer, 1999.

[93] C. Soutar, D. Roberge, A. Stoianov, R. Gilroy, and B. V. K. Vijaya Kumar, "Optimal trade-off filter for the correlation of fingerprints," *Optical Engineering*, **38**, 1999, 108–113.

[94] A. Mahalanobis and H. Singh, "Application of correlation filters for texture recognition," *Applied Optics*, **33**, 1994, 2173–2179.

[95] A. Mahalanobis, B. V. K. Vijaya Kumar, and D. W. Carlson, "Evaluation of MACH and DCCF correlation filters for SAR ATR using the MSTAR Public Data Base," *Algorithms for Synthetic Aperture Radar Imagery* V, Proceedings of SPIE, **3370**, 1998, 460–468.

[96] M. Turk and A. Pentland, "Eigenfaces for recognition," *Journal of Cognitive Neuroscience*, **3**, 1991, 71–86.

[97] Advanced Multimedia Processing Lab of Carnegie Mellon University Electrical and Computer Engineering Department, http://amp.ece.cmu.edu.

[98] F. J. Huang and T. Chen, "Tracking of multiple faces for human-computer interfaces and virtual environments," *IEEE International Conference on Multimedia and Expo.*, New York, IEEE, 2000, 1563–1566.

Index

DATE DUE

Demco, Inc. 38-293